Mechanical Stretch and Cytokines

Mechanosensitivity in Cells and Tissues

Volume 5

Series Editors

A. Kamkin
Department of Fundamental and Applied Physiology, Russian State Medical University, Ostrovitjanova Str. 1, 117997 Moscow, Russia

I. Kiseleva
Department of Fundamental and Applied Physiology, Russian State Medical University, Ostrovitjanova Str. 1, 117997 Moscow, Russia

For further volumes:
http://www.springer.com/series/7878

Andre Kamkin · Irina Kiseleva
Editors

Mechanical Stretch and Cytokines

Foreword by Leonid V. Rosenshtraukh

Springer

Editors
Prof. Andre Kamkin
Department of Fundamental and Applied Physiology
Russian State Medical University
Ostrovitjanova Str. 1
117997 Moscow
Russia
Kamkin.A@g23.relcom.ru;
andre.kamkin@gmail.com

Prof. Irina Kiseleva
Department of Fundamental and Applied Physiology
Russian State Medical University
Ostrovitjanova Str. 1
117997 Moscow
Russia

Editorial Assistant
Dr. Natalia E. Lapina
Division of Neurosurgical Research
Medical Faculty Mannheim
Ruprecht-Karls-University Heidelberg
Theodor-Kutzer-Ufer 1-3
D-68167 Mannheim
Germany
Natalia.Lapina@medma.uni-heidelberg.de

ISBN 978-94-007-2003-9 e-ISBN 978-94-007-2004-6
DOI 10.1007/978-94-007-2004-6
Springer Dordrecht Heidelberg London New York

Library of Congress Control Number: 2011936153

Printed on acid-free paper

Springer is part of Springer Science+Business Media (www.springer.com)

Foreword

The book "Mechanosensitivity in Cells and Tissues: Mechanical Stretch and Cytokines" edited by Andre Kamkin and Irina Kiseleva impressively demonstrates the diversity of cellular effects of cytokines depending on mechanical stimuli. Clinical and fundamental investigation of biological effects of cytokines in regulation of functioning of cells, tissues and organs is a relatively new, interesting and fast growing field, which draws a lot of attention recently. A proposed link between cytokines, their production and mechanisms of action, and their impact on effects of cellular and tissue stretch given a new momentum to this field. The role of cytokines as regulators of stretch related mechanisms is of special importance since mechanosensitivity plays an important role in a wide variety of biological processes. It is known that cytokines can be marked out as a new separate regulatory system which, along with nervous and endocrine systems, helps to maintain homeostasis, and these three systems are tightly interconnected and interdependent. Presently two major approaches to investigation of cytokines and mechanosensitivity exist. The first one focuses on the effects of interplay of mechanical stimulation of tissue or cells and cytokine production. The second approach investigates the influence of cytokines application on mechanosensitivity and mechanotransduction. This Volume is devoted to those two issues.

To my knowledge Andre Kamkin and Irina Kiseleva present the first book on this topic, which consists of a collection of publications, written by leading scientists in this field, which guaranties that this Volume will remain of great interest to readers for a long period of time. Present book is devoted to discussion of the latest findings in the filled of cytokines and mechanical stretch of different tissues research. It is evident that the data, available on this topic is scarce, and this Volume is especially valuable, since its 10 Chapters provide a complete description of the modern vision of the field, which is rapidly developing. Volume begins with an overview of cytokines, addressed to those readers, who would like to get acquainted with this topic. In following Chapters authors summarize the recent findings about the cellular and molecular effect of mechanical stretch, which induce cytokine production in cardiomyocytes, fibroblasts, vascular smooth muscle cells, endothelial cells, lung and chondrocytes. Following description of mechanosensitive proinflammatory gene expression is of special interest. Chapters describing the role of major proinflammatory cytokines in regulation of cardiac bioelectrical activity, and in particular the influence of cytokines on mechanoelectrical feedback present latest data from this field. Following chapters describe the interplay between mechanical stretch and cytokine synthesis in pulmonary endothelial cells. Then readers will find discussion of implications of cytokines in cochlear pathophysiology and in primary cilia. As a mechanosensory organelle, primary cilium is capable of initiation of the release of various cytokines-like substances, involving nitric oxide and purinergic agonists.

In general discussion which is present in the Volume in detail covers the research devoted to interplay between mechanical stretch and cytokines. I think that this book will be of great interest to a wide audience of readers, including scientists and doctors from this and neighboring fields.

Moscow, Russia Leonid V. Rosenshtraukh

Editorial

Cytokines and Mechanical Stretch of Different Tissues

Andre Kamkin and Irina Kiseleva

The research of the role of cytokines, which began approximately in the middle of the last century, lead to appearance of a huge amount of data in this field. Presently purification of the natural molecules is achieved and a wide spectrum of biological effects of cytokines is reported. However recently it has been shown that biological effects of cytokines span far beyond the immune system and influence functioning of many organs and tissues in previously unexpected ways. For example in the last decade the interplay between cytokines effects and mechanical stretch of cells, tissues and organs was shown. Primarily it relates to elevation of cytokine production in response to cellular cyclic stretching and related underlying mechanisms and signaling cascades. Physiological and pathophysiological processes, related to this issue drawn a lot of attention recently. But even now this field remains relatively unexplored and only few reviews are devoted to this topic. The other aspect of this field deals with the effects of cytokines on naturally or artificially stretched organs, tissues and especially cells remains completely uncovered. Moreover even effects of cytokines on action potentials and ion currents, flowing through ion channels have been investigated only in few studies. Possible relationship between cytokines and mechanical stretch is of such interest that we decided to devote this Volume to this topic.

We begin description of the modern vision of cytokines and mechanical stretch of different tissues from general review covering cytokine system (Simbirtsev and Kozlov, 2012). Authors present the general characteristics of cytokines, classification of cytokines, the profile of basic cytokine families. Then authors discuss the role of cytokine in pathology development and prospects of their clinical application. Cytokines can be marked out as a new separate regulatory system which, along with nervous and endocrine systems, helps to maintain homeostasis, and these three systems are tightly interconnected and interdependent. Authors draw attention to the fact that cytokines are the most universal regulatory system, because they can possess both distant biological action after secretion by cell-producer (locally and systemically), and action by intercellular contact as membrane-bound biologically active form.

The second article is devoted to mechanical stretch of cells of different tissues and the role of mediators of innate immunity (Kovalchuk et al., 2012). Authors show that mechanical deformation of the cells may lead to the release of cytokines and proinflammatory mediators, resulting in excessive immune system activation. *In vivo* and *in vitro*, it has been shown that the stretch of cells can cause the secretion of cytokines. Authors summarize the recent findings about the cellular and molecular effect of mechanical stretch on cardiomyocytes, fibroblasts, vascular smooth muscle cells, endothelial cells and lung.

In the next Chapter authors discuss mechanosensitive pro-inflammatory gene expression in vascular cells (Cattaruzza et al., 2012). In particular authors describe molecular mechanisms governing circumferential wall tension and fluid shear stress-mediated signal transduction in vascular cells, the differential and variable impact of haemodynamically-induced pro-inflammatory gene expression on these remodelling processes.

Next Chapter is devoted to cytokines, heart and calcium current in sepsis (Stengl et al., 2012). In this review the recent experimental work was summarized with emphasis on cardiac function. The major cytokines are characterized with special emphasis on cardiac function and on regulation of cardiac calcium current.

In the next review (Kuzmin et al., 2012) the role of proinflammatory cytokines is considered in regulation of cardiac bioelectrical activity. Authors discuss ionic current alternation as a possible mechanism of cytokines action in heart. They consider TNF-α as a possible player in this signaling cascade. It was shown that TNF-α induced alternation of transmembrane action potentials. Influence of TNF-α on transient outward current (I_{to}), I_{Kur}, I_{Kr}, I_{Ks}, I_{K1} is also reported. Authors discuss the interplay between TNF-α and Ca^{2+} current, influence of TNF-α on SERCA; influence of IL-1 on action potentials, I_{Na}, I_{Ca}, I_K. Finally using TNF-α and IL-6 as an example authors discuss the effects of cytokines on mechanoelectric feedback. The circulating levels of TNF-α and IL-6 were found to be significantly higher in patients suffering from atrial fibrillation. It seems that there is a positive feedback between inflammation and atrial fibrillation and proinflammatory cytokines are believed to be markers of atrial fibrillation, and more over, the key element in this positive feedback system.

Next Chapter describes physiological role of interleukin-13 in the heart and mechanoelectrical feedback (Abramochkin et al., 2012). This review briefly summarizes existing evidences of involvement of anti-inflammatory cytokine interleukin-13 in the control of heart functioning and presents latest findings on IL-13 influence on cardiomyocytes activity. According to this data, application of the IL-13 led to moderate acute changes in electrical activity of cardiomyocytes while did not cause any electrical abnormalities contrary to inflammatory cytokines. Application of IL-13 reduced the effect of the mechanical stretch on electrical activity of cardiomyocytes. Negative inotropic effect of anti-inflammatory IL-13 contrasts with positive inotropic effect of most pro-inflammatory cytokines. Special attention is given to possible mechanisms of IL-13-signalling and its influence on cardiac function in norm and pathology.

Next review describes the link between mechanical stretch and cytokine synthesis in pulmonary endothelial cells (Ito and Hasegawa, 2012). This chapter focuses on recent evidence regarding regulation of mechanical stress-induced cytokine/chemokine production by pulmonary endothelial cells and the role of pulmonary endothelial cells in the pathogenesis of pulmonary diseases related to mechanical stress.

Next Chapter discusses implications of cytokines in cochlear pathophysiology (Tabuchi and Hara, 2012). Recent reports have clarified the involvements of cytokines in the death of cochlear cells in various cochlear injuries. On the other hand, neurotrophic factors play key roles in the development and maintenance of spiral ganglion neurons. This review summarizes what is currently known about the involvement of cytokines in cochlear pathophysiology.

Next Chapter is devoted to primary cilia (Muntean et al., 2012). As a mechanosensory organelle, primary cilium could initiate the release of various cytokines-like substances, involving nitric oxide and purinergic agonists. As a chemosensory organelle, primary cilium also contains various receptors and signaling systems. In this chapter, authors describe cilia as newly recognized communication devices in response to agonist to regulate cell cycle and cellular development.

The final Chapter is devoted to identification of mechanosensitive genes in chondrocytes and osteoblasts and their role in OA pathogenesis (Henrotin et al., 2012). Authors describe responses of chondrocytes and osteoblasts to mechanical stimuli.

In general this Volume contains a throughout description of the modern vision of the link between mechanical stretch and cytokines. Knowledge of the mechanisms which underlie these processes is necessary for understanding of the normal functioning of different living organs and tissues and allows to predict changes, which arise due to alterations of their environment, and possibly will allow to develop new methods of artificial intervention. The book brings up the problem closer to the experts in related medical and biological sciences as well as practicing doctors besides just presenting the latest achievements in the field.

References

Abramochkin DV, Makarenko EYu, Mitrochin VM, Kalugin L, Kovalchuk LV, Khoreva MV, Nikonova A, Lozinsky I, Kamkin A (2012) An anti-inflammatory cytokine interleukin-13: Physiological role in the heart and mechanoelectrical feedback. In: Kamkin A, Kiseleva I (eds) Mechanosensitivity in cells and tissues 5. Mechanical stretch and cytokines. Springer, New York, NY, pp 155–164

Cattaruzza M, Wagner AH, Hecker M (2012) Mechanosensitive pro-inflammatory gene expression in vascular cells. In: Kamkin A, Kiseleva I (eds) Mechanosensitivity in cells and tissues 5. Mechanical stretch and cytokines. Springer, New York, NY, pp 59–86

Henrotin Y, Pesesse L, Sanchez Ch (2012) Identification of mechanosensitive genes in chondrocytes and osteoblasts and their role in OA pathogenesis. In: Kamkin A, Kiseleva I (eds) Mechanosensitivity in cells and tissues 5. Mechanical stretch and cytokines. Springer, New York, NY, pp 223–233

Ito S, Hasegawa Y (2012) Mechanical stretch and cytokine synthesis in pulmonary endothelial cells. In: Kamkin A, Kiseleva I (eds) Mechanosensitivity in cells and tissues 5. Mechanical stretch and cytokines. Springer, New York, NY, pp 165–187

Kovalchuk LV, Khoreva MV, Nikonova A, Kazanski VE, Lozinsky I, Diakonova IN, Kamkin A (2012) Mechanical stretching of cells of different tissues: The role of mediators of innate immunity. In: Kamkin A, Kiseleva I (eds) Mechanosensitivity in cells and tissues 5. Mechanical stretch and cytokines. Springer, New York, NY, pp 35–58

Kuzmin VS, Abramochkin DV, Mitrochin VM, Tian B, Makarenko EYu, Kovalchuk LV, Khoreva MV, Nikonova A, Kalugin L, Lysenko NN, Lozinsky I, Rozanov A, Arutyunov G, Kiseleva I, Kamkin A (2012) The role of proinflammatory cytokines in regulation of cardiac bioelectrical activity: Link to mechanoelectrical feedback. In: Kamkin A, Kiseleva I (eds) Mechanosensitivity in cells and tissues 5. Mechanical stretch and cytokines. Springer, New York, NY, pp 107–153

Muntean BS, Jin X, Nauli SM (2012) Primary cilia are mechanosensory organelles with chemosensory roles. In: Kamkin A, Kiseleva I (eds) Mechanosensitivity in cells and tissues 5. Mechanical stretch and cytokines. Springer, New York, NY, pp 201–222

Simbirtsev AS, Kozlov IG (2012) Cytokine system. In: Andre Kamkin, Irina Kiseleva (eds) Mechanosensitivity in cells and tissues 5. Mechanical stretch and cytokines. Springer, New York, NY, pp 1–33

Stengl M, Prucha M, Matejovic M (2012) Cytokines, heart and calcium current in sepsis. In: Kamkin A, Kiseleva I (eds) Mechanosensitivity in cells and tissues 5. Mechanical stretch and cytokines. Springer, New York, NY, pp 87–106

Tabuchi K, Hara A (2012) Implications of cytokines in cochlear pathophysiology. In: Kamkin A, Kiseleva I (eds) Mechanosensitivity in cells and tissues 5. Mechanical stretch and cytokines. Springer, New York, NY, pp 189–199

Contents

Contributors

Denis V. Abramochkin Department of Fundamental and Applied Physiology, Russian State Medical University, 117997 Moscow, Russia, abram340@mail.ru

Grigory Arutyunov Department of Internal Diseases, Russian State Medical University, 117997 Moscow, Russia, arutyunov_gp@rsmu.ru

Marco Cattaruzza Division of Cardiovascular Physiology, Institute of Physiology and Pathophysiology, University of Heidelberg, 69120 Heidelberg, Germany, cattaruzza@physiologie.uni-heidelberg.de

Irina N. D'iakonova Department of Fundamental and Applied Physiology, Russian State Medical University, 117997 Moscow, Russia, i-dyak@mail.ru

Akira Hara Department of Otolaryngology, Graduate School of Comprehensive Human Sciences, University of Tsukuba, Tsukuba 305-8575, Japan, haraakir@md.tsukuba.ac.jp

Yoshinori Hasegawa Department of Respiratory Medicine, Nagoya University Graduate School of Medicine, Showa-ku, Nagoya 466-8550, Japan, yhasega@med.nagoya-u.ac.jp

Markus Hecker Division of Cardiovascular Physiology, Institute of Physiology and Pathophysiology, University of Heidelberg, 69120 Heidelberg, Germany, hecker@physiologie.uni-hd.de

Yves Henrotin Bone and Cartilage Research Unit, Institute of Pathology Level 5, University of Liège, 4000 Liège, Belgium, yhenrotin@ulg.ac.be

Satoru Ito Department of Respiratory Medicine, Nagoya University Graduate School of Medicine, Showa-ku, Nagoya 466-8550, Japan, itori@med.nagoya-u.ac.jp

Xingjian Jin Department of Pharmacology, The University of Toledo, Toledo, OH 43614, USA, Xingjian.Jin@utoledo.edu

Leonid Kalugin Department of Fundamental and Applied Physiology, Russian State Medical University, 117997 Moscow, Russia, leonidkumpf@gmail.com

Andre Kamkin Department of Fundamental and Applied Physiology, Russian State Medical University, 117997 Moscow, Russia, Kamkin.A@g23.relcom.ru; andre.kamkin@gmail.com

Viktor E. Kazanski Department of Fundamental and Applied Physiology, Russian State Medical University, 117997 Moscow, Russia, vkazanski@mail.ru

Marina V. Khoreva Department of Immunology, Russian State Medical University, 117997 Moscow, Russia, mv@rsmu.ru

Irina Kiseleva Department of Fundamental and Applied Physiology, Russian State Medical University, 117997 Moscow, Russia, Kamkin.A@g23.relcom.ru

Leonid V. Kovalchuk Department of Immunology & Department of Fundamental and Applied Physiology, Russian State Medical University, 117997 Moscow, Russia, lkov@rsmu.ru

Ivan G. Kozlov Department of Pharmacology, Russian State Medical University, 117997 Moscow, Russia; Laboratory of Experimental Immunology and Immunopharmacology, Federal Scientific and Clinical Center for Pediatrics Hematology, Oncology and Immunology, 117997 Moscow, Russia, immunopharmacology@yandex.ru

Vlad S. Kuzmin Department of Fundamental and Applied Physiology, Russian State Medical University, 117997 Moscow, Russia, ku290381@gmail.com

Natalia E. Lapina Division of Neurosurgical Research, Medical Faculty Mannheim, Ruprecht-Karls-University Heidelberg, D-68167 Mannheim, Germany, Natalia.Lapina@medma.uni-heidelberg.de

Ilya Lozinsky Department of Fundamental and Applied Physiology, Russian State Medical University, 117997 Moscow, Russia, Ilya.Lozinsky@gmx.net

Natalia N. Lysenko Department of Fundamental and Applied Physiology, Russian State Medical University, 117997 Moscow, Russia, fraunata@yandex.ru

Ekaterina Yu. Makarenko Department of Fundamental and Applied Physiology, Russian State Medical University, 117997 Moscow, Russia, kmakarenko@yandex.ru

Martin Matejovic 1st Medical Department, Faculty of Medicine in Plzen, University Hospital Plzen, Charles University in Prague, Plzen, Czech Republic, MATEJOVIC@fnplzen.cz

Vadim M. Mitrochin Department of Fundamental and Applied Physiology, Russian State Medical University, 117997 Moscow, Russia, vmitrohin@gmail.com

Brian S. Muntean Department of Pharmacology, The University of Toledo, Toledo, OH 43614, USA, Brian.Muntean@rockets.utoledo.edu

Surya M. Nauli Department of Pharmacology, The University of Toledo, Toledo, OH 43614, USA, Surya.Nauli@UToledo.Edu

Anna Nikonova Department of Immunology, Russian State Medical University, 117997 Moscow, Russia, annesha@mail.ru

Laurence Pesesse Bone and Cartilage Research Unit, Institute of Pathology Level 5, University of Liège, 4000 Liège, Belgium, l.pesesse@ulg.ac.be

Miroslav Prucha Department of Clinical Biochemistry, Hematology and Immunology, Na Homolce Hospital, Prague, Czech Republic, miroslav.prucha@homolka.cz

Leonid V. Rosenshtraukh Laboratory of Heart Electrophysiology, Institute of Experimental Cardiology, Cardiology Research Center, 121552 Moscow, Russia, Lvros@cardio.ru

Alexander Rozanov Department of Internal Diseases, Russian State Medical University, 117997 Moscow, Russia, rozanov_av@rsmu.ru

Christelle Sanchez Bone and Cartilage Research Unit, Institute of Pathology Level 5, University of Liège, 4000 Liège, Belgium, christelle.sanchez@ulg.ac.be

Andrey S. Simbirtsev Department of Immunopharmacology, State Scientific Institute of High Pure Biologicals, St. Petersburg 197110, Russia, simbirtsev@hpb-spb.com

Milan Stengl Department of Physiology, Faculty of Medicine in Plzen, Charles University in Prague, 301 66 Plzen, Czech Republic, milan.stengl@lfp.cuni.cz

Keiji Tabuchi Department of Otolaryngology, Graduate School of Comprehensive Human Sciences, University of Tsukuba, Tsukuba 305-8575, Japan, ktabuchi@md.tsukuba.ac.jp

Bo Tian Department of Fundamental and Applied Physiology, Russian State Medical University, 117997 Moscow, Russia, Kamkin.A@g23.relcom.ru

Andreas H. Wagner Division of Cardiovascular Physiology, Institute of Physiology and Pathophysiology, University of Heidelberg, 69120 Heidelberg, Germany, a.wagner@physiologie.uni-heidelberg.de

Chapter 1
Cytokine System

Andrey S. Simbirtsev and Ivan G. Kozlov

Abstract Cytokines can be marked out as a new separate regulatory system which, along with nervous and endocrine systems, helps to maintain homeostasis, and these three systems are tightly interconnected and interdependent. Cytokines are the most universal regulatory system, because they can possess both distant biological action after secretion by cell-producer (locally and systemically), and action by intercellular contact as membrane-bound biologically active form. Cytokines include interferons, interleukins, growth and colony-stimulating factors, chemokines, mediators from tumor necrosis factor group, transforming growth factors and some other molecules. Cytokines are not primary mediators of pathology. However cytokines can participate in immunopathologic processes formation and function as diagnostic markers in some diseases. On the other hand the promising perspective of cytokine clinical usage for treatment of widespread diseases, including infectious and cancer, always was the driving force for cytokine study.

Keywords Cytokines · Cytokine receptors · Interferons · Interleukins · Growth factors · Colony-stimulating factors · Chemokines

1.1 Introduction

Cytokines are the family of endogenous polypeptide mediators ensuring intercellular interaction. Cytokines regulate embryonic development, some normal physiological functions of an organism, defense reactions in case of pathogens invasion and tumor growth, ensure allergic, autoimmune and other immunopathological processes and reparation of damaged tissues after injury (Thomson and Lotze, 2003). Cytokines can be marked out as a new separate regulatory system which, along with nervous and endocrine systems, helps to maintain homeostasis, and these three systems are tightly interconnected and interdependent (Irwin, 2008). At present about 250 individual substances belonging to cytokine system are known. Cytokines include interferons, interleukins, growth and colony-stimulating

I.G. Kozlov (✉)
Department of Pharmacology, Russian State Medical University, 117997 Moscow, Russia;
Laboratory of Experimental Immunology and Immunopharmacology, Federal Scientific and Clinical Center for Pediatrics Hematology, Oncology and Immunology, 117997 Moscow, Russia
e-mail: immunopharmacology@yandex.ru

A. Kamkin, I. Kiseleva (eds.), *Mechanical Stretch and Cytokines*,
Mechanosensitivity in Cells and Tissues 5, DOI 10.1007/978-94-007-2004-6_1,

factors, chemokines, mediators from tumor necrosis factor group, transforming growth factors and some other molecules (Dumonde, 1986).

The history of cytokine studying began in the middle of XX century (Cohen et al., 1974; Henney et al., 1974; Ward et al., 1969). At that time some single biological effect, which was found out during the study, gave an occasion to name of the relevant mediator. Thus in the 1950s the interferon (IFN) was named by its ability to interfere or enhance resistance in cases of recurrent viral infection (Isaacs and Lindermann, 1957). Interleukin-1 initially was named as endogenous pyrogen in opposition to bacterial lipopolysacharides considered as exogenous pyrogens (Dinarello and Wolff, 1982). The next period of cytokine studying, referring to 1960–1970s, was associated with purification of the natural molecules and detailed characterization of their biological action (Dumonde et al., 1969). In 1979 the term "interleukins", i.e. leukocyte connecting mediators, was suggested for their designation and classification. However it was found out in short term, that biological effects of cytokines are extended well over the limits of immune system, and thus the earlier term "cytokines" (Cohen, 2004) became more acceptable and persist up till now.

Cytokines include several types of above-listed mediators, but even now the most of newly discovered molecules are named as interleukins. Interleukins, historically having serial numbers from one, do not form the single cytokine group associated with similar functions. They, in turn, can be divided in proinflammatory cytokines, lymphocytic growth and differentiation factors, certain regulatory cytokines.

The name "interleukin" is given to the newly discovered mediator in case of compliance of the following criteria, generated by Nomenclatural Committee of International Union if Immunological Societies (Paul et al., 1992): molecular cloning and expression of studying factor's gene, existence of unique nucleotide and corresponding amino acid sequence, obtaining of neutralizing monoclonal antibodies. Furthermore, the new molecule must be produced by immune cells (lymphocytes, monocytes and other leukocyte types), has important biological function in immune response regulation, consequently it cannot have a functional name. Finally, specified properties of the new interleukin must be published in the reviewed scientific publication.

It seems that formation of the cytokine regulatory system in the evolutionary process passed together with the development of multicellular organisms, and was determined by necessity in formation of intercellular messengers, which can include hormones, neuropeptides, adhesion molecules and some more substances (Bazan, 1990; Huising et al., 2006). In this respect, cytokines are the most universal regulatory system, because they can possess both distant biological action after secretion by cell-producer (locally and systemically), and action by intercellular contact as membrane-bound biologically active form (Fan et al., 1996; Kostyal et al., 1991; Musso et al., 1999; Schäfer et al., 2004; Tuck et al., 1994; Yasukawa et al., 1993). This characteristic distinguishes the cytokine system from adhesion molecules which act only in direct cell-to-cell contact (Gumbiner, 1996). At the same time, the cytokine system differs from hormones which basically are synthesized by specialized organs and spread with blood circulation through the whole organism to realize their action.

1.2 The General Characteristics of Cytokines

In spite of a lot of mediators included in the cytokine system, they have a number of general characteristics which allow integrating them in the independent regulation system. These characteristics include (Cytokines Online Pathfinder Encyclopaedia (http://www.copewithcytokines.de/); Frankenstein et al., 2006; Gonda and D'Andrea, 1997; Oppenheim et al., 2000; Taga and Kishimoto, 1992; The Cytokines Web (http://cmbi.bjmu.edu.cn/cmbidata/cgf/CGF_Database/cytweb/targets/index.html); Thomson and Lotze, 2003; Tieri et al., 2005; Zlotnik and Yoshie, 2000):

1. Cytokines are polypeptides or proteins, which are usually glycosylated, majority of them have molecular mass from 5 to 50 kDa. Biologically active cytokine molecules may consist of one, two, three or more identical or different subunits.
2. The biological action of cytokines is not antigen-specific. They affect the functional activity of cells assisting in innate and adaptive immune reactions. However cytokines can stimulate antigen-induced processes in immune system by affecting T- and B-lymphocytes.
3. There are three variants of cytokine gene expression: (a) stage-specific expression at certain stages of embryonic development, (b) constitutive expression for the regulation of some normal physiological functions, (c) inducible type of expression which is characteristic for the majority of cytokines. In fact, most cytokines are not synthesized by cells in the absence of inflammatory reaction or immune response. Cytokine gene expression begins in response to the pathogen invasion, antigenic stimulation or tissue injury.
4. The majority of cytokines are inducible mediators. In case of inducible expression, cytokines are synthesized in response to the short-time stimulation. Cytokines has an important characteristic of self-limiting synthesis which is ceased by various autoregulation mechanisms such as enhanced RNA instability and existence of negative feedbacks including following mechanisms: (a) induction of prostaglandin synthesis, (b) induction of corticosteroid hormone synthesis, (c) increased synthesis of soluble receptors and expression of membrane trap receptors, (d) induction of intracellular transcription factors blocking signal transduction form cytokine receptors and cytokine synthesis (e.g. SOC-2).
5. One cytokine can be produced in organism by histogenetically different cell types in different organs.
6. Some cytokines can be associated with membranes of producing cells, possessing the whole spectrum of biological activity in membrane-bound form and acting by intercellular contact.
7. Cytokine biological effects are mediated by specific cellular receptor complexes binding cytokines with very high affinity, and certain cytokines can use common receptor's subunits. Cytokine receptors can exist in soluble form retaining ability to bind ligands.

8. Cytokines have pleiotropic biological action. The same cytokine can affect several cell types with different effects depending on the type of target cells. Pleiotropy of cytokine action is provided by expression of cytokine receptors in cells with different origin or functions, and by signal transduction using several different intracellular messengers and transcription factors.
9. Cells can synthesize several cytokines simultaneously (multiple inducibility) participating in the formation of the cytokine network, in response to single activation signal. Biological effects in tissues and in the whole organism depend on presence and concentrations of other cytokines with synergistic, additive or opposite type of action.
10. Cytokines are characterized by redundancy of their biological action. Several different cytokines induce or suppress their own synthesis, synthesis of other cytokines and their receptors (cross-inducibility). Cytokines have overlapping biological activity. However there is no one cytokine which has the same function or duplicate action of the other cytokine.
11. Cytokines can affect proliferation, differentiation and functional activity of target cells.
12. Cytokines can affect cells by different mechanisms, such as autocrine (affecting cell synthesizing and secreting this cytokine); paracrine (affecting cells located close to cell-producer, e.g. in the inflammation site or in lymphoid organ), or endocrine (distantly, affecting cells of every organ and tissue after release in circulation). In the last case cytokine action resembles hormone action.

The cytokine role in regulation of organism functions can be divided in four main components:

1. Regulation of fertilization, embryogenesis, anlage and organ development including organs of immune system.
2. Regulation of certain normal physiological function.
3. Regulation of organism's local and systemic defense reactions including protection from all types of pathogens, maintenance of homeostasis under influence of external and internal disturbing factors including tumor growth.
4. Regulation of tissue regeneration process.

1.3 Classification of Cytokines

Cytokines can be classified by their biochemical and biological characteristics, and also by receptor types by which cytokines realize their biological functions (Bodmer et al., 2002; Lata and Raghava, 2008; Oppenheim et al., 2000; Thomson and Lotze, 2003; Zlotnik and Yoshie, 2000). The integral structural-functional classification in which all cytokines are divided into groups, in the first place by their biological action, and also by the structural patterns of cytokine and their receptor molecules, is presented in Table 1.1.

Table 1.1 The structural-functional classification of cytokines

No	Cytokine family	Subfamilies and ligands	Main biological functions
1	**2**	**3**	**4**
1	Interferons (IFN)	Interferons type I: IFNα, β, δ, κ, ω, τ	Antiviral, antiproliferative, immunomodulating action
		Interferon type II: IFNγ	
		Interferons type III: IFNλ1 (IL-29), IFNλ2 (IL-28A), IFNλ3 (IL-28B)	
2	Hemopoietic cell growth factors	Stem cells factor (kit-ligand, steel factor), Flt-3 ligand, G-CSF, M-CSF, IL-7, IL-11	Stimulation of different cell precursor types proliferation and differentiation in bone marrow, hematopoiesis activation
		Ligands of gp140 receptor subunit: IL-3, IL-5, GM-CSF	
		Erythropoietin, thrombopoietin	
3	Superfamily of interleukin-1 (IL-1) and fibroblast growth factor (FGF)	FGF family: acid FGF, basic FGF, FGF3-FGF23	Activation of fibroblast and epithelial cell proliferation
		IL-1 family (F1–11): IL-1α, IL-1β, IL-1 receptor antagonist, IL-18, IL-33, IL-37 and others	Proinflammatory effect, activation of specific immunity
4	Tumor necrosis factor (TNF) family	TNF, lymphotoxin α and β, Fas-ligand and others	Proinflammatory action, apoptosis regulation and immune cell intercellular interaction
5	Interleukin-6 family	Ligands of gp130: IL-6, IL-11, IL-31, oncostatin-M, cardiotropin-1, leukemia inhibitory factor, ciliary neurotrophic factor	Proinflammatory and immunoregulatory action

Table 1.1 (continued)

No	Cytokine family	Subfamilies and ligands	Main biological functions
6	Chemokines	Subclasses: CC, CXC, CX3C, C	Different leukocyte types chemotaxis regulation
7	Interleukin-10 family	IL-10, -19, -20, -22, -24, -26	Immunosupressive action, inflammation and tumor growth regulation
8	Interleukin-12 family	IL-12, -23, -27, -35	Regulation of T-helper differentiation
9	Cytokines of T-helper clones and lymphocyte function regulators	T-helpers type I: IL-2, IL-15, IL-21, IFNγ, TNF	Activation of cell immunity
		T-helpers type II: IL-4, IL-5, IL-10, IL-13	Activation of humoral immunity, immunomodulation effects
		Ligands of IL-2 receptor γ-chain: IL-2; IL-4 | IL-13; IL-7 | TSLP; IL-9; IL-15; IL-21	Differentiation, proliferation and functional activity stimulation of different types of lymphocytes, DC, NK-cells, macrophages and others
10	Interleukin-17 family	IL-17A, B, C, D, E, F	Proinflammatory cytokine synthesis activation
11	Superfamily of nerve growth factor (NGF), platelet deriving growth factor (PDGF) and transforming growth factor (TGF)	NGF, brain neurotrophic factor	Regulation of inflammation, angiogenesis, neuron functioning, embryonic development and tissue regeneration
		PDGF, vascular endothelial growth factors (VEGF)	

Table 1.1 (continued)

No	Cytokine family	Subfamilies and ligands	Main biological functions
		TGFβ, activins, inhibins, nodal, bone morphogenic proteins, mullerian inhibitory substance	
12	Epidermal growth factor (EGF) family	EGF, TGFα and others	Stimulation of different cell type proliferation
13	Insulin-like growth factor (IGF) family	IGF-I, IGF-II	Stimulation of different cell type proliferation

Cytokine biological functions are pleiotropic, every molecule has several activities manifesting to various cell types participating in different immune reactions (Ozaki and Leonard, 2002). Therefore it is impossible to classify cytokines simply by their biological characteristics. On the other hand, previous classifications based on structure or receptor type only did not take into account variability of biological characteristics of these unique molecules. An attempt to consider both biological and biochemical characteristics was made in the given classification. However in the given classification some cytokines are artificially included in different families at the same time because of their biological activity features. Active studying of intracellular transcriptional factors and signal transduction systems from cytokine receptors (Gadina et al., 2001; Ihle, 1995; Leonard and Lin, 2000), and updating of ligand-receptor interaction variants and biological characteristics is going at present that should allow finishing cytokine classification.

1.4 The Profile of Basic Cytokine Families

The profile of biochemical and biological characteristics of basic cytokine families and certain cytokines according to the present classification is given below.

1.4.1 Interferons

At present interferons (IFN) are classified into three families: types I, II and III. Type I IFN family has the simplest organization because all included molecules have the similar structure and mostly the same functions ensuring antiviral defense, and all interferons possess species-specific action (Schultz et al., 2004; Sen, 2001). All type I IFN genes are localized beside each other in one gene cluster 9p22 in 9 chromosome (Gray and Goeddel, 1982; Gresser, 2007; Hardy et al., 2004). Type I IFNs have molecular mass about 20–22 kDa, possess similar biological characteristics and bind to the receptors of the same type called Type I IFN Receptors (Uzé et al., 2007).

This is due to high degree of homology between interferon molecules reaching 80% for IFN-alpha group. The main function of type I IFNs is the antiviral action. IFN synthesis begins when viruses entry in organism, after interaction of viral proteins and nucleic acids (viral pathogen-associated molecular patterns) with membrane and cytoplasmic pattern-recognizing receptors in epithelial cells, different leukocyte types, macrophages and dendritic cells (Liu, 2005).

Type I IFNs possess four principal biological characteristics (Vilcek, 2007):

1. Direct antiviral action due to virus transcription inhibition.
2. Suppression of cell proliferation necessary for viral distribution.
3. Activation of NK-cells functions that can kill virus-infected cells in organism.
4. Enhancing of class I major histocompatibility complex molecules necessary for increasing effectiveness of viral antigen presentation by infected cells to cytotoxic T-lymphocytes. This results in activation of virus-infected cell specific recognition by T-lymphocytes – the first stage of virus-infected target cells lysis.

Consequently besides direct antiviral action both mechanisms of innate (NK-cells) and adaptive (T-lymphocytes) immunity are activated. Therefore all type I IFNs besides antiviral action also possess immunoregulation effect. This is an example of how one small cytokine molecule with molecular mass almost in 10 fold less than antibodies is able by its pleiotropic action to activate absolutely different mechanisms of defense reactions aimed at one purpose – removal of virus entered in an organism.

There is an exception of quite recently discovered IL-28 and IL-29 otherwise called IFNλ or type III IFNs (Bartlett et al., 2005; Vilcek, 2003) which possess also other functions due to interaction of these molecules with certain subunits of IL-10 family cytokine receptors. Moreover, there is one more interferon type, IFNγ, which possess low antiviral activity but have some more important immunoregulative functions (Schroder et al., 2004). IFNγ, called type II IFN, interacts with another receptor type, type II IFN Receptors (Bach et al., 1997), regulates lymphocyte functions and classified as T-helper I cytokine (Table 1.2).

1.4.2 Hematopoietic Cell Growth Factors

The second group includes growth and differentiation factors of hematopoietic cells stimulating hematopoietic precursor cells development from stem cell. This group includes cytokines stimulating predominantly early stages of hemopoiesis at stem cells and early precursors level (kit-ligand (Bowie et al., 2007; Geissler et al., 1991), Flt-3 ligand (Antonysamy and Thomson, 1999; Brasel et al., 1995; Shortman and Naik, 2007)); cytokines with high specificity to certain cell line differentiation (erythropoietin (Jelkmann, 2007), thrombopoietin (Kaushansky, 2006) and IL-7 (Appasamy, 1992; Aspinall et al., 2004; Dittel and LeBien, 1995) affecting T- and B-lymphocyte precursors); and also cytokines with more wide biological activity

Table 1.2 Classification and biological characteristics of interferons

IFN types	Molecule names	Number of genes	Receptor types	The main biological function
Type I	IFNα IFNβ IFNδ IFNκ IFNω IFNτ	23 1 1 1 1 1	Type I	High antivirus activity, antiproliferative and immunomodulation effect
Type II	IFNγ	1	Type II	Low antivirus activity, activation of macrophages and cell immunity
Type III	IL-29 (IFNλ1) IL-28A (IFNλ2) IL-28B (IFNλ3)	1 1 1	Type I + IL-10 family receptors	Antivirus activity, immunomodulating effect

spectrum such as IL-3 (Martinez-Moczygemba and Huston, 2003; Morris et al., 1990), IL-11 (Bhatia et al., 2007; Curti et al., 2002), colony-stimulating factors (G-CSF, M-CSF, GM-CSF) (Nemunaitis, 1997; Throm, 2008). In this cytokine group gp140 ligands with common receptor subunit and also thrombopoietin and erythropoietin are marked out because of structural organization similarity. Hematopoietic system is very complex and has extremely important significance for immune system functioning and vital activity of the whole organism.

1.4.3 Fibroblast Growth Factor (FGF) and Interleukin-1 (IL-1) Superfamily

Cytokines from FGF and IL-1 superfamily have high degree of homology and similar structure of proteins which confirms general origin. However FGF superfamily members significantly differ from IL-1 family agonists by biological activity manifestations (Krejci et al., 2009; Smallwood et al., 1996). At present 2 polypeptides with molecular mass about 18 kDa named "IL-1" are well-known, IL-1α and IL-1β (Dinarello et al., 2010). Both cytokines are encoded by different genes but have 26% homology in amino acid sequence, nearly the same biological activity spectrum and compete for binding with the same receptors. Moreover, there is third protein with similar structure which has almost the same homology with IL-1α and IL-1β as they have with each other, and can bind specifically with IL-1 receptor without biological action (Arend, 1993). It blocks biological activity of IL-1 competing with it for the same receptor, and due to these characteristics it was named as IL-1 receptor antagonist (IL-1RA) (Arend, 2003). The fourth member of this family, IL-18, which

Table 1.3 Cytokines, IL-1 family members

New classification name	Previous name	Receptors	Biological functions
IL-1F1	IL-1α	IL-1R I, IL-1R II, IL-1 Ac	Proinflammatory cytokine
IL-1F2	IL-1β	IL-1R I, IL-1R II, IL-1 Ac	Proinflammatory cytokine
IL-1F3	IL-1RA	IL-1R I, IL-1R II, IL-1 Ac	IL-1 antagonist
IL-1F4	IL-18	IL-18Rα, IL-18Rβ	Proinflammatory cytokine
IL-1F5	IL-1δ	?	Unknown functions, 47% homology with IL-1RA, does not possess IL-1, IL-18 or IL-1RA activities
IL-1F6	IL-1ε	IL-1Rrp2	Unknown functions
IL-1F7	IL-1ζ, IL-37	IL-18Rα, IL-18Rβ	Unknown functions, 36% homology with IL-1RA, binds with IL-18 receptor without action
IL-1F8	IL-1H2	IL-1Rrp2	Unknown functions
IL-1F9	IL-1H1	IL-1Rrp2	Unknown, 37% homology with IL-1RA
IL-1F10	ILHy2	?	Unknown functions
IL-1F11	IL-33	ST2	Tx2 function and allergy stimulation

? - unknown.

got its serial number before discovering its similarity with other cytokines from IL-1 family, is also quite well-studied (Huising et al., 2004; Lebel-Binay et al., 2000). At present the IL-1 molecules family besides of functional names has names F1-F11 (F – family), where F1 corresponds to IL-1α, F2 – IL-1β, F3 – IL-1 receptor agonist, F4 – IL-18 (Sims et al., 2001). Other members of IL-1 family were discovered after gene analysis and has high homology with IL-1 molecules but their biological functions are not completely elucidated (Table 1.3). Recently two more members of IL-1 family, IL-33 and IL-37, were discovered and now are under active study (Lingel et al., 2009; Schmitz et al., 2005).

The most human IL-1 family genes are located in 2 chromosome. IL-1α, IL-1β and IL-18 are synthesized as precursors with molecular mass about 30 kDa, without signal peptides or hydrophobic fragments that are necessary for association with cellular membrane and common mechanism of secretion through endoplasmic reticulum. IL-1β and IL-18, unlike IL-1α, are actively secreted by human cells to surroundings. By now the IL-1 converting enzyme (ICE) converting IL-1β precursor and IL-18 precursor in mature biologically active secretor forms with molecular mass about 18 kDa has been discovered. ICE or caspase-1 is the serine protease, heterodimer consisting from two different polypeptide chains with molecular mass 10 and 20 kDa (Dinarello, 1998; Mariathasan et al., 2004; Wilson et al., 1994).

After pathogen invasion IL-1 production begins in the zone of the first contact of producing cells with microorganisms, i.e. locally, in the area of skin and mucosa injury and in regional lymphoid tissue (Glauser, 1997). Therefore the first manifestation of IL-1 biological action is activation of local defense reactions. IL-1 synthesis results in local hyperemia, oedema, tissue leukocyte infiltration, simulating inflammation development with all its classic features. Generalization of inflammation due to inadequate local defense mechanisms functioning leads to release IL-1 in circulation and its systemic action provided by acute phase response at the organism level, and different organs and systems functions stimulation, needed for ensuring defense reactions. Thus, due to wide spectrum of IL-1 biological activity, it is the basic mediator of local inflammatory reaction development and of acute phase response at the organism level (Dinarello, 2009).

1.4.4 Tumor Necrosis Factor (TNF) Family

Historically the first member of this family is the TNF itself (Aggarwal et al., 1985; Carswell et al., 1975). In the middle of XX century experiments were performed demonstrated the existence of endogenous mediator which caused necrosis of tumor node after injection in animals with induced tumor growth. These experiments gave the hope of obtaining the effective antineoplastic drug named tumor necrosis factor (TNF), but unfortunately it was discovered that TNF caused necrosis of solid tumors due to massive induction of inflammation in the area of tumor growth, endothelial activation and thrombosis of vessels supplying the tumor, that cause tumor necrosis. According to contemporary point of view, TNF can cause apoptosis of some malignant cell types, but this fact was not used in clinical practice. TNF is one of the most important proinflammatory cytokine that determines its general biological function (Aggarwal and Natarajan, 1996; Alimzhanov et al., 1997).

TNF gene is located in 6 chromosome between major histocompatibility complex molecule genes. As IL-1, TNF is synthesized by macrophages, NK-cells, T-lymphocytes and other cells after pathogen invasion in organism. Secreted TNF has molecular mass 17 kDa, but monomeric polypeptide does not possess biological activity. Biologically active TNF is formed after noncovalent association of three monomers in the molecule with molecular mass 51 kDa. This structure of biologically active molecules are characteristic for all cytokines from TNF family and provide interaction with relevant three-dimensional cellular receptors (Hehlgans and Pfeffer, 2005; Tang et al., 1996). Furthermore majority of cytokines from TNF family can exist both in soluble and in membrane form. In the last case ligand-receptor interaction occurs by intercellular contact (Bodmer et al., 2002).

TNF can bind with two types of cellular receptors called type I and type II receptors that determines the type of cellular response to its action (Naismith and Sprang, 1996). Type I receptors with molecular mass of 55 kDa have the so-called death domain in their cytoplasmic fragment. TNF binding to these receptors affects the conformation of death domains and results in activation of cytoplasmic transcriptional factor TRADD (TNF receptor-associated death domain), leading to cellular apoptosis (Bodmer et al., 1997; Chinnaiyan et al., 1996). From the other hand, TNF interaction with type 2 receptors with molecular mass 75 kDa leads to activation of well-known transcriptional factor NF-κB and induces expression of quite a number of genes participating in stimulation of inflammatory process and enhancing cell viability. Predominance of one or another response depends on type of cells affected by TNF in specific tissue environment, and on proportion of two receptor type expression by cells (Aggarwal, 2003).

TNF biological action as the proinflammatory cytokine depends to a large extent of its concentration in tissues. Therefore, the role of TNF in inflammation development can be considered in two ways. In low doses it is the important physiologic regulator of inflammation, and is necessary for development of normal defense reactions after pathogen invasion. In conditions of local inflammatory process TNF causes leukocyte, fibroblast and endothelial cell activation, enhance phagocytosis and cytokine synthesis by macrophages. However in high concentrations TNF acts as a pathological inflammatory mediator, and TNF in combination with other cytokines cause hyperactivation of endothelium affecting its permeability and resulting in microcirculation disturbance in organs. At the same time decrease of myocardial and vessel wall smooth muscular contractility and drop of blood pressure are observed, and some other symptoms of septic shock are developed (Croft, 2009; Kwon et al., 2003; So et al., 2006).

TNF family has been formed depending on similarity in ligand and their receptor structure, consisting of three noncovalent binded similar subunits forming biologically active homotrimeric molecules. At the same time, talking about biological activity, this family includes cytokines with quite different activities. For example, TNF is one of the most impressive proinflammatory cytokines, Fas-ligand cause apoptosis of target cells, and CD40-ligand provides stimulating signal in intercellular interaction of T- and B-lymphocytes. These distinctions in biological activities of structurally close molecules are determined firstly by patterns of their receptor expression and structure, e.g. intercellular "death" domain determining cellular apoptosis presence of absence.

1.4.5 Interleukin-6 (IL-6) Family

IL-6 was discovered as a mediator of T- and B-lymphocyte intercellular interaction which is synthesizing by T-lymphocytes and enhancing proliferation and antibodies synthesis by activated B-cells (Kishimoto, 2010; Kishimoto et al., 1995). During the study of IL-6 following characteristics of its receptors were discovered: they have two receptor transmembrane subunits, one of them specifically binds to IL-6, and another (named gp130 according to its molecular weight) is the common subunit for several cytokines: IL-6 itself, IL-11, IL-31, oncostatin M (OSM), leukemia inhibitory factor (LIF), ciliary neurotrophic factor (CNTF) and cardiotropin 1 (CT-1) (Schuster et al., 2003; Schwantner et al., 2004). All specified cytokines are encoding by their own genes, but have similarity in structure and can interact with receptor subunit gp130 ensuring the same pattern of intercellular activation signal transduction and partial overlap of biological characteristic. Therefore these factors are included in IL-6 cytokine family or, differently, cytokines ligands gp130 (Kishimoto, 2010). However, all these cytokines besides gp130 have other unique and specific receptor subunits, co-expression of which together with gp130 in different cell types results in appearance of partially different own biological functions of each cytokine.

The mature secretory form of IL-6 has 184 amino acid residue and has molecular mass 21 kDa. IL-6 gene is located in 7p21 chromosome, includes 4 exons and 5 introns and is structurally close to G-CSF gene (Bowcock et al., 1988). IL-6 is synthesized by many cell types participating in initiation and regulation of inflammation and immune response, such as T-lymphocytes, monocytes/macrophages, endothelial and smooth muscle cells, fibrobl .sts, keratinocytes and some other cells. IL-6 gene expresses under influence of viruses, bacteria entering in organism and their products, and also proinflammatory cytokines. IL-6 is the typical early inducible cytokine rapidly accumulating in circulation after encountering with pathogens (Jones, 2005; Vgontzas et al., 2005).

General manifestations of IL-6 biological activity in organism are: antigen activation of B-lymphocyte proliferation and enhancing of antibodies synthesis without selective action on certain immunoglobulin classes; T-lymphocyte proliferation activation due to induction of IL-2 receptor expression and enhancing IL-2 production; activating of T-killers functional activity; granulocytic hemopoiesis stimulation, and multi-CSF activity together with IL-3; proinflammatory action due to activation of adhesion molecules expression on endothelium and some leukocyte types chemotaxis, enhancing of fibroblast and osteoclast functional activity which can result in cartilage degradation in joint area; acute-phase response activation by induction of C-reactive protein, serum amyloid A and fibrinogen synthesis in liver, and also pyrogenic action (Kishimoto, 2006). Comparing with other cytokines, IL-6 is the main activator of majority acute-phase proteins synthesis in liver; whereas proinflammatory cytokines IL-1 and TNF stimulate only certain proteins synthesis and their action can be mediated by IL-6.

Probably IL-6, besides the main function of defense reaction activator, acts as a regulator of immune system development. Data obtained in experimental models of inflammation showed two opposite characteristics of IL-6 exhibited depending on immunopathology development stage or type. In experimental chronic inflammatory processes IL-6 enhances pathology development showing proinflammatory action. On the other hand, in some models of acute inflammation IL-6 shows immunoregulatory or even anti-inflammatory behavior that can be explained by its ability to transfer acute phase inflammation to chronic phase, with mononuclear attraction and mononuclear granuloma formation. Therefore one of the biological functions of IL-6 can be connected with switch from defense reactions development from initial developing inflammation and innate immune reactions to adaptive immune reactions (Ishihara and Hirano, 2003; Jones, 2005; Nishimoto, 2006).

1.4.6 Chemokines

Studying of inflammatory process regulation has resulted in discovery of large group of mediators with similar biological activity and high structural homology, integrated in the separate family of chemokines (chemotactic cytokines) under the main biological action manifestation – different leukocyte type chemotaxis (directed motion) stimulation (Luster, 1998; Zlotnik and Yoshie, 2000; Zlotnik et al., 2006). One of the first chemokines described was the most well-known IL-8, which has got its serial number as typical interleukin (Walz et al., 1987; Yoshimura et al., 1987). After discovering of the whole family consisting of similar molecules IL-8 was placed in chemokine family, including by now over 50 individual substances with specific structural and biological features.

Chemokines are 8–12-kDa polypeptides with 20–80% homology in the primary amino acid sequence. The characteristic feature of chemokine structure is the presence of disulfide bridges between cysteine residues. These bridges form the unique spatial configuration of molecules necessary for interaction with specific receptors and manifestation of biological activity (Fernandez and Lolis, 2002). Basing on cysteine residues position in molecules all chemokines are classified into 4 basic groups: CXC or α-chemokines, CC (β-chemokines), C (γ-chemokines) and CX3C (δ-chemokines). In CXC chemokines there is one any amino acid residue named "X" between the first two cysteine residues (of 4 in the molecule), whereas in CC chemokines the first two cysteine residues are located next to each other. In CX3C chemokines cysteines are separated by three different residues, and in C chemokine molecules there are only two cysteine residues instead of four, which are located away from each other, closing one disulfide bridge (Zlotnik and Yoshie, 2000).

Indicated structural features determine biological action of chemokines. CXC chemokines including IL-8 affect predominantly neutrophilic granulocytes and certain types of mononuclear cells, but not monocytes, whereas CC chemokines act as chemoattractants predominantly for monocytes, lymphocytes and some other cells. In general molecules from chemokine family affect almost all leukocyte types, however each separate chemokine has its own individual characteristics (Colobran et al., 2007). Both chemokines with short range action and chemokines with broad spectrum of functions exist, and several chemokines often affect the same cells.

The most numerous chemokine subclass is CC. The main function of these chemokines is chemotactic activation of predominantly monocytes/macrophages, lymphocytes, basophiles and eosinophils. Two members of the third subclass, C chemokines, are chemoattractive for T-lymphocytes. They are synthesized by T-lymphocytes, NK-cells, mast cells, and their unique characteristic is quite selective action on lymphocytes only. The exclusive representative of CX3C chemokines is fractalkin. It is the only one membrane-bound polypeptide from all chemokine family which is synthesized by endothelial cells and expressed on their surface, and biologically active peptide domain is located on the long mucine "peduncle". Fractalkin action is directed at attraction of monocytes and T-lymphocytes to the inflammation site.

The most well-studied CXC chemokine regulating tissue granulocyte migration is IL-8. The main producers of IL-8 appear to be activated by pathogens monocytes/macrophages and endothelial cells, however it can be produced by many other cell types such as lymphocytes, neutrophilic granulocytes, different epithelial cells, fibroblasts, hepatocytes and other cells. Endothelial cells and neutrophilic granulocytes can accumulate pre-produced IL-8 and some other cytokines in granules ensuring very rapid secretion after cellular activation comparing with de novo protein synthesis demanding 2–3 h.

At present about 20 different chemokine receptors are known interacting with various affinity with one or several ligands of chemokine family (Allen et al., 2007; Murdoch and Finn, 2000). For example 2 types of high-affinity receptors for IL-8 exist, type I and II which have 77% homology to each other. According to the new uniform nomenclature for chemokine receptors, they are designated as CXCR1 and CXCR2. Chemokine receptors belong to rhodopsin receptor superfamily with 7 transmembrane domains, and conduct signal via G-proteins; their genes in humans are located in 2q34–q35 chromosome.

The main biological characteristic of chemokines is the regulation of different cell type migration. This function became apparent as early as in embryonic development. The second manifestation of cell migration regulation by chemokines is conditioned by their homeostatic function ensured by regulation of lymphocyte and dendritic cell migration in lymphoid organs and mucosa, that is necessary for cellular content renewal and occur in organism during all its life. This function is ensured by constant (constitutive) expression of certain chemokine genes. Finally, inducible expression of chemokine genes occurs in defense reactions after pathogen invasion and immunopathological processes development, and also in tumor growth. In this case the main goal of cytokine action is the stimulation of different leukocyte types and other cells migration for ensuring the optimal quantitative and qualitative composition of all cell types without exception assisting in innate and adaptive immunity development. According to two alternatives of chemokine genes expression in postnatal period they are also classified on proinflammatory (inducible), homeostatic and chemokines with mixed characteristics (Ransohoff et al., 2007; Moser and Loetscher, 2001). Homeostatic chemokines predominantly regulate T-lymphocyte, B-lymphocyte and DC migration and are synthesized in tissues where different cells are attracted by concentration gradient interacting with specific receptors. Proinflammatory chemokines are expressed in inflammation and immune response development and activate migration of many cell types to zone of defense reactions development, particularly in inflammation site.

1.4.7 Interleukin-10 (IL-10) Family

Interleukin-10 family includes several polypeptides with similar structure: IL-10, IL-19, IL-20, IL-22, IL-24 and IL-26. Three recently discovered interferons lambda, IL-28A, IL-28B and IL-29, are also referred to IL-10 family. IL-10 was discovered first, other cytokines were included in this family under genetic analysis results and homologous genes search. In spite of similar structure, no one from specified cytokines has absolutely similar characteristics with IL-10, because all of them use different combinations of receptor subunits for signal transduction and activate different transcriptional factors (Dumoutier and Renauld, 2002; Moore et al., 1993; Pestka et al., 2004).

Human IL-10 is the polypeptide of 160 amino acids, existing in solution of noncovalent binded homodimer. Human IL-10 gene is encoded by one exon of 3.5 kb and is located in 1 chromosome (Eskdale et al., 1997; Zdanov et al., 1996). In mouse IL-10 is synthesized by Th2 but not Th1 clones. In human IL-10 is synthesized by Th0, Th1 and Th2 clones, cytotoxic T-lymphocytes, activated B-lymphocytes, mast cells and macrophages. IL-10 differs from many other cytokines by its biological action, directed on inhibition of inflammation and immune response (Pestka et al., 2004). Mechanism of anti-inflammatory and immunosuppressive action is predominantly realized through inhibition of antigen-presenting cells functions and suppression of proinflammatory cytokines (IL-1, IL-6, IL-8, TNF) production by macrophages and DC, and also of immunoregulatory cytokines by T-helper lymphocyte clones (Beebe et al., 2003; Bogdan et al., 1991; Bonder et al., 1999). It appears that biological role of IL-10 in organism consists from restriction of innate and adaptive anti-infectious immune reactions that can result in tissue injury (Asadullah et al., 2004).

Biochemical and biological characteristic of other cytokines – IL-10 family members – shows that, in spite of structural similarity, all these cytokines possess absolutely different biological activities, quite different from IL-10 itself. Evolutional divergence of biological characteristics of IL-10 family cytokines, that have probably the same progenitor, is fixed in genome, indicating the importance of each member in vital activity of an organism.

1.4.8 Interleukin-12 (IL-12) Family

Cytokines from IL-12 family are heterodimeric molecules consisting of two different polypeptide chains (Hamza et al., 2010). IL-12 consists of p40 and p35 polypeptides with molecular mass 40 and 35 kDa, respectively, bound to each other by disulfide bonds. The cumulative heterodimeric cytokine is named IL-12p70. Each polypeptide chain is encoded by its own gene, their expression is regulated separately (Trinchieri, 2003). Moreover, it was shown that p40 subunit can bind to absolutely different polypeptide, p19 with molecular mass 19 kDa. This heterodimeric complex has different biological characteristics and was named IL-23. The third family member, IL-27 is derived from p28 polypeptide and a protein with previously unknown function synthesized after Epstein-Barr virus infection – EB13 (Epstein–Barr virus-induced gene 3) (Beadling and Slifka, 2006; Brombacher et al., 2003). Further studies have shown two more variants of association between different described polypeptides. The first is formed if two p40 polypeptides are bound and homodimeric variant IL-12p80 is produced that have similar to IL-12p70 biological activity but is rarely found. The second heterodimeric cytokine is formed after binding between p35 and EB13 subunits and was named IL-35 (Collison et al., 2007; Niedbala et al., 2007). Its biological functions include stimulation of T-regulatory lymphocytes (Treg) differentiation and restriction of IL-17 induced inflammation development.

Gene expression and all cytokines from IL-12 family production predominantly occur in dendritic cells and macrophages in response to different pathogens after interaction between pathogen-associated molecular patterns with cellular Toll-like receptors (Trinchieri, 2003). The main function of IL-12 family cytokines is regulation of T-helper lymphocyte clones differentiation activated after meeting with an antigen. At present it appears that IL-12 causes Th1 differentiation, IL-23 – Th17 differentiation, IL-27 – inducible Treg, and IL-35 participates in formation of native CD4+CD25+ Treg (Goriely et al., 2008). Moreover, cytokines from this family can maintain proliferation of some T-lymphocyte clones, including memory T-lymphocytes, and also activate NK-cells.

1.4.9 Interleukin 17 (IL-17) Family

Initially IL-17 was described as T-cellular cytokine stimulating inflammation development by activation of known proinflammatory cytokines synthesis (Yu and Gaffen, 2008). From 2000 5 more cytokines with high structural homology, up to 55%, with IL-17 were opened. Consequently the new cytokine family was formed including following members: IL-17A, IL-17B, IL-17C, IL-17D, IL-17E and IL-17F. IL-17 family includes cytokine, previously known as IL-25, named as IL-17E. As with some other cytokine families, in spite of structural similarity, cytokines from IL-17 family had significantly different biological functions except IL-17A and IL-17F with nearly similar characteristics (Goriely et al., 2008).

Human IL-17 (previously named simply IL-17) is the glycoprotein from 155 amino acids with molecular mass 17.5 kDa; its gene is localized in 6p12 chromosome. All cytokines from IL-17 family are homodimeric molecules consisting from two identical subunits. It was considered previously that IL-17 is synthesized exclusively by T-helper lymphocytes, and in general Th17 clones, but after that IL-17A gene expression was discovered in many tissues, in lymphoid and non-lymphoid cells (Aarvak et al., 1999). The same was shown during the study of other cytokines from this family, and every cytokine from IL-17 family has its unique pattern of tissue distribution. Among T-helper lymphocytes activated after meeting with antigen, IL-17A synthesis is performed by specific T-helper clone type, named Th17 (Moseley et al., 2003). This clone type is differentiated under action of IL-23 produced by activated DC. Additional factors of Th17 differentiation are proinflammatory cytokines, IL-1β and IL-6, synthesized by activated macrophages in the course of the inflammatory reaction (Harrington et al., 2005).

IL-17A receptors consist from two subunits that are typical transmembrane polypeptides interacting for signal transduction after binding with cytokine. After IL-17A binding with receptors intracellular molecular cascade is activated which leads to activation of transcriptional factors NF-κB, AP-1, and also some kinases (ERK, c-Jun N-terminal kinase and p38 MAP kinase). Consequently, transcription of proinflammatory cytokine genes and other molecules regulating inflammation starts, depending on cell type (Dong, 2008; Toy et al., 2006).

IL-17A is synthesized predominantly by antigen-activated T-lymphocytes and acts on innate immune cells: monocytes/macrophages and neutrophils, stimulating IL-1, TNF, IL-6, G-CSF, chemokine and antimicrobial peptide production. Experimental deletion of IL-17A gene leads to significant decrease of anti-infectious resistance, first of all due to neutrophilic granulocyte functions disorder (Stockinger and Veldhoen, 2007). Therefore IL-17A is the link between activation of innate and adaptive immunity. However enhanced IL-17 family cytokine production results in development of immune system hyperactivity, pathological inflammation and can underlie immunopathology development (Tesmer et al., 2008).

1.4.10 T-Helper Clone Cytokines

This cytokine group is the most heterogenous and difficult for classification. Molecules considered here do not belong to any single cytokine family by their structure, as it was accepted for other cytokines, but are integrated mostly by functional characteristics. Inclusion in this group is based on two essential principles: (1) belonging to Th1 or Th2 synthesized cytokines that determines basic cellular or humoral immune reaction type, (2) presence of common receptor subunit – the gamma chain of IL-2 receptor complex. Among gamma chain ligands, IL-4 is additionally marked out, due to common receptor subunits with IL-13 that determines to a large extent partially overlapped biological activity of these cytokines. Similarly IL-7 is marked out due to its receptor structure close to TSLP (thymic stromal lymphopoietin). Cytokines stimulating cellular immunity and synthesized by human Th1 include IL-2, IL-15, IL-21 and IFNγ, and basic cytokines of Th2 clones regulating humoral immune response development are IL-4, IL-5, IL-13 (Agnello et al., 2003; Feuerer et al., 2009; Mosmann and Coffman, 1989; Murphy and Reiner, 2002).

1.4.10.1 Interleukin-2 (IL-2)

IL-2 is synthesized by cells as polypeptide consisting from 133 amino acid residues with molecular mass about 15 kDa. It has 3 cysteine, 2 of which in 58 and 105 position form disulfide bridge absolutely necessary for generation of biologically active molecular conformation. IL-2 molecule can undergo posttranslational glycosylation in N-terminal region by threonine residue in position 3 (Dr. Kendall Smith's Immunology Resource Site (http://www.kendallasmith.us/); Smith, 1988).

Human IL-2 gene is located in 4q chromosome. Control of IL-2 gene transcription predominantly depends on certain zones in 5′ region where binding sites of transcription enhancers NFAT-1, NF-κB, AP-1 and others are located. mRNA transcripts appear already after 1 h after lymphocyte activation, and their amount reaches maximal levels in 4–6 h. Secreted IL-2 is detected in culture medium of stimulated lymphocytes after 6–10 h and reaches maximal levels in 24 h. The main cells-producers of IL-2 are T-lymphocytes with T-helper phenotype (CD3+CD4+) with functional characteristics of T-helpers type 1. IL-2 is produced in inducible manner, resting lymphocytes do not express IL-2 gene. Consequent stages of gene transcription, mRNA translation and protein secretion occur after antigen interaction with T-cell receptor (Waldmann, 2006).

IL-2 interacts with specific membrane receptors expressed in some cells of immune system. IL-2 receptor complex consists of 3 subunits that are polypeptides of different size, named as α-, β- and γ-chains, each encoded by its own gene (Almeida et al., 2002; Ozaki et al., 2000; Smith, 2006).

Target cells for IL-2 action are T- and B-lymphocytes, NK-cells, monocytes and tissue macrophage-like cells, e.g. olygodendrocytes. The main biological effect of IL-2 is stimulation of different lymphocyte type proliferation, which determined its title of lymphocytic growth factor. IL-2 stimulates cell division of both T-helper lymphocytes synthesizing it after antigenic stimulation and T-killer lymphocytes, acting by autocrine and paracrine mechanisms. Furthermore IL-2 leads to functional activation of all specified cells, in each case directed on their basic functions performance for defense reaction development in organism (Gaffen and Liu, 2004; Malek and Bayer, 2004).

In IL-2 gene knockout mice, in spite of absence of one lymphocyte growth factor, hyperplasia of lymphoid organs, increased level of CD4+ T-lymphocytes, multiorganic inflammatory reaction, autoimmune hemolytic anemia and autoimmune ulcerative colitis development were shown indicating of immune system activation. Experiments on animals with deficit of certain IL-2 receptor chain genes confirm these observations. At present it appears that described phenotypic changes in IL-2 gene or IL-2 receptor alpha chain gene knockout mice are connected with Treg functions disturbance (Almeida et al., 2002). These data indicate that the unique role of IL-2 in immune regulation is rather conditioned by control of immune system hyperactivity or tolerance induction by stimulation of Treg differentiation in vivo than with growth factor functions which is duplicated by other cytokines (Malek and Bayer A, 2004).

1.4.10.2 Interleukin-15 (IL-15)

IL-15 gene is located in 4q31 chromosome, as IL-2 and some other growth factor genes. IL-2 and IL-15 genes have very close structural organization. IL-15 molecule with molecular mass just above 14 kDa consists from 114 residues and has two disulfide bridges between 35–85 and 42–88 residues, and also sites for N-glycozilation (Asn-79 and Asn-112). IL-15 receptor complex consists from 3 subunits. Two of them are similar with IL-2 receptor complex – beta and gamma chains, and third is receptor subunit specific for IL-15 only – receptor alpha chain. IL-15 receptors are expressed in T-lymphocytes, NK-cells, monocytes, endothelial and some other cells (Lodolce et al., 2003).

Unlike IL-2, IL-15 is synthesized by different cell types in multiple organs. In humans IL-15 is synthesized by monocytes/macrophages, epithelial cells of bone marrow stroma, and other epithelial cells, but IL-15 mRNA was not found in activated T- and B-lymphocytes. IL-15 synthesis is activated under action of type I interferons, IFNγ, double stranded RNA, lipopolysaccharide and some other pathogen's derivates. IL-15 mRNA expression was detected in several cell types, but in cell culture supernatants or blood plasma it was found in very rare cases. Much of IL-15 can exist in biologically active membrane from (Budagian et al., 2004; Waldmann, 2006).

Due to similar receptor structure, IL-15 possess many biological activities of IL-2: it enhances T- and B-lymphocyte and NK-cell proliferation, cytotoxic T-lymphocyte generation, etc. IL-15 is synthesized in bone marrow stroma, and probably participates in maturation and leaving to periphery of functionally mature NK-cells, and consequently maintains their viability and stimulates functional activity (Cooper et al., 2002).

1.4.10.3 Interleukin-21 (IL-21)

IL-21 gene is located in 4q26-q27 chromosome, only 180 kb away from IL-2 gene. IL-2 and IL-21 genes have very similar structural organization, and it cannot be excluded that evolutionally they were formed as a result of duplication. IL-21 protein itself with molecular mass about 15 kDa has homology with IL-2, IL-4 and IL-15, and homology between mouse and human IL-21 reaches 57% (Parrish-Novak et al., 2000). Receptor complex for IL-21 consists from two subunits: specific for IL-21 subunit named as IL-21 receptor, and IL-2 receptor complex gamma chain, which is the common subunit for IL-2, -4, -7, -9, -15 and -21 receptors. IL-21 interaction with heterodimeric receptor leads to standard signal transduction mechanism with Jak1, Jak3 and intracellular messengers STAT1 and STAT2 (O'Sullivan et al., 2007).

IL-21 is produced predominantly by activated CD4+ T-lymphocytes, but affects wide range of cells of different origin. IL-21 receptors are expressed in T- and B-lymphocytes, NK-cells, some myeloid cell populations and keratinocytes. IL-21 synthesized by T-lymphocytes regulates naïve and memory CD8 T-cell functions. IL-21 synergistically with IL-7 enhances proliferation of these cells, thereby participating both in initiation and maintenance of T-killer lymphocyte subpopulation response. Experiments with knockout mice show that in absence of IL-21 CD8 T-lymphocytes normally develop, but their functional activity is defective, e.g. there is reduction of antiviral response (Zeng et al., 2005). IL-21 can be used for anticancer and anti-infectious immunity enhancement due to its ability to maintain cytotoxic T-lymphocyte functional activity. It has been shown in experimental models of cancer development in mice studying IL-2, IL-15 and IL-21 anticancer action, that IL-21 was the most active, and in some tumor types only IL-21 possesses anticancer action among all specified cytokines (Moroz et al., 2004).

1.4.10.4 Interferon gamma (IFNγ)

IFNγ firstly was found in cultural medium of antigen-activated leukocyte suspension. This factor possess antiviral activity but its biochemical characteristics were different from other known IFNs because of sensitivity to acid action. Therefore it was named "acid-labile" interferon or type II IFN, and later IFNγ in contrast to other acid-resistant type I IFNs. It was found that antiviral activity is not the main characteristic of this molecule, IFNγ act rather as immunity regulator and has been also named as "immune" interferon (Sen, 2001).

Human IFNγ consists of 4 exons and is located on 12 chromosome. The typical signal 23-member polypeptide is removed during synthesis with formation of mature polypeptide IFNγ consisting from 143 amino acid residues (Gray and Goeddel, 1982). Human IFNγ receptor complex consists from two subunits named as alpha and beta receptor chains which do not relate to IL-2 receptor complex. There is very rapid activation of intracellular tyrosine kinases (Jak1 and Jak2) and the transcriptional factor STAT1 after IFNγ binding with both receptor chains. In spite of type I IFNs synthesized by almost all cells in organism, IFNγ is synthesized predominantly by antigen-activated T-lymphocytes and NK-cells. Among T-lymphocytes cytotoxic T-cells and predominantly CD4+ T-helpers type I synthesize and secrete IFNγ, but there are some examples of IFNγ production by different T-helper clones in humans. Also it was shown that small amounts of IFNγ can be synthesized by other cell types: γδT-lymphocytes, NK-T-cells, macrophages and dendritic cells (Bach et al., 1997).

Biological activity of IFNγ, as other interferons, is species-specific. All biological characteristics of IFNγ can be reduced to 5 basic functions: 1) macrophage activation, 2) NK-cell activation, 3) stimulation of class II histocompatibility antigen expression, 4) regulation of immunoglobulin isotype synthesis by B-lymphocytes, 5) direct antiviral action. As regards the latter, specific antiviral action of IFNγ is lower than in type I IFNs, in which this activity appears to be the main biological function. Other specified IFNγ characteristics related to cell-mediated immunity activation, performing by T-lymphocytes, NK-cells and macrophages (Schroder et al., 2004).

Two mutations were described in human IFNγ receptor gamma chain gene, which lead to shortened protein and receptor functioning disturbance. All these genetic defects are exhibited firstly by enhanced susceptibility to mycobacterium infection and especially severe tuberculosis course, which confirms the principal role of IFNγ in defense from intracellular pathogens. Among 140 patients with genetic defects of IFNγ receptors only 4 had serious viral infections (Novelli and Casanova, 2004). It appears that IFNγ can enhance antiviral immunity, but does not play the key role in antiviral defense reactions.

1.4.10.5 Interleukin-4 (IL-4) and Interleukin-13 (IL-13)

IL-4 was initially described in 1982 as a factor stimulating proliferation of B-lymphocytes activated by surface IgM. Later it was found that this mediator also stimulates proliferation and functional activity of some T-lymphocyte types. In 1986 after gene cloning and production of purified protein possessing all specified characteristics, this cytokine received its final name, IL-4 (Brown and Hural, 1997). Human IL-13 gene cloning and in-depth study of its biological functions show some principal differences between two cytokines. IL-13, like IL-4, stimulate B-lymphocyte and monocyte/macrophage functions, however, IL-13 absolutely does not affect T-lymphocytes because T-cells do not express its receptors (de Vries and Zurawski, 1995).

Human IL-4 and IL-13 genes are located in 5q23–31 chromosome in cluster also containing IL-3, IL-5, IL-9 and GM-CSF genes. Both genes consist of 4 exons and 3 introns, are situated only 25 kb away from each other and are situated in the same orientation, which lead to hypothesis of emergence of two gene due to duplication during evolutional process (Wills-Karp, 2000). IL-4 is biologically active in monomeric form with molecular mass 15 kDa, pI 10.5, and has 6 cysteine residues forming 3 disulfide bridges in 3–127, 24–65 and 46–99 positions, maintaining biologically active conformation of globular structure. IL-13 consists of 112 amino acid residues with molecular mass 12.3 kDa; glycosylated IL-13 molecular mass raises up to 17,000 Da. Molecule has two disulfide bridges between cysteine residues in 28–56 and 44–70 positions (Puri, 1995).

Activated type II T-helper lymphocytes, basophils and mast cells are main producers of IL-4, in less extent it is produced by cytotoxic T-lymphocytes, $\gamma\delta$T-lymphocytes, eosinophils and some other cells. Antigenic activation via T-cellular antigenic receptor is a signal for gene expression and IL-4 synthesis in T-lymphocytes. Basophils, eosinophils and mast cells start to produce IL-4 after interaction of IgE with high-affinity membrane Fcε-receptors and after activation by bacterial cell wall components and C5a. Human IL-13 is produced equally by antigen- or polyclonal inductors-activated Th0, Th1 and Th2 clones of CD4+ T-lymphocytes, CD8+ T-lymphocytes, and NK-T-cells. IL-13 is also synthesized by activated B-lymphocytes and mast cells, basophils and eosinophils and smooth muscle cells of human high respiratory tract (Izuhara et al., 2003).

IL-4 receptor consists from two subunits: one is specific for alpha chain IL-4 with molecular mass about 140 kDa, and second is common for several cytokines of IL-2 receptor complex gamma chain. IL-4 binds directly with receptor alpha chain with rather high affinity ($K_d = 0.5 \times 10^{-10}$M), then it binds with gamma chain, and affinity rises in 2–3 times more. One IL-4 molecule interacts simultaneously with alpha and gamma receptor chains resulting in their dimerization. There are two types of IL-13 receptors. The first type of functional receptor ensuring signal transduction from IL-13 consists of two polypeptide chains: (1) specific for IL-13 subunit or IL-13 receptor alpha chain (CDw124) and (2) IL-4 receptor alpha chain, which is common for IL-4 and IL-13 receptors (Chatila, 2004; Puri, 1995).

IL-4 biological action is species-specific. One of the main biological characteristic of IL-4 is its ability to activate B-lymphocyte proliferation and functional activity. However IL-4 itself is not able to activate the whole program of resting B-lymphocyte activation without specific antigenic activation. IL-4 acts as costimulator for B-lymphocytes activated by specific antigen. Besides proliferation stimulation IL-4 induces production of certain antibody types, IgE and IgG4 by activated B-lymphocytes. The mechanism of IL-4 action is related to switch of synthesis of immunoglobulin molecule heavy chains from IgM to IgE or IgG4. IL-4 can also regulate proliferation and differentiation of antigen-activated T-lymphocytes, acting as not only B-cellular, but also T-cellular growth factor (Fallon et al., 2002).

Talking about biological activity spectrum, IL-13 affects B-lymphocytes similarly as IL-4 does, stimulating proliferation of antigen-activated B-cells, expression of several surface markers, and synthesis of IgE and IgG4 antibodies, but in all cases response intensity is significantly lower than response for IL-4 action. As IL-4, IL-13 enhances expression of several membrane antigens and inhibits proinflammatory cytokine synthesis. In general IL-13 enhances humoral immune response manifestations and allergic inflammation, acting as cytokine duplicating IL-4 action in this way. Two cytokines differ because IL-13 cannot activate T-lymphocytes and has significantly more pronounced action on fibroblast activation and connective tissue metabolism (Dessein et al., 2005).

1.4.10.6 Interleukin-5 (IL-5)

Th2 synthesize some cytokines regulating functions of B-lymphocytes and other cells participating in humoral immune reactions. IL-5 is one of such cytokines; it acts as B-lymphocyte growth factor, enhances immunoglobulin synthesis by B-cells and stimulates T-lymphocyte and eosinophil functions (Tavernier et al., 1995).

Human IL-5 gene is located in 5q chromosome in cluster, containing IL-4, IL-13 and GM-CSF genes (Lee et al., 1989). IL-5 is synthesized by classic mechanism, after signal peptide removal mature cytokine consists from 115 amino acid residues and has molecular mass about 13 kDa. IL-5 is biologically active only as 27-kDa homodimer generated after formation of disulfide bridges between molecules. IL-5 is synthesized by antigen-activated T-lymphocytes (Th1), usually in parallel with IL-4, but expression of these genes is regulated independently. Additionally IL-5 is produced by B-lymphocytes, basophils, mast cells and eosinophils in allergic inflammation sites.

Human IL-5 receptor consists of two subunits designated as receptor complex alpha and beta chains. Alpha chain is the single-stranded transmembrane protein with molecular mass 45 kDa, which specifically binds with IL-5 only. Receptor complex beta chain is the common receptor subunit of IL-5, IL-3 and GM-CSF receptors (Tominaga et al., 1992).

Biological action of IL-5 in allergy consists of enhancing of bone marrow eosinophil precursors and increasing of mature eosinophil number, eosinophil chemotaxis stimulation, their functional activation and apoptosis delay. Consequently IL-5 appears to be the one of basic cytokines participating in allergy immunopathogenesis, and can be used as a target for anticytokine therapy in the treatment of allergic diseases (Simon et al., 2004).

1.5 Role of Cytokine in Pathology Development and Prospect of Their Clinical Application

The significant role of cytokines in pathogenesis of several diseases and clinical syndromes serves a basis for a theory of cytokine-mediated diseases. Conventional view of immune system formed in age of infectious immunology indicated that immunity in all its manifestations is useful for organism, because it defends organism from pathogens. It was difficult to imagine that organism itself can destroy its own cells or produce molecules provoking pathological alterations in organs and tissues, or even induce death. Study of cytokine role in sepsis proved this possibility and radically changed the established views (Proulx et al., 2009).

As long ago as 1960s the supposition was stated that significant systemic reaction of patient in septic condition is not related with pathogen itself but rather with response of an organism mediated by endogenous factors and connected with immune system activation. These results gave the stimulus to search of sepsis and septic shock mediators. Initially it was demonstrated that condition with clinical symptoms reminding septic shock can be induced in mice by injection of gram-negative bacteria cell wall components – LPS, without presence of living bacteria (Aird, 2003). Consequently it has been shown that LPS is the potent inducer of endogenous inflammatory mediators, including cytokines, and TNF appeared to be the principal player of such cytokines. Experiments of Bruce Beutler, demonstrated the principal role of TNF in animal death from experimental septic shock, gave the key information for understanding of the whole endogenous reaction cascade in infection resulting in septic shock development. In these experiments LPS injection in mice resulted in septic shock development and death, however if neutralizing antibodies were injected after LPS, manifestations of septic shock were much less, and animals did not die (Beutler et al., 1985).

The character of proinflammatory cytokine action in organism depends on their levels. In low concentrations cytokines are necessary for correct course of local inflammation, in higher doses they cause the development of protective systemic inflammatory reaction, but in pathologically high concentrations they can lead to septic shock development and death of an organism (Table 1.4). If cytokine levels exceed physiological concentrations, they act as mediators of pathology development and, in some cases, as "death" mediators, rather than protective mediators.

Cytokine theory of disease development (Dinarello, 1996; Tracey, 2007) runs that endogenous cytokines can cause symptoms of pathological changes in organs, damage tissues, as it does TNF in sepsis, and other cytokines do in autoimmune and septic inflammation. However this fashionable theory somewhat displaces accents in correct understanding of true cytokine role in regulation of basic physiological processes in organism, particularly, in immunopathology development. Cytokines are not primary mediators of pathology. Cytokine action is only the final accord in complex immunopathogenesis in the majority of diseases, excluding hereditary disorders of cytokine regulation.

Table 1.4 Dose dependent biological effects of proinflammatory cytokines

Cytokine production types	Manifestations of inflammatory reaction	Biological effects of cytokines	Clinical significance
Local synthesis in site of inflammation	Local inflammation	Leukocyte activation, phagocytosis and oxygen radical production enhancement. Enhancement of adhesion molecules expression in endothelium. Stimulation of cytokine and chemokine synthesis. Metabolic activation in connective tissue.	Local inflammatory reaction. Removal of pathogen.
Cytokines enter in circulation	Systemic inflammatory reaction	Fever. Increase of steroid hormones level. Leukocytosis. Increase of acute phase protein synthesis.	Systemic protective reaction. Removal of pathogen and recovery.
Pathologically high cytokine levels in circulation	Septic shock	Reduced contractility of myocardium and smooth muscle cells in vessels. Increased permeability of endothelium. Dysfunction of microcirculation in organs. Blood pressure drop. Hypoglycemia.	Pathological inflammation. Death of an organism.

Thus cytokine theory of pathology development in human diseases is just the description of one important part but not all immunopathogenesis. Certainly it is necessary to remember that cytokines can participate in immunopathologic processes formation and function as diagnostic markers in some diseases. But most likely they should be considered as the instrument of immune system functioning rather than underlying course of pathology development, and use them as a basis for development of individual approaches to cytokine or anticytokine treatment prescription depending of immunopathogenesis study.

Cytokines have huge potential of formation and regulation of defense reactions, and pretend to the role of the most effective medical products affecting all sides of non-specific resistance and non-specific immunity development without exception. The promising perspective of cytokine clinical usage for treatment of widespread diseases, including infectious and cancer, always was the driving force for cytokine study. There are some cytokine products registered in the world, including interferons, colony-stimulating factors, interleukins, their antagonists and some other cytokines (Debets and Savelkoul, 1994; Rouveix, 1997; Smith, 2001). Undoubtedly, the future of cytokine therapy is connected with genetically engineered products, synthesizing using last achievements of biotechnology.

In some cases, if the endogenous cytokine production is not sufficient, recombinant cytokine products can be used for replacement therapy, hemopoiesis activation or for enhancement of defense reactions. But in cases of cytokine hyperproduction their excess synthesis should be inhibited. In this connection there are two principal variants of cytokine usage in clinical practice:

1. Cytokine therapy directed at lack of endogenous cytokine replacement, or enhancement of their biological action.
2. Anticytokine therapy for inhibition of biological activity of endogenous cytokines synthesizing in excess during acute and chronic inflammatory processes, including autoimmune and allergic diseases.

References

Aarvak T, Chabaud M, Miossec P, Natvig JB (1999) IL-17 is produced by some proinflammatory Th1/Th0 cells but not by Th2 cells. J Immunol 162:1246–1251

Aggarwal BB (2003) Signalling pathways of the TNF superfamily: a double-edged sword. Nat Rev Immunol 3(9):745–756

Aggarwal BB, Natarajan K (1996) Tumor necrosis factors: developments during the last decade. Eur Cytokine Netw 7:93–124

Aggarwal BB, Kohr WJ, Hass PE, Moffat B, Spencer SA et al (1985) Human tumor necrosis factor. Production, purification, and characterization. J Biol Chem 260:2345–2354

Agnello D, Lankford CSR, Bream J et al (2003) Cytokines and transcription factors that regulate T helper cell differentiation: new players and new insights. J Clin Immunol 23(3):147–161

Aird WC (2003) The role of the endothelium in severe sepsis and multiple organ dysfunction syndrome. Blood 101(10):3765–3777

Alimzhanov MB, Kuprash DV, Kosco-Vilbois MH, Luz A, Turetskaya RL et al (1997) Abnormal development of secondary lymphoid tissues in lymphotoxin beta-deficient mice. Proc Natl Acad Sci USA 94:9302–9307

Allen SJ, Crown SE, Handel TM (2007) Chemokine: receptor structure, interactions, and antagonism. Annu Rev Immunol 25:787–820

Almeida A, Legrand N, Papiernik M et al (2002) Homeostasis of peripheral CD4+ T cells: IL-2R alpha and IL-2 shape a population of regulatory cells that controls CD4+ T cell numbers. J Immunol 169:4850–4860

Antonysamy MA, Thomson AW (1999) FLT3 ligand (FL) and its influence on immune reactivity. Cytokine 12:87–100

Appasamy PM (1992) IL 7-induced T cell receptor-gamma gene expression by pre-T cells in murine fetal liver cultures. J Immunol 149:1649–1656

Arend WP (1993) Interleukin-1 receptor antagonist. Adv Immunol 54:167–227

Arend WP (2003) The balance between IL-1 and IL-1Ra in disease. Cytokine Growth Factor Rev 13(4–5):323–340

Asadullah K, Sabat R, Friedrich M et al (2004) Interleukin-10: an important immunoregulatory cytokine with major impact on psoriasis. Curr Drug Targets Inflamm Allergy 3(2):185–192

Aspinall R, Henson S, Pido-Lopez J, Ngom PT (2004) Interleukin-7: an interleukin for rejuvenating the immune system. Ann NY Acad Sci 1019:116–122

Bach EA, Aguet M, Schreiber RD (1997) The IFN gamma receptor: a paradigm for cytokine receptor signaling. Annu Rev Immunol 15:563–591

Bartlett NW, Buttigieg K, Kotenko SV, Smith GL (2005) Murine interferon lambdas (type III interferons) exhibit potent antiviral activity in vivo in a poxvirus infection model. J General Virol 86:1589–1596

Bazan JF (1990) Structural design and molecular evolution of a cytokine receptor superfamily. Proc Natl Acad Sci USA 87(18):6934–6938

Beadling C, Slifka MK (2006) Regulation of innate and adaptive immune responses by the related cytokines IL-12, IL-23, and IL-27. Arch Immunol Ther Exp 54:15–24

Beebe AM, Cua DJ, de Waal Malefyt R (2003) The role of interleukin-10 in autoimmune disease: systemic lupus erythematosus (SLE) and multiple sclerosis (MS). Cytokine Growth Factor Rev 13(4–5):403–412

Beutler B, Milsark IW, Cerami AC (1985) Passive immunization against cachectin/tumor necrosis factor protects mice from lethal effect of endotoxin. Science 229(4716):869–871

Bhatia M, Davenport V, Cairo MS (2007) The role of interleukin-11 to prevent chemotherapy-induced thrombocytopenia in patients with solid tumors, lymphoma, acute myeloid leukemia and bone marrow failure syndromes. Leuk Lymphoma 48(1):9–15

Bodmer JL, Burns K, Schneider P, Hofmann K, Steiner V et al (1997) TRAMP, a novel apoptosis-mediating receptor with sequence homology to tumor necrosis factor receptor 1 and Fas(Apo-1/CD95). Immunity 6:79–88

Bodmer JL, Schneider P, Tschopp J (2002) The molecular architecture of the TNF superfamily. Trends Biochem Sci 27:19–26

Bogdan C, Vodovotz Y, Nathan C (1991) Macrophage deactivation by interleukin 10. J Exp Med 174:1549–1555

Bonder CS, Finlay-Jones JJ, Hart PH (1999) Interleukin-4 regulation of human monocyte and macrophage interleukin-10 and interleukin-12 production. Role of a functional interleukin-2 receptor gamma-chain. Immunology 96:529–536

Bowcock AM, Kidd JR, Lathrop GM, Daneshvar L, May LT et al (1988) The human interferon-beta 2/hepatocyte stimulating factor/interleukin-6 gene: DNA polymorphism studies and localization to chromosome 7p21. Genomics 3:8–16

Bowie MB, Kent DG, Copley MR, Eaves CJ (2007) Steel factor responsiveness regulates the high self-renewal phenotype of fetal hematopoietic stem cells. Blood 109(11):5043–5048

Brasel K, Escobar S, Anderberg R, de Vries P, Gruss H-J, Lyman SD (1995) Expression of the flt3 receptor and its ligand on hematopoietic cells. Leukemia 9:1212–1218

Brombacher F, Kastelein RA, Alber G (2003) Novel IL-12 family members shed light on the orchestration of Th1 responses. Trends Immunol 24:207–212

Brown MA, Hural J (1997) Functions of IL-4 and control of its expression. Crit Rev Immunol 17:1–32

Budagian V, Bulanova E, Orinska Z et al (2004) Reverse signaling through membrane-bound interleukin-15. J Biol Chem 279:42192–42201

Carswell EA, Old LJ, Fiore N, Schwartz MK (1975) An endotoxin-induced serum factor that causes necrosis of tumors. Proc Natl Acad Sci USA 72:3666–3670

Chatila TA (2004) Interleukin-4 receptor signaling pathways in asthma pathogenesis. Trends Mol Med 10(10):493–499

Chinnaiyan AM, O'Rourke K, Yu GL, Lyons RH, Garg M et al (1996) Signal transduction by DR3, a death domain-containing receptor related to TNFR-1 and CD95. Science 274: 990–992

Cohen S, Bigazzi PE, Yoshida T (1974) Similarities of T cell function in cell-mediated immunity and antibody production. Cell Immunol 12:150–159

Collison LW, Workman CJ, Kuo TT, Boyd K, Wang Y et al (2007) The inhibitory cytokine IL-35 contributes to regulatory T cell function. Nature 450:566–569

Colobran R, Pujol-Borrell R, Armengol MP, Juan M (2007) The chemokine network. I. How the genomic organization of chemokines contains clues for deciphering their functional complexity. Clin Exp Immunol 148:208–217

Cooper M, Bush J, Fehniger T et al (2002) In vivo evidence for a dependence on interleukin-15 for natural killer cell survival. Blood 100:3633–3638

Croft M (2009) The role of TNF superfamily members in T-cell function and diseases. Nat Rev Immunol 9(4):271–285

Curti A, Tafuri A, Ricciardi MR et al (2002) Interleukin-11 induces proliferation of human T-cells and its activity is associated with downregulation of p27(kip1). Haematologica 87(4):373–380

Cytokine: more than a new word, a new concept proposed by Stanley Cohen thirty years ago (2004) Cytokine 28:242–247

de Vries JE, Zurawski G (1995) Immunoregulatory properties of IL-13: its potential role in atopic disease. Int Arch Allergy Immunol 106(3):175–179

Debets R, Savelkoul HF (1994) Cytokine antagonists and their potential therapeutic use. Immunol Today 15(10):455–458

Dessein A, Kouriba B, Eboumbou C et al (2005) Interleukin-13 in the skin and interferon-gamma in the liver are key players in immune protection in human schistosomiasis. Immunol Rev 201:180–190

Dinarello C (1996) Biological basis for interleukin-1 in disease. Blood 87:2095–2147

Dinarello CA (1998) Interleukin-1 beta, interleukin-18 and the interleukin-1 beta converting enzyme. Ann NY Acad Sci 856:1–11

Dinarello CA (2009) Immunological and inflammatory functions of the interleukin-1 family. Annu Rev Immunol 27:519–550

Dinarello CA, Wolff SM (1982) Molecular basis of fever in humans. Am J Med 72(5): 799–819

Dinarello C, Arend W, Sims J, Smith D, Blumberg H et al (2010) IL-1 family nomenclature. Nat Immunol 11(11):973

Dittel BN, LeBien TW (1995) The growth response to IL-7 during normal human B cell ontogeny is restricted to B-lineage cells expressing CD34. J Immunol 154:58–67

Dong C (2008) Th17 cells in development: an updated view of their molecular identity and genetic programming. Nat Rev Immunol 8(5):337–348

Dumonde DC (1986) Work of the Lymphokine Standardization Subcommittee of the International Union of Immunological Societies. Lymphokine Res 5(Suppl 1):S1–S5

Dumonde DC, Wolstencroft RA, Panayi GS, Matthew M, Morley J, Howson WT (1969) "Lymphokines": non-antibody mediators of cellular immunity generated by lymphocyte activation. Nature 224(5214):38–42

Dumoutier L, Renauld J (2002) Viral and cellular interleukin-10 (IL-10)-related cytokines: from structures to functions. Eur Cytokine Netw 13:5–15

Eskdale J, Kube D, Tesch H, Gallagher G (1997) Mapping of the human IL10 gene and further characterization of the 5 flanking sequence. Immunogenetics 46:120–128

Fallon PG, Jolin HE, Smith P, Emson CL, Townsend MJ et al (2002) IL-4 induces characteristic Th2 responses even in the combined absence of IL-5, IL-9 and IL-13. Immunity 17:7–17

Fan X, Sibalic V, Niederer E, Wüthrich RP (1996) The proinflammatory cytokine interleukin-12 occurs as a cell membrane-bound form on macrophages. Biochem Biophys Res Commun 225(3):1063–1067

Fernandez E, Lolis E (2002) Structure, function, and inhibition of chemokines. Annu Rev Pharmacol Toxicol 42:469–499

Feuerer M, Hill JA, Mathis D, Benoist C (2009) Foxp3+ regulatory T cells: differentiation, specification, subphenotypes. Nat Immunol 10(7):689–695

Frankenstein Z, Alon U, Cohen IR (2006) The immune-body cytokine network defines a social architecture of cell interactions. Biol Direct 1:32. doi:10.1186/1745–6150–1–32

Gadina M, Hilton D, Johnston JA, Morinobu A, Lighvani A et al (2001) Signaling by type I and II cytokine receptors: ten years after. Curr Opin Immunol 13(3):363–373

Gaffen SL, Liu KD (2004) Overview of interleukin-2 function, production and clinical applications. Cytokine 28(3):109–123

Geissler EN, Liao M, Brook JD, Martin FH, Zsebo KM et al (1991) Stem cell factor (SCF), a novel hematopoietic growth factor and ligand for c-kit tyrosine kinase receptor, maps on human chromosome 12 between 12q14.3 and 12qter. Somat Cell Mol Genet 17(2):207–214

Glauser MP (1997) Cytokines in severe infections: from pathophysiology to clinical applications. Eur Cytokine Netw 8(3):317–318

Gonda TJ, D'Andrea RJ (1997) Activating mutations in cytokine receptors: implications for receptor function and role in disease. Blood 89(2):355–369
Goriely S, Neurath M, Goldman M (2008) How microorganisms tip the balance between interleukin-12 family members. Nat Rev Immunol 8:81–86
Gray PW, Goeddel DV (1982) Structure of the human immune interferon gene. Nature 298(5877):859–863
Gresser I (2007) Interferon: an unfolding tale. J Interferon Cytokine Res 27(6):447–452
Gumbiner BM (1996) Cell adhesion: the molecular basis of tissue architecture and morphogenesis. Cell 84:345–357
Hamza T, Barnett JB, Li B (2010) Interleukin 12 a key immunoregulatory cytokine in infection applications. Int J Mol Sci 11:789–806
Hardy MP, Owczarek CM, Jermiin LS, Ejdebäck M, Hertzog PJ (2004) Characterization of the type I interferon locus and identification of novel genes. Genomics 84:331–345
Harrington LE, Hatton RD, Mangan PR et al (2005) Interleukin 17-producing CD4+ effector T cells develop via a lineage distinct from the T helper type 1 and 2 lineages. Nature Immunol 6(11):1123–1132
Hehlgans T, Pfeffer K (2005) The intriguing biology of the tumour necrosis factor/tumour necrosis factor receptor superfamily: players, rules and the games. Immunology 115(1):1–20
Henney CS, Gaffney J, Bloom BR (1974) On the relation of products of activated lymphocytes to cell-mediated cytolysis. J Exp Med 140(3):837–852
Huising MO, Stet RJ, Savelkoul HF, Verburg-van Kemenade BM (2004) The molecular evolution of the interleukin-1 family of cytokines; IL-18 in teleost fish. Dev Comp Immunol 28(5): 395–413
Huising MO, Kruiswijk CP, Flik G (2006) Phylogeny and evolution of class-I helical cytokines. J Endocrinol 189:1–25
Ihle JN (1995) Cytokine receptor signalling. Nature 377(6550):591–594
Irwin MR (2008) Human psychoneuroimmunology: 20 years of discovery. Brain Behav Immun 22(2):129–139
Isaacs A, Lindermann J (1957) Virus interference. I. The interferon. Proc R Soc Lond B Biol Sci 147(927):258–267
Ishihara K, Hirano T (2003) IL-6 in autoimmune disease and chronic inflammatory proliferative disease. Cytokine Growth Factor Rev 13(4–5):357–368
Izuhara K, Arima K, Yasunaga S (2003) IL-4 and IL-13: their pathological roles in allergic diseases and their potential in developing new therapies. Curr Drug Targets Inflamm Allergy 1(3): 263–269
Jelkmann W (2007) Erythropoietin after a century of research: younger than ever. Eur J Haematol 78(3):183–205
Jones SA (2005) Directing transition from innate to acquired immunity: defining a role for IL-6. J Immunol 175(6):3463–3468
Kaushansky K (2006) Lineage-specific hematopoietic growth factors. N Engl J Med 354(19):2034–2045
Kishimoto T (2006) Interleukin-6: discovery of a pleiotropic cytokine. Arthritis Res Ther 8(Suppl2):2–14
Kishimoto T (2010) IL-6: from its discovery to clinical applications. Int Immunol 22(5):347–352
Kishimoto T, Akira S, Narazaki M, Taga T (1995) Interleukin-6 family of cytokines and gp130. Blood 86:1243–1254
Kostyal DA, Beezhold DH, Hall RE (1991) Differentiation-inducing cytokine P48 exists in a membrane-associated form. J Immunol 147(3):893–898
Krejci P, Prochazkova J, Bryja V, Kozubik A, Wilcox WR (2009) Molecular pathology of the fibroblast growth factor family. Hum Mutat 30(9):1245–1255
Kwon B, Kim BS, Cho HR, Park JE, Kwon BS (2003) Involvement of tumor necrosis factor receptor superfamily(TNFRSF) members in the pathogenesis of inflammatory diseases. Exp Mol Med 35(1):8–16

Lata S, Raghava GPS (2008) CytoPred: a server for prediction and classification of cytokines. Prot Engin, Design Select 21(4):279–282

Lebel-Binay S, Berger A, Zinzindohoué F, Cugnenc P-H, Thiounn N et al (2000) Interleukin-18: biological properties and clinical implications. Eur Cytokine Netw 11(1):15–26

Lee JS, Campbell HD, Kozak CA, Young IG (1989) The IL-4 and IL-5 genes are closely linked and are part of a cytokine gene cluster on mouse chromosome 11. Somat Cell Mol Genet 15(2):143–152

Leonard WJ, Lin JX (2000) Cytokine receptor signaling pathways. J Allergy Clin Immunol 105(5):877–888

Lingel A, Weiss TM, Niebuhr M, Pan B, Appleton BA et al (2009) Structure of IL-33 and its interaction with the ST2 and IL-1RAcP receptors-insight into heterotrimeric IL-1 signaling complexes. Structure 17(10):1398–1410

Liu YJ (2005) IPC: professional type 1 interferon-producing cells and plasmacytoid dendritic cell precursors. Annu Rev Immunol 23:275–306

Lodolce JP, Burkett PR, Koka RM et al (2003) Regulation of lymphoid homeostasis by interleukin-15. Cytokine Growth Factor Rev 13(6):429–439

Luster AD (1998) Chemokines – chemotactic cytokines that mediate inflammation. N Engl J Med 338:436–445

Malek T, Bayer A (2004) Tolerance, not immunity, crucially depends on IL-2. Nat Rev Immunol 4:665–674

Mariathasan S, Newton K, Monack D, Vucic D, French D et al (2004) Differential activation of the inflammasome by caspase-1 adaptors ASC and Ipaf. Nature 430(6996):213–218

Martinez-Moczygemba M, Huston DP (2003) Biology of common beta receptor-signaling cytokines: IL-3, IL-5, and GM-CSF. J Allergy Clin Immunol 112(4):653–665

Moore KW, O'Garra A, de Waal Malefyt R, Vieira P, Mosmann TR (1993) Interleukin-10. Annu Rev Immunol 11:165–190

Moroz A, Eppolito C, Li Q et al (2004) IL-21 enhances and sustains CD8+ T cell responses to achieve durable tumor immunity: comparative evaluation of IL-2, IL-15, and IL-21. J Immunol 173:900–909

Morris CF, Young IG, Hapel AJ (1990) Molecular and cellular biology of interleukin-3. Immunol Ser 49:177–214

Moseley TA, Haudenschild DR, Rose L, Reddi AH (2003) Interleukin-17 family and IL-17 receptors. Cytokine Growth Factor Rev 14:155–174

Moser B, Loetscher P (2001) Lymphocyte traffic control by chemokines. Nat Immunol 2:123–128

Mosmann TR, Coffman RL (1989) TH1 and TH2 cells: different patterns of cytokine secretion lead to different functional properties. Annu Rev Immunol 7:145–173

Murdoch C, Finn A (2000) Chemokine receptors and the role in inflammation and infectious disease. J Am Soc Hematol 95(10):3032–3043

Murphy KM, Reiner SL (2002) The lineage decisions of helper T cells. Nat Rev Immunol 2: 933–944

Musso T, Colosso L, Zucca M et al (1999) Human monocytes constitutively express membrane-bound, biologically active, and interferon-gamma-upregulated interleukin-15. Blood 93: 3531–3539

Naismith JH, Sprang SR (1996) Tumor necrosis factor receptor superfamily. J Inflamm 47(1–2): 1–7

Nemunaitis J (1997) A comparative review of colony-stimulating factors. Drugs 54(5):709–729

Niedbala W, Wei XQ, Cai B, Hueber AJ, Leung BP et al (2007) IL-35 is a novel cytokine with therapeutic effects against collagen-induced arthritis through the expansion of regulatory T cells and suppression of Th17 cells. Eur J Immunol 37(11):3021–3029

Nishimoto N (2006) Interleukin-6 in rheumatoid arthritis. Curr Opin Rheumatol 18(3): 277–281

Novelli F, Casanova J-L (2004) The role of IL-12, IL-23 and IFN-γ in immunity to viruses. Cytokine Growth Factor Rev 15:367–377

Oppenheim JJ, Feldmann M, Durum SK, Hirano T, Vilcek J, Nicola NA (2000) The Online Cytokine Reference Database. http://apresslp.gvpi.net/apcyto/lpext.dll?f=templates&fn=main-h.htm&2.0

O'Sullivan LA, Liongue C, Lewis RS, Stephenson SE, Ward AC (2007) Cytokine receptor signaling through the Jak-Stat-Socs pathway in disease. Mol Immunol 44(10):2497–2506

Ozaki K, Leonard WJ (2002) Cytokine and cytokine receptor pleiotropy and redundancy. J Biol Chem 277(33):29355–29358

Ozaki K, Kikly K, Michalovich D, Young PR, Leonard WJ (2000) Cloning of a type I cytokine receptor most related to the IL-2 receptor beta chain. Proc Natl Acad Sci USA 97(21):11439–11444

Parrish-Novak J, Dillon S, Nelson A et al (2000) Interleukin 21 and its receptor are involved in NK cell expansion and regulation of lymphocyte function. Nature 408:57–63

Paul WE, Kishimoto T, Melchers E et al (1992) Nomenclature for secreted regulatory proteins of the immune system (interleukins). WHO-IUIS Nomenclature Subcommittee on Interleukin Designation. Clin Exp Immunol 88:367

Pestka S, Krause CD, Sarkar D, Walter MR, Shi Y, Fisher PB (2004) Interleukin-10 and related cytokines and receptors. Annu Rev Immunol 22:929–979

Proulx F, Joyal JS, Mariscalco MM, Leteurtre S, Leclerc F, Lacroix J (2009) The pediatric multiple organ dysfunction syndrome. Ped Crit Care Med 10(1):2–22

Puri RK (1995) Structure and function of interleukin-4 and its receptor. Cancer Treat Res 80: 143–185

Ransohoff RM, Liu L, Cardona AE (2007) Chemokines and chemokine receptors: multipurpose players in neuroinflammation. Int Rev Neurobiol 82:187–204

Rouveix B (1997) Clinical pharmacology of cytokines. Eur Cytokine Netw 8(3):291–293

Schäfer A, Schulz C, Eigenthaler M, Fraccarollo D, Kobsar A et al (2004) Novel role of the membrane-bound chemokine fractalkine in platelet activation and adhesion. Blood 103: 407–412

Schmitz J, Owyang A, Oldham E, Song Y, Murphy E et al (2005) IL-33, an interleukin-1-like cytokine that signals via the IL-1 receptor-related protein ST2 and induces T helper type 2-associated cytokines. Immunity 23(5):479–490

Schroder K, Hertzog PJ, Ravasi T, Hume DA (2004) Interferon-gamma: an overview of signals, mechanisms and functions. J Leukoc Biol 75(2):163–189

Schultz U, Kaspers B, Staeheli P (2004) The interferon system of non-mammalian vertebrates. Dev Comp Immunol 28:499–508

Schuster B, Kovaleva M, Sun Y, Regenhard P, Matthews V et al (2003) Signaling of human ciliary neurotrophic factor (CNTF) revisited. The interleukin-6 receptor can serve as an alpha-receptor for CTNF. J Biol Chem (US) 278(11):9528–9535

Schwantner A, Dingley AJ, Ozbek S, Rose-John S, Grötzinger J (2004) Direct determination of the interleukin-6 binding epitope of the interleukin-6 receptor by NMR spectroscopy. J Biol Chem (US) 279(1):571–576

Sen GC (2001) Viruses and interferons. Ann Rev Microbiol 55:255–281

Shortman K, Naik SH (2007) Steady-stat and inflammatory dendritic-cell development. Nat Rev Immunol 7(1):19–30

Simon D, Braathen L, Simon H (2004) Eosinophils and atopic dermatitis. Allergy 59:561–570

Sims J, Nicklin M, Bazan J et al (2001) A new nomenclature for IL-1-family genes. Trends Immunol 22:536–537

Smallwood PM, Munozsanjuan I, Tong P, Macle JP, Hendry SHC et al (1996) Fibroblast growth factor (FGF) homologous factors – new members of the FGF family implicated in nervous-system development. Proc Natl Acad Sci USA 93(18):9850–9857

Smith KA (1988) Interleukin-2: inception, impact, and implications. Science 240(4856):1169–1176

Smith K (2001) Interleukin 2 immunotherapy. In: Austen F, Burakoff S, Rosen F, Strom T (eds) Therapeutic immunology, 2nd edn. Blackwell Science, Malden, MA, pp 240–250

Smith KA (2006) The structure of IL2 bound to the three chains of the IL2 receptor and how signaling occurs. Med Immunol 5(3):1–5

So T, Lee SW, Croft M (2006) Tumor necrosis factor/tumor necrosis factor receptor family members that positively regulate immunity. Int J Hematol 83(1):1–11

Stockinger B, Veldhoen M (2007) Differentiation and function of Th17 T cells. Curr Opin Immunol 19(3):281–286

Taga T, Kishimoto T (1992) Cytokine receptors and signal transduction. FASEB J 6:3387–3396

Tang P, Hung MC, Klostergaard J (1996) Human pro-tumor necrosis factor is a homotrimer. Biochemistry 35(25):8216–8225

Tavernier J, Cornelis S, Devos R, Guisez Y, Plaetinck G, Van der Heyden J (1995) Structure/function analysis of human interleukin 5 and its receptor. Agents Actions Suppl 46:23–34

Tesmer LA, Lundy SK, Sarkar S, Fox DA (2008) Th17 cells in human disease. Immunol Rev 223:87–113

The Cytokines Web. http://cmbi.bjmu.edu.cn/cmbidata/cgf/CGF_Database/cytweb/targets/index.html

Thomson AW, Lotze MT (2003) The cytokine handbook, 4th edn. Academic, Amsterdam

Throm MJ (2008) A review of the pharmacology and pharmacotherapy of colony-stimulating factors. J Infus Nurs 31(5):295–306

Tieri P, Valensin S, Latora V, Castellani GC, Marchiori M et al (2005) Quantifying the relevance of different mediators in the human immune cell network. Bioinformatics 21:1639–1643

Tominaga A, Takaki S, Hitoshi Y, Takatsu K (1992) Role of the interleukin 5 receptor system in hematopoiesis: molecular basis for overlapping function of cytokines. Bioessays 14(8): 527–533

Toy D, Kugler D, Wolfson M, Vanden Bos T, Gurgel J et al (2006) Cutting edge: interleukin 17 signals through a heteromeric receptor complex. J Immunol 177(1):36–39

Tracey K (2007) Physiology and immunology of the cholinergic antiinflammatory pathway. J Clin Invest 117:289–296

Trinchieri G (2003) Interleukin-12 and the regulation of innate resistance and adaptive immunity. Nat Rev Immunol 3:133–146

Tuck DP, Cerretti DP, Hand A, Guha A, Sorba S, Dainiak N (1994) Human macrophage colony-stimulating factor is expressed at and shed from the cell surface. Blood 84:2182–2188

Uzé G, Schreiber G, Piehler J, Pellegrini S (2007) The receptor of the type I interferon family. Curr Top Microbiol Immunol 316:71–95

Vgontzas AN, Bixler EO, Lin HM et al (2005) IL-6 and its circadian secretion in humans. Neuroimmunomodulation 12(3):131–140

Vilcek J (2003) Novel interferons. Nat Immunol 4:8–9

Vilcek J (2007) My fifty years with interferon. J Interferon Cytokine Res 27(7):535–542

Waldmann TA (2006) The biology of interleukin-2 and interleukin-15: implications for cancer therapy and vaccine design. Nat Rev Immun 6(8):595–601

Walz A, Peveri P, Aschauer H et al (1987) Purification and amino acid sequencing of NAF, a novel neutrophil-activating factor produced by monocytes. Biochem Biophys Res Commun 149:755–761

Ward PA, Remold HG, David JR (1969) Leukotactic factor produced by sensitized lymphocytes. Science 163(871):1079–1081

Wills-Karp M (2000) The gene encoding interleukin-13: a susceptibility locus for asthma and related traits. Respir Res 1:19–23

Wilson KP, Black JA, Thomson JA et al (1994) Structure and mechanism of interleukin-1 beta converting enzyme. Nature 370:270–273

Yasukawa M, Yakushijin Y, Hasegawa H, Miyake M, Hitsumoto Y et al (1993) Expression of perforin and membrane-bound lymphotoxin (tumor necrosis factor-beta) in virus-specific CD4+ human cytotoxic T-cell clones. Blood 81:1527–1534

Yoshimura T, Matsuhima K, Tanaka S et al (1987) Purification of a human monocyte-derived neutrophil chemotactic factor that has peptide sequence similarity to other host defence cytokines. Proc Natl Acad Sci USA 94:9233–9237

Yu J, Gaffen S (2008) Interleukin-17: a novel inflammatory cytokine that bridges innate and adaptive immunity. Front Biosci 13:170–177

Zdanov A, Schalkhini C, Wlodawer A (1996) Crystal-structure of human interleukin-10 at 1.6 angstrom resolution and a model of a complex with its soluble receptor. Prot Sci 5(10):1955–1962

Zeng R, Spolski R, Finkelstein SE, Oh S, Kovanen PE et al (2005) Synergy of IL-21 and IL-15 in regulating CD8+ T cell expansion and function. J Exp Med 201:139–148

Zlotnik A, Yoshie O (2000) Chemokines: a new classification system and their role in immunity. Immunity 12(2):121–127

Zlotnik A, Yoshie O, Nomiyama H (2006) The chemokine and chemokine receptor superfamilies and their molecular evolution. Genome Biol 7:243

Chapter 2
Mechanical Stretching of Cells of Different Tissues: The Role of Mediators of Innate Immunity

Leonid V. Kovalchuk, Marina V. Khoreva, Anna Nikonova, Viktor E. Kazanski, Ilya Lozinsky, Irina N. D'iakonova, and Andre Kamkin

Abstract The current review describes the modern conce of how the mechanical stretch (MS) affects cytokine and chemokine production by the cells of different tissues (cardiomyocytes, fibroblasts, smooth muscle cells, endothelial cells and pulmonary cells). Released mediators regulate cell functions such as synthesis of the extracellular matrix proteins, proliferation, apoptosis and others, in autocrine or paracrine manner. Endogenous cytokines (tumor necrosis factor α (TNFα), insulin-like growth factor 1 (IGF-1), vascular endothelial growth factor (VEGF), interleukin 6 (IL-6) and others) produced in myocardium in response to mechanical stretch (MS) may trigger pathological processes resulting in myocyte growth, apoptosis and formation of reactive fibrosis. Mechanical load is associated with increase in tissue volume and tissue remodeling. This review provides data about changes in expression of cytokine receptors expression, as well as receptors of innate immunity (TLRs), in response to MS. TLR4 is expressed on the surface of cells of the heart, including cardiomyocytes, smooth muscle cells and endothelial cells. Cyclic MS enhances expression of TLR4 in cultured neonatal rat cardiomyocytes. Excessive MS may result in alterations of cell structure and functions, composition of extracellular matrix (ECM), and promote development of pathological conditions such as hypertrophy, fibrosis, atherosclerosis, osteoporosis, etc. Searching for drugs with targeted action working at the extracellular, membrane and intracellular levels and which will improve the consequences of excessive MS is of undoubted interest and is actual for the treatment of many human pathologies.

Keywords Mechanical stretch · Cytokines · Toll-like receptors · Cardiomyocites · Fibroblasts · Smooth muscle cell · Endothelial cell · Lung

L.V. Kovalchuk (✉)
Department of Immunology & Department of Fundamental and Applied Physiology, Russian State Medical University, 117997 Moscow, Russia
e-mail: lkov@rsmu.ru

A. Kamkin, I. Kiseleva (eds.), *Mechanical Stretch and Cytokines*, Mechanosensitivity in Cells and Tissues 5, DOI 10.1007/978-94-007-2004-6_2,

2.1 Introduction

Both mechanosensitivity and mechanotransduction are fundamental physiological processes which are responsible for sensing of mechanical forces and their transformation into electrical or (and) biochemical signals. The cell membrane interfaces with the external medium or with neighbouring cells and gets mechanical stress mainly in the form of stretch or compression due to general tissue deformation. Mechanical stimuli trigger different electrophysiological and biochemical responses. In living tissues, mechanical sensing and stress-induced responses in cells are defined by several cellular components including extracellular matrix, integrins, the cytoskeleton, and mechanically gated ion channels. The linkage between these cellular components undoubtedly plays a critical role. Moreover, mechanical deformation of the cells may lead to the release of cytokines and pro-inflammatory mediators, resulting in excessive immune system activation. In vivo and vitro, it has been shown that the stretch of cells can cause the secretion of cytokines. In this book chapter, we summarize the recent findings about the cellular and molecular effect of mechanical stretch on *cardiomyocites*, *fibroblasts*, *vascular smooth muscle cells*, *endothelial cells* and *lung*.

However for other cells production of cytokines during cell stretch is shown (Blain, 2011). For example, compression activated an integrin-dependent interleukin-4 (IL-4) autocrine/paracrine loop in normal *chondrocytes*, whereas mechanotransduction was mediated via the activation of an IL-1β autocrine/paracrine loop in osteoarthritis cells (Salter et al., 2002). IL-1 has been shown to increase F-actin amounts in chondrocytes (Pritchard and Guilak, 2006) suggesting that a finely balanced interplay of soluble factors e.g. cytokines may exist to regulate cytoskeletal element dynamics and hence organization. A whole genome array demonstrated that IL-1β inhibited the mRNA expression of both tubulin and vimentin, as well as the LIM protein FHL2 which is associated with the actin cytoskeleton (Joos et al., 2008).

An increase of immunostaining of interleukin (IL)-1beta and tumor necrosis factor (TNF) alpha in *cartilage* (see for review Sanchez et al., 2011) is observed after joint impact of canine patellae (Pickvance et al., 1993). In vitro studies show that exposure of cartilage explants to mechanical strain of high amplitude leads to the synthesis of inflammatory mediators. Intermittent compression and shear stress stimulates nitric oxide (NO), prostaglandin (PG) E_2 and IL-6 productions through the activation of respectively inducible NO synthase (iNOS), cyclooxygenase (COX)-2 and IL-6 gene expression (Mohtai et al., 1996; Das et al., 1997; Fermor et al., 2002). In contrast, biochemical signals could have anti-inflammatory effects (see for review Sanchez et al., 2011). Cyclic tensile stress counteracts the IL-1 dependent chondrocyte catabolic response. Cyclic tensile stress reduce the IL-1beta dependent synthesis of IL-6 and PGE_2 (Mathy-Hartert et al., 2008) in chondrocytes. The IL-1beta dependent stimulation of NO/iNOS, PGE_2/COX-2 protein and gene expression is decreased by cyclic tensile stress application (Agarwal et al., 2004; Gassner et al., 1999). The inhibition of Agg synthesis due to IL-1 is also reversed. These effects are mediated by Nuclear Factor (NF)-κB. Its translocation to the nucleus is prevented by cyclic tensile stress. When magnitude of the strain is increased to 15% of deformation, these anti-inflammatory effects are nullified (Agarwal et al., 2004). IL-1 dependent mRNA expression of MMP-3, 7, 8, 9, 13, 16, 17 and 19 of chondrocytes are also decreased by cyclic tensile stress. Expression of MMP-2, 11, 14 and Tissue Inhibitor of Metalloprotease -1, 2 and 3 are not affected (Deschner et al., 2006).

It should be noted that in the literature of the last years, considerable attention is given to a detailed study of various signaling pathways, membrane changes, mechanosensitivity and other intracellular changes during MS (Chien, 2007; Lehoux et al., 2006; Li and Xu, 2007; Ruwhof et al., 2000; Shyu, 2009; Wang and Thampatty, 2006). However, to date the large number of the facts has been accumulated confirming that innate immunity factors may influence on mechanobiological properties of the cells under MS conditions. Autocrine and paracrine effects of soluble mediators in the process of cardiomyocyte MR detected more than 15 years ago (Sadoshima and Izumo, 1993). The most important scientific achievement of the recent years was the discovery of membrane proteins, called Toll-like receptors (TLRs). Their function is tightly connected with such immune system mediated processes as production of cytokines (predominantly pro-inflammatory), triggering of inflammation, activation of adaptive immunity, tissue regeneration, etc. TLRs are a family of molecules that play a critical role in innate immunity. Among TLRs, TLR4, the receptor for lipopolysaccharide, is the most frequently observed and best-characterized receptor. In addition to playing a role in heart failure and myocardial ischemia/reperfusion injury, TLR4 has been found to play a role in neointimal formation, atherosclerosis and stroke. The ability of cells to respond via activation of TLRs is critical in innate immune sensing in most tissues, but also extends to more general danger sensing, e.g. of oxidative stress, in cardiomyocytes. Of particular importance factors of innate immunity (TLR4, receptors TNFα, expression of TNFα and others) are in the heart tissue in situ, because in addition to cardiomyocytes in the structure of the body consists of many satellite cells, as fibroblasts, macrophages, and others. These cells are also able to participate in the development of innate immune responses and inflammation by expressing TLRs, receptors, cytokines, chemokines, etc.

2.2 Cardiomyocytes and Mechanical Stretching

Physiological and pathological mechanisms of mechanical stretching (MS) action are of particular interest in understanding of cardiovascular system functioning under normal and pathological conditions. Prolonged hemodynamic stimulation by high pressure and volume overload leads to functional and structural changes in ventricular cardiomyocytes what in turn results in cardiac hypertrophy. The process of cardiac hypertrophy is often accompanied by remodeling characterized by cardiomyocytes loss, interstitial fibrosis and collagen deposition, leading to decreased elasticity and increased risk for heart failure (Paradis et al., 2000).

Cardiac hypertrophy might be induced by both mechanical loading and action of humoral factors such as angiotensin II (AngII), endothelin 1 (ET-1) and others. In recent years, a growing body of work on mediators of innate immunity supports their action on cardiovascular system.

MS is one of the most important stimulus in the development of cardiovascular system. To exclude the involvement of neurohumoral effects on cardiomyocytes to study the cellular and molecular mechanisms of signal transduction and gene expression there were developed different methods of biomechanical load. At least three models of mechanical stress have been used for cardiomyocytes: mechanical stretch (Shyu et al., 1995), aortocaval shunt (McNicholas-Bevensee et al., 2006) and baloon dilation of the left ventricle in the isolated heart (Otani et al., 2006).

It was proved by in vitro experiments for cardiomyocytes that MS optimally modulate cell parameters such as growth, mechanosensitivity, apoptosis, electric remodeling, and alterations in gene expression, autocrine-paracrine and number of other important effects. Under physiological conditions mechanical cell stretching provides optimal function of vital organs (heart, lungs, blood vessels, muscles, etc.). Mechanobiological investigations indicate a significant role of MS in the regulation of key cellular functions including proliferation, differentiation, apoptosis, gene induction and synthesis of the proteins that are necessary for maintenance of tissue homeostasis.

However, excessive MS may result in alterations of cell structure and functions, composition of extracellular matrix (ECM), and promote development of pathological conditions such as hypertrophy, fibrosis, atherosclerosis, osteoporosis, etc. Mechanical load is associated with increase in tissue volume and tissue remodeling. In response to MS cellular metabolism changes and ECM synthesis increases.

MS causes hypertrophy of atrial (Saygili et al., 2007) and ventricular (Yamazaki et al., 1996) cardiomyocytes. The hypertrophic effect caused by mechanical stretching is characterized by increase in cell size, enhanced sarcomeric organization and expression of hypertrophy markers, e.g. atrial natriuretic factor (ANF), brain natriuretic peptide (BNP) and myosin heavy chain (β-MHC) (Frank et al., 2008).

Torsoni et al. (2005) reported that Ras homolog gene family, member A (RhoA) and focal adhesion kinase (FAK) play a key role in the regulation of hypertrophic genetic program induced by MS. ERK1/2 kinase mediates the effect of RhoA in maintenance of cardiac hypertrophy by regulating activation of GATA4 transcription factor, expressed in heart. Both, PKC-α and -δ are important regulators that mediate activation of Rho GTPases and MAPKs in stretching-induced hypertrophic processes (Pan et al., 2005). MS-induced cardiac hypertrophy is comparable to that induced by α-adrenergic agonist, phenylephrine. Hypertrophy induced by MS also implies activation of the early genes, c-jun, c-fos, c-myc and skeletal muscle α-actin (Liang and Gardner, 1999; Sadoshima and Izumo, 1993).

During MS only a total 185 genes of more than 28,000 analyzed genes are significantly regulated (Frank et al., 2008). Specific genes induced by 24 h MS of neonatal cardiomyocytes were identified by DNA microarray method. Genes that are activated include ANF, BNP, skeletal muscle α-actin, HSP70, proto-oncogene c-myc, CKS-2 (cyclin-dependent kinase regulatory subunit), and also cardioprotective genes, such as metallthionein-1 and HO-1 (haem oxidase-1), cytokines, growth and differentiation factors. At the same time expression of lipocalin, Cx40, cell adhesion molecules and phospholipase genes is suppressed.

MS induced gene expression may be significantly attenuated by AngII receptor blocker irbesartan, what matches with well-known key role of AngII in induction of gene expression during MS (Sadoshima et al., 1993). Zou et al. (2004) have demonstrated in their work that short-time mechanical stretching (5–8 min) activates AT1 receptor (AngII receptor type 1) without involving of AngII. To prove involvement of AngII in MS induced expression of genes in cardiomyocytes they use more than 4 h MS period. Different duration of MS triggers different mechanisms of cardiac hypertrophy and may influence on pattern and type of gene expression in cardiomyocytes. MS at a physiological level does not induce cell damage of cardiac myocytes; however, pathological stretch or stretch elongation of more than 20% may induce myocyte apoptosis. Apoptosis of cardiomyocytes is an important pathological factor in the transition from hypertensive heart disease to heart failure. Low amplitude MS of 5% induces cardiomyocyte hypertrophy, whereas a high amplitude MS of 25% induces cardiomyocyte apoptosis (Pimentel et al., 2001). Cardiomyocyte apoptosis and the cell death signal gene bax are significantly induced by high amplitude MS through ROS production and MAPK activation. ROS production is also stretch-amplitude-dependent. This MS-induced apoptosis is reproduced in an isolated heart model by balloon dilation of the left ventricle during the early phase of reperfusion (Otani et al., 2006).

Although the exact mechanism of how mechanical potential is converted into biochemical signals is not fully understood, different possible mechanisms are proposed. One of the most well studied mechanisms involves ECM, cell surface integrins and cytoskeleton. Cell adhesion to the ECM via integrins that transduce the siganal to the cytoskeleton provides a structural basis for mechanical signal transmission from the ECM to the cell (Juliano and Haskill, 1993; Maniotis et al., 1997). The major cellular components involved in the mechanotransduction mechanisms are the integrins, cytoskeleton, G proteins, receptor tyrosine kinases (RTKs), mitogen-activated protein kinases (MAPKs), and stretch-activated ion channels (SAC) (Davies, 1995; Haga et al., 2007; Ingber, 1997; Katsumi et al., 2004; Orr et al., 2006; Shyy and Chien, 2002). ROS together with the other soluble factors capable to regulate production of ECM proteins in paracrine and autocrine manner play an important role in the signaling process. These soluble factors include cytokines, chemokines, growth factors and other mediators (Ali and Schumacker, 2002; Waters, 2004).

It should be noted that in the literature of the last years, considerable attention is given to a detailed study of various signaling pathways, membrane changes, mechanosensitivity and other intracellular changes during MS. However, to date the large number of the facts has been accumulated confirming that innate immunity factors may influence on mechanobiological properties of the cells under MS conditions. The most demonstrative are the studies using the MS models on cardiomyocytes in vitro. It is known that cardiomyocytes of human and experimental animals (mice, rats) express components of the innate immunity including Toll-like receptors (TLRs), e.g. TLR4, tumor necrosis factor α receptors (TNFR) and can produce TNFα and other cytokines in response to various stimuli (Al-Lamki et al., 2009; Boyd et al., 2006). Studies conducted by our scientific group demonstrated the important role of TNFα and interleukin 6 (IL-6) in the formation of mechanoinduced electric abnormalities in the model of the MS in rat cardiomyocytes.

Pattern recognition receptors (PRR) are the major component of the innate immunity system that respond to conservative molecular structures of microorganisms, pathogen-associated molecular patterns (PAMP). TLRs are signaling PRR and considered to be the key receptors of innate immunity (Takeda et al., 2003). TLRs belong to the family of transmembrane receptors that are characterized structurally by the presence of a leucine-rich repeat (LRR) domain in their extracellular domain which is involved in ligand recognition and TLR dimerization. Cytoplasmic portion of TLR, a TIR (Toll/IL-1 receptor) domain has a high homology with the IL-1 receptor (IL-1R) family and mediates interaction between receptors and transduction molecules (Armant and Fenton, 2002).

TLR ligand binding and recognition trigger intracellular signaling pathways resulting in production of numerous effector molecules: proinflammatory cytokines, chemokines, co-stimulatory molecules, etc. (Takeda and Akira, 2005). Amount and pattern of released cytokines depend on the type of stimulated TLR and inducing ligand. Ligands recognized by TLRs are divided to several groups: various components of microorganisms (lipopolysaccharide, peptidoglycan, zymosan and others), artificially synthetized molecules with structure close to microbial molecules and endogenous ligands (heat shock proteins HSP60, HSP70, hyaluronan, heparan sulphate, fibrinogen, fibronectin, β2 defensin). Leukocytes providing the first line of defense – neutrophils, macrophages, dendritic cells, endothelial and epithelial cells of the mucosal tissues, express TLRs. However, many of the non-hematopoietic cells of the organism express TLRs (cardiomyocytes, fibroblasts, hepatocytes and others). Part of TLRs is localized on the cell surface (TLR1, TLR2, TLR4, TLR5, TLR6 and TLR11), and the rest of TLR are localized intracellular to the endosomal membranes (TLR3, TLR7, TLR8 and TLR9) (Sandor and Buc, 2005).

Endogenous ligands ("danger signals") produced as the result of cell destruction and inflammation (HSP, minimally modified (oxidized) LDL, etc.) are the markers of tissue damage and can activate TLRs. Induction of proinflammatory cytokine (TNFα and others) production mediated by TLRs may contribute to the development of chronic inflammation, pathogenesis of various diseases, e.g. Crohn's disease, atherosclerosis (Chen et al., 2007; Frantz et al., 2007; Guillot et al., 2002; Li and Sun, 2007; Misch and Hawn, 2008; Sitrin et al., 1998; Tsan and Gao, 2004).

It was demonstrated that not only specific ligands of TLRs, but cyclic MS can upregulate expression of TLR4 in cultured neonatal rat cardiomyocytes (Shyu et al., 2010). TLR4 is expressed on the surface of cells of the heart, including cardiomyocytes, smooth muscle cells and endothelial cells. Increased expression of TLR4 is observed in human and animal cardiomyocytes in heart failure. TLR4 can modulate contractility of myocytes and plays an important role in myocardium disfunction during bacterial sepsis and heart hypertrophy. TNFα and IL-1 are linked to TLR4 to mediate postishemic cardiac disfunction. Production of TNFα mediated by TLR4 ligands is increased in patients with acute myocardial infarction (Koval'chuk et al., 2008). Cyclic MS enhances expression of TLR4, receptors of AngII and TNFα. Expression of TLR4 is mediated by AngII, TNFα, p38 kinase and NF-κB signaling pathways. At the same time TLR4 mediated adhesion of monocytes to cardiomyocytes increases in conditions of MS. Authors suppose that expression of TLR4 induced by cyclic MS may have the protective effect under hemodynamic overload of cardiomyocytes (Shyu et al., 2010).

Factors of innate immunity are especially important in heart tissue in situ as far as not only cardiomyocytes but also satellite cells, e.g. fibroblasts, macrophages and other cells form the organ structure. These cells are also involved in the development of innate immune and inflammatory reactions expressing TLRs, cytokine and chemokine receptors, etc.

Since MS induce secretion and synthesis of bioactive molecules in cardiomyocytes, these molecules may act on the adjacent cells as well as on the secreting cells. Autocrine and paracrine effects of soluble mediators during MS were described more than 15 years ago (Sadoshima et al., 1993).

Released mediator proteins while interacting with specific receptors activate intracellular signaling pathways and induce cellular and molecular effects in cardiomyocytes. Among capable mediators during MS of cardiomyocytes cytokines have particular significance. Different pathological processes resulting in myocyte growth, apoptosis and formation of reactive fibrosis are triggered by endogenous cytokines produced in myocardium in response to MS. The increasing attention is paid to TNFα, IGF-1, VEGF, IL-6 and other mediators. (Pimenteal et al., 2002; Shyu et al., 2005a; Wang et al., 2007a).

The special place in regulation of functional activity of cardiomyocytes is given to TNFα. TNFα is a pleiotropic cytokine that plays as known an important role in tissue damage. At physiological levels, TNFα exerts cytoprotective action within myocardium, mainly antioxidant and antiapoptotic effects. At pathological concentrations TNFα stimulates apoptosis of myocytes, reactive fibrosis, and contribute to involvement of cardiomyocytes in inflammatory process. It is important that levels of proinflammatory cytokines in circulation (plasma levels) may be elevated in case of many cardiovascular diseases. Cardiomyocytes are considered to be the major target of TNFα (Kaser et al., 2003). TNFα reduce contractile function of myocardium, decreasing cardiac output. Action of TNFα is mediated by activation of metalloproteinases responsible for the remodeling of ECM. Then increased synthesis of collagen is observed, loss of its supportive function and development of hypertrophy, with the following development of ventricular dilatation and decreased myocardium contractility, declaring by diastolic disfunction. Thus, in heart failure TNFα induces various effects: myocardium remodeling, negative ionotropic action and vascular disfunction.

Cultured rat neonatal cardiomyocytes in conditions of cyclic MS produce not only TNFα but also resistin that can act in heart as autocrine regulator factor. (Wang et al., 2007a) Induction of resistin is mediated by TNFα through ERK/MAP kinase and NF-κB signaling pathways. Upregulation of resistin results in decreased glucose uptake in cardiomyocytes.

The moderate 20% mechanical stretch resulted in the elevation of IL-18 mediated by endothelin A receptor and angiotensin subtype 1 receptor (Naka et al., 2008). Maximal level of cytokine expression achieved 36 h after the stretch. IL-18 as a proinflammatory cytokine with multiple biological functions produced by immune and non-immune cells plays a great role in pathogenesis of different diseases, including myocardial infarction, ischemia and myocarditis. Increased circulatory level of IL-18 is one of the risk factors for the development of cardiovascular diseases. Hence, IL-18 induction in cardiomyocytes may result in disruption of cardiac functions in autocrine and paracrine manner.

Anti-inflammatory cytokine transforming growth factor β1 (TGF-β) may play a central role in heart defense during hypertrophic response supporting the recovery of normal myocardium functions (van Wamel et al., 2002). Cardiomyocytes and fibroblasts in vitro can produce TGF-β. In cardiomyocytes TGF-β production increased after 4 h MS, while in fibroblasts increase of TGF-β was observed only after 24 h of stretching. Secreted TGF-β acting either in autocrine or paracrine manner upregulates it's own mRNA in cardiomyocytes. TGF-beta mRNA expression in stationary, non-stretched, cardiomyocytes is increased by AngII release from cardiomyocytes that had been stretched for 30–60 min (Ruwhof et al., 2000).

It is known that IL-13 promotes development of fibrosis and apoptosis. Elevated plasma levels of IL-13 is revealed in patients with acute myocardial infarction and chronic heart failure. High affinity IL-13 receptor, IL-13R2, acts as trap-receptor. Cyclic MS of cardiomyocytes leads to increased expression of IL-13 mRNA, and also to a more than 5 fold increase in expression of IL13R2 mRNA in comparison with control (Nishimura et al., 2008).

Induced by mechanical stretch cytokines, chemokines and other growth factors realize autocrine and paracrine effects acting directly on cardiomyocytes and other cells and regulate their functions.

2.3 Fibroblasts and Mechanical Stretching

Influence of MS on fibroblast of various origins is relatively well studied. In many respects MS effects on fibroblasts are similar in spite of different tissue localization. It is well known that human fibroblasts express a wide variety of cytokines, including IL-1β, IL-4, IL-6, IL-8, IL-10, IL-12, TNFα, interferon γ (IFN-γ), CCL20 and macrophage colony stimulating factor (M-CSF) (Ruiz et al., 2007).

Cardiac fibroblasts (CF) play a central role in regulation of the ECM metabolism in the heart. They regulate synthesis of the ECM components (collagen and others), enzymes involved in both degradation of the ECM and inhibition of the ECM degradation. CF are usually stay in quiescent condition with long life cycle thus cell population of CF stays rather stable. However, under pathological conditions, e.g. heart failure or myocardial infarction, CF proliferate actively, their number increases disrupting the balance between cell populations in the heart. Proliferation of CF and synthesis of the ECM play an important role in the cardiac wound healing after myocardial infarction caused by acute coronary occlusion. Cardiac fibroblasts have been termed "sentinel" cells, because they can sense changes in chemical, mechanical, and electrical signals in the heart and develop the appropriate response. One of the major functions of the CF is the synthesis and degradation of the ECM to provide a 3D network for cardiomyocytes and other cells of the heart to ensure proper cardiac structure and function.

The major part of stimuli function as activators of CF, but chronic stimulation can lead to pathological remodeling. Various studies have demonstrated that mechanical stimulation of fibroblasts results in increased production of ECM components, expression of receptors for ECM as well as cytokines, chemokines, growth factors and other signaling molecules. Particular attention is paid to AngII, TGF-β, ET1, TNFα that could regulate synthesis of ECM and its deposition (MacKena et al., 2000).

CFs are the main source of IL-1β in the damaged heart. IL-1β acting through its receptor, IL-1R, suppresses fibroblast proliferation, mediates changes in the ECM by decreasing synthesis of type I and II collagen, and upregulating production of matrix metalloproteinases (MMP). IL-1β can induce expression of IL-6 and angiotensin subtype 1 receptor as it shown in rat model.

Recent studies have indicated that IL-6 is a critical component of cell-cell interactions that occur between cardiomyocytes and fibroblasts (Banerjee et al., 2009). The biological activity of IL-6 is regulated by binding to the receptor complex composed of α chain of IL-6 receptor and gp130 subunit and following activation of several signaling pathways (Szabo-Fresnais et al., 2010). IL-6 and the gp130/JAK/STAT signaling are crucial for proper cardiac function and cardioprotection. However, continuous activation of gp130 results in myocardial hypertrophy and increased expression of IL-6 (Hirota et al., 1995). The majority of IL-6 appears to be secreted by cardiac fibroblasts and to a lesser extent by cardiomyocytes, but both cell types can respond to IL-6 stimulation. IL-6 production can be regulated by IL-1β, TNF-α and AngII.

In vitro studies of rat CF showed that 20% MS at 1.5 Hz during 48 h leads to the enhanced response of the cells to TGF-β and IGF-1 stimulating procollagen mRNA levels by 4.3 and three-fold, respectively, in comparison to cells treated only with growth factors (Butt and Bishop, 1997). Higher magnitude of tension (stretching and compression) essentially increases TGF-β1 level in rat CF (Lee et al., 1999). Studies on the model of tendon fibroblasts have shown that cyclic biaxial stretch at 1 Hz and amplitude of 5% increases secretion of TGF-β1, basic fibroblast growth factor (bFGF), IL-6, TNF-α, and platelet-derived growth factor (PDGF), which are known as stimulators of cell proliferation, differentiation and formation of the ECM (Skutek et al., 2001). Similar studies of MS on smooth muscle cells have demonstrated time-dependent increase in expression of TGF-β1 and PDGF mRNAs that mediate induction of cell proliferation and the ECM proteins production, such as type I and IV collagen and fibronectin (Li et al., 1998; Joki et al., 2000). Moreover, cyclic biaxial MS (20%, 1.5 Hz) increased the stimulatory effect of PDGF on procollagen production of pulmonary fibroblasts by 96% (Bishop et al., 1998). One possible mechanism by which MS enhance the stimulatory effects of soluble factors such as PDGF may be that mechanical loads induce the overexpression of cell surface receptors as well as the augmentation of tyrosine phosphorylation of adhesive proteins (Tanabe et al., 2000). However, the presence of growth factors is crucial for cell proliferation in response to mechanical load (Banes et al., 1995).

Thus, intimate intracellular interactions regulated by receptor and mediatory components of innate immunity are established in the myocardium and in many other tissues. As a result of excess MS the major cellular elements (e.g., cardiomyocytes) and satellite cells (e.g., fibroblasts) may form pathological lesions, the nature of which depends on the predominant destruction of interacting cells.

2.4 Vascular Smooth Muscle Cells and Mechanical Stretch

The pulsatile nature of blood pressure and flow creates hemodynamic stimuli in the forms of cyclic stretch and shear stress, which exert continuous influences on the constituents of the blood vessel wall. Vascular smooth muscle cells (VSMCs) use multiple sensing mechanisms to detect the mechanical stimulus resulting from pulsatile stretch and transduce it into intracellular signals. Cyclic stretch and shear stress lead to modulations of gene expression and cellular functions, e.g., proliferation, differentiation, apoptosis, migration, and vascular remodeling.

In various studies different kinds of VSMCs were used: human venous or arterial VSMCs or cells from different species of animals (mouse, rat, rabbits, swine and others), and this resulted in sometimes controversial findings. Most of the studies used in vitro models. According to Shyu (2009) the cellular functions induced by in vitro stretch may not really represent cellular functions in vivo. Further and detailed studies are needed to elucidate the real effect and mechanism of mechanical stress on VSMCs functions.

Multiple modulatory effects of MS on synthesis and secretion of biologically active molecules by VSMCs were demonstrated in different models. These molecules may influence on nearby cells or act in autocrine manner. In particular, PDGF have an autocrine function in VSMCs under stretching conditions (Wilson et al., 1993). Mechanical stretch using portal vein SMCs induces ET-1 release and promotes the synthesis of smooth-muscle-specific proteins by a mechanism requiring an intact cytoskeleton (Zeidan et al., 2003). Mechanical stretch also stimulates autocrine IGF-1 production from arterial and venous VSMCs (Cheng and Du, 2007; Standley et al., 1999). MS stimulates proliferation of venous SMCs mediated by activation of IGF-1 and IGF-1 receptor. Inactivation of IGF-1 receptor blocks MS-induced proliferation of VSMCs. The IGF-1 receptor level is increased in the neointima of vein grafts, and IGF-1 receptor deletion reduces neointima formation in vein grafts (Cheng and Du, 2007).

Transforming growth factor α (TGF-α) has been shown to modulate the NF-κB activation and vascular remodeling under stress (Lemarie et al., 2006). The intracellular signaling pathways of the autocrine and paracrine effects of TGF-α under mechanical stress involve ROS (reactive oxygen species) and NF-κB. Cyclic stretch enhanced GADD153 expression in cultured rat VSMCs. The stretch-induced GADD153 is mediated by TNF-α, at least in part, through the JNK and AP-1 (activator protein-1) pathway (Cheng et al., 2008). These findings suggest that GADD153 plays a role in stretch-induced VSMC apoptosis.

AngII and TGF-β1 have autocrine and paracrine actions on DDR-2 (discoidin domain receptor-2) expression in VSMCs under MS (Shyu et al., 2005b). DDR-2 can regulate cell proliferation and ECM remodelling mediated by MMP activities. The intracellular signaling pathways of stretch-induced DDR-2 involve the p38 MAPK and Myc pathways.

In the endothelial cells (ECs)–smooth muscle cells (SMCs) co-culture model, SMCs secrete IL-1β and IL-6 after application of shear stress, resulting in the inhibition of E-selectin expression via the inhibition in SMC activation of IRAK (IL-1-receptor-associated kinase)/Gp-130 (glycoprotein-130), JNK/p38 MAPK and NF-κB (Lehoux, 2006; Haga et al., 2007).

MS plays a role in the vascular remodeling. The signaling pathways that modulate vascular remodeling in response to MS include ROS, NO, NF-κB, epidermal growth factor receptor (EGFR), MAPK and PKC (Kouri and Eickelberg, 2006). Mechanical force can be transduced via ROS-dependent autocrine and paracrine EGFR activation, and may regulate VSMC proliferation and synthetic activity through the NF-κB pathway. One of potential specific targets for vascular remodeling induced by MS is supposed to be a TGF-α. In vivo, increased hemodynamic forces in a model of hypertension induced by AngII infusion, activation of NF-κB and associated cell proliferation and wall thickening are reduced in TGF-α-mutant mice compared with wild-type controls (Lemarie et al., 2006). According to Albinsson and Hellstrand (2007) remodeling of SMCs to stretch requires a dynamic cytoskeleton. The stabilization of actin filaments is essential for the growth and synthesis of contractile proteins in response to physiological levels of MS. MS enhances expression of hypoxia-inducible factor-1 (HIF-1)α in VSMCs. This is associated with an increase in the expression of vascular endothelial growth factor (VEGF) through transcriptional regulation (Chang et al., 2003). The transient increase in VEGF and HIF-1α gene expression induced by MS may be linked to the blood vessel pathology, including atherosclerosis and hypertension. The induction of VEGF and the HIF-1α genes by MS may contribute in vascular remodeling.

SMC are the major components of the blood vessel wall and play an important role in the formation of atherosclerotic lesions. SMCs are able to sense changes in hemodynamic forces and MS induces phosphorylation and activation of PDGF receptor and subsequent triggering of intracellular signaling pathways associated with cell proliferation (Wernig et al., 2003). MS can lead to the apoptosis induction in SMC and this process is mediated by activation of pathways that involve integrin β1 receptor, PKC-δ and p38 MAPK kinase (Li et al., 2000).

Cyclic MS, the increased mechanical forces sensed by SMCs caused by altered hemodynamic blood flow contribute to the development of atherosclerosis. Elevation of mRNA and protein levels of IL-6 is observed in the SMCs as the result of activation of Ras/Rac/p38 MAPK/NF-κB/NF-IL6 pathways. Activation of PKC-δ downregulates production of IL-6. Secreted IL-6 accumulates in intracellular space and initiates inflammatory response what can lead to the formation of early atherosclerotic lesions (Zampetaki et al., 2005). These findings provide a link between mechanical forces and the inflammatory response in the vessel wall and thus highlight the molecular mechanisms of biomechanical stress that contribute to the initiation of atherosclerosis.

2.5 Endothelial Cells and Mechanical Stretch

Vascular endothelial cells (ECs) play important role in regulating circulatory functions. Mechanical stimuli, including the stretch and shear stress resulting from circulatory pressure and blood flow, modulate EC functions by activating mechanosensors, signaling pathway and gene and protein expressions. Alterations of normal hemodynamic load can contribute to cardiovascular diseases, such as hypertension, intimal hyperplasia, vascular restenosis and atherosclerosis. (Cummins et al., 2007; Paravicini and Touyz, 2006). ECs are continually exposed to shear stress. Laminar shear stress, the frictional force created by blood flow, exerts different effects on structure and function of ECs: it changes morphology, metabolism, gene expression and can change the phenotype of ECs. Data characterizing response of ECs to shear stress and cyclic stretch have been accumulated from different in vivo, ex vivo, and in vitro experiments. Studies addressing the effect of laminar shear stress on ECs have shown that it influences significantly on the expression of genes associated with vascular remodeling and cellular functions such as proliferation, apoptosis, migration (Kakisis et al., 2004; Lehoux et al., 2006) DNA mircoarrays have been used to analyse a large number of genes in ECs exposed to shear stress (Andersson et al., 2005; Wasserman et al., 2002). The analysis showed that about 3% (600 out of 20,000) of genes respond to shear stress (Ohura et al., 2003). Proteomic analysis shows that a broad spectrum of proteins is altered by shear stress (Wang et al., 2007b). These proteins included transcriptional regulators, enzymes, protein kinases, GPCRs (G-protein-coupled receptors), cytokines, adhesion molecules (Shyu, 2009).

ECs produce various growth factors and cytokines and shear stress leads to the elevation in the synthesis of PDGF, heparin binding-epidermal growth factor (HB-EGF), basic fibroblast growth factor (bFGF), TGFβ, IL-1, IL-6, granulocyte-macrofage colony stimulating factor (GM-CSF) (Hsieh et al., 1991; Kosaki et al., 1998; Mitsumata et al., 1993; Morita et al., 1993; Malek et al., 1993; Ohno et al., 1995; Sterpetti et al., 1993).

The shear stress can cause vasculoprotective effects that have been reported more then 10 years ago (Traub and Berk, 1998). Recently it was revealed that bone morphogenetic protein (BMP) is a member of TGFβ cytokine family, it induces anti-inflammatory effect in ECs and is involved in atherogenesis. BMP is activated in atherosclerosis prone lesion areas of the vessels and may contribute to vascular calcification and the development of atherosclerotic plaques (Csiszar et al., 2006). It is demonstrated that laminar shear stress activates cAMP and PKA kinase and supresses expression of BMP type 4 in coronary athery ECs (Csiszar et al., 2007). These data support the anti-atherogenic and vascularprotective effects of shear stress because BMP-4 triggers activation of endothelium, disfunction, hypertension and vascular calcification.

Shear stress leads to the activation of transcriptional factors AP-1, NF-κB, GATA-6 and Krüppel-like factor (KLF)-2 in ECs (Huddleson et al., 2006; Khachigian et al., 1997; Korenaga et al., 2001; Lin et al., 1997; Shyy et al., 1995; Sokabe et al., 2004). The transcription factor KLF-2 is an important mediator of anti-inflammatory and antitrombotic properties of the endothelium. In vitro experiments prolonged shear stress stabilizes KLF-2 mRNA and induces KLF-2 protein expression, especially in the presence of proinflammatory cytokine TNFα. In vitro studies on ECs showhs that shear stress in the presence of proinflammatory cytokine has more pronounced atheroprotective effect than statins, well known lipid-lowering agents (van Thienen et al., 2006). This finding also supports the atheroprotective effect of prolonged shear stress on ECs. Complement-inhibitory protein CD59 can be activated through KLF-2 in venous and arterial ECs indicating vascular protective action of shear stress during complement-mediated damage (Kinderlerer et al., 2008).

Endothelial inflammation is one of the major reasons for the development of atherosclerotic lesions in the vessels. ECs exposed to disturbed flow experience oxidative stress, increasing expression of markers of inflammation and attraction of monocytes as early features of atherosclerosis (Harrison et al., 2006). Anti-inflammatory and anti-oxidant defense are crucial for the protection of cellular macromolecules and prevention of atherosclerosis progression. Reactive oxygen species (ROS), including superoxide anion (O^{2-}), hydrogen peroxide (H_2O_2), are produced by ECs and act as secondary messengers and also mediate development of different diseases, e.g. atherosclerosis (Finkel, 1998; Laurindo et al., 1994). Laminar shear stress upregulates expression of PRXs (peroxiredoxins) critical antioxidants in the ECs (Mowbray et al., 2008). Laminar shear stress induces transcriptional activity of ERK-5 and PPAR-γ and downregulates expression of adhesion molecules on ECs. Laminar shear stress increases NO production activating eNOS (endothelial NO-synthase) expression through ERK-5 and thus inhibiting formations of ROS (Woo et al., 2008). Shear stress also induces PKA in the ECs promoting high activity of eNOS and subsequent production of NO (Zhang et al., 2006), activates antioxidant genes and transcriptional factor Nrf-2 (nuclear factor-erythroid 2 p45 subunit related factor 2), which is a major transcriptional factor for EC redox homeostasis (Dai et al., 2007).

Cyclic stretch is regarded as a repetitive mechanical deformation of vascular cells, as it rhythmically distends and relaxes with the cardiac cycle. Multiple intracellular events triggered by distension similar to those triggered by shear stress, e.g. increase of intracellular Ca^{2+}, triphosphoinositol, diglyceride and activation of PKC (Naruse and Sokabe, 1993; Rosales and Sumpio, 1992; Cohen et al., 1997). Big amount of biologically active mediators regulated by cyclic stretch are expressed in ECs, including IL-8, TGFβ, VEGF, MCP-1 (monocyte chemotactic protein-1) (Ali and Schumacker, 2002; Wung et al., 2001; Zheng et al., 2001). In vitro studies demonstrate that cyclic stretch increases expression of NOS, ET-1, ICAM-1 (Awolesi et al., 1995; Cheng et al., 1996; Wang et al., 1995). Expression of gene products in the ECs depends on the physiological and pathological (elevated) level of mechanical stress. Recently, DNA microarray technology were used to assess the profile of gene expression in human lung ECs depending on the time and amplitude of cyclic stretch (Birukov et al., 2003). The results showed that the cyclic MS under physiological or pathological conditions (at 5 and 18% elongation, respectively) causes various patterns of gene expression involved in signal transduction, cytoskeleton remodeling, cell adhesion, inflammatory reactions and regulation of endothelial barrier function. Significant increase in production of IL-8 observed in ECs under cyclic tension at 15%, while stretching to 6% does not affect the levels of cytokine. (Birukov, 2005).

Biomechanical forces promote development of atherosclerosis by inducing formation of ROS in ECs and activating expression of proatherogenic cytokines. Cyclic stretch enhances TNFα production and MMP expression in HUVECs (human umbilical vein ECs) (Wang et al., 2003). These findings indicate that hemodynamic forces acting in atherosclerosis-resistant and -susceptible areas of the vasculature induce different responses in the vascular wall. Atherosclerotic lesions frequently develop in areas of the vasculature exposed to disturbed flow, whereas areas that experience pulsatile laminar flow are relatively protected from plaque formation

2.6 Cytokine Production by Pulmonary Cells and Mechanical Stretch

Cells of the lung are exposed to different types of the mechanical stress. For example, endothelial cells are exposed to shear stress in conditions of flow, and epithelial cells of the airways and alveoli are subjected to tension and compression during the respiratory cycle. Therefore, many researchers in recent years studied the effects of mechanical stretch on the function of different cell of the airways wall and distal lung, including epithelial, endothelial, smooth muscle cells and fibroblasts (Liu et al., 1999). Cells respond to mechanical stress by changes in intracellular ion concentration, cytoskeletal reorganization and changes in gene expression. Abnormal pulmonary overdistension may cause inflammation due to mechanical activation of macrophages, epithelial and endothelial cells, which may lead to alveolar and endothelial barrier dysfunction, resulting in ventilator-induced lung injury (VILI) syndrome or pulmonary edema (Dos Santos and Slutsky, 2000; Uhlig, 2002). The growing body of experimental evidence confirms that proinflammatory cytokines serve as mediators of VILI (Dreyfuss and Saumon, 1998; Tremblay and Slutsky, 1998), which can cause or burden lung injury by initiating inflammatory reaction (Liu and Slutsky, 1997). The level of TNFα mRNA was increased in the cells isolated from bronchoalveolar lavage fluid (BAL) after conventional ventilation (Takata et al., 1997). Using an ex vivo rat lung ventilation model, Tremblay et al. (1997) demonstrated that injurious ventilation regimens increased BAL concentrations of several cytokines, including TNFα, IL-1β, IL-6, IL-10, macrophage inflammatory protein 2 (MIP-2) and IFNγ. Blockade of the function of pro-inflammatory mediators such as IL-1 or TNFα, attenuated the severity of VILI in animals (Imai et al., 1999; Narimanbekov and Rozycki, 1995).

However, other researchers argue the concept that hyperventilation induces production of proinflammatory cytokines, reporting that ventilation, which severely damages the lungs, does not lead to the release of significant amount of TNFα or IL-1β in the absence of stimulation by lipopolysaccharide (LPS) (Ricard et al., 2001). These controversial observations suggest that type of the model system used to study the cytokine production is very important (Simon, 2001).

To study cellular and molecular mechanisms of ventilation-induced cytokine production in the lung several research groups have used cell culture models to examine the stretch-induced inflammatory response. Pugin et al. (1998) have shown that production of IL-8 by human macrophages increased after 8–32 h of stretch with a frequency of 20 cycles/min; the same regimen enhanced LPS-induced production of TNFα and IL-6 by macrophages. The same stretch regimen applied to human lung epithelial A549 cells did not result in increased IL-8 production. Comparing different stretch regimens influence on IL-8 production by A549 cells revealed that 30% (not 20%) MS at 20 or 40 cycles/min upregulated the level of IL-8 after 12–48 h (Vlahakis et al., 1999). Using different devices for MS, Tsuda et al. (1999) observed that stretch alone did not affect IL-8 production after 8 h; however, in the presence of glass fibers or crocidolite asbestos, stretch significantly increases IL-8 production in cultured A549 cells. MIP-2 is a homologue of human IL-8 in rodents and acts as mediator of neutrophil activation and recruitment (Luster, 1998). To simulate ventilation of preterm newborns and study the effect of MS on MIP-2 production the stretch of organotypic cultured primary fetal rat lung cells was used. Only LPS stimulation, not MS increased the level of MIP-2 mRNA. Pretreatment of the cells with cycloheximide, an inhibitor of protein synthesis, did not inhibit the MS-induced release of MIP-2 (Mourgeon et al., 2000). Effect similar to additive effect of LPS and MS on MIP-2 production in primary rat embryonic lung cells was also observed in the model of IL-8 production by A549 cells using stretch and TNFα (Vlahakis et al., 1999). In clinical practice mechanical ventilation is commonly used in patients with sepsis, bacterial infections or other pathological conditions of respiratory failure. The combined effect of the inflammatory response and mechanical ventilation can influence cytokine production through different mechanisms.

In most of the studies mentioned above mechanical stretch alone did not have a great effect on cytokine gene expression. Most of the studied cytokines were chosen on the basis of animal studies and clinical observations and, thus, the choice was limited to several molecules. Several research groups have used microarray technique combined with bioinformatics analysis to analyze expression of gene related to cytokine, chemokine, inflammatory mediators, or proteins that are involved in intracellular signaling of acute inflammatory responses. Preliminary results have demonstrated that mechanical stretch alone has a profound effect on mRNA levels of multiple genes in human lung epithelial cells. After 1 h of stretch (18% elongation, 30 cycles/min), mRNA levels of more then 35 genes were significantly increased, while mRNA levels of more then 300 genes decreased. Interestingly, most of the commonly studied cytokine genes, such as IL-1β, IL-6, IL-1Rα, and TNF-R1 were not affected by stretch, whereas mRNA levels of IL-1Rβ, IL-1RA, TNF-α and TNF-R2 were decreased. Genes encoding several novel cytokines and inflammatory mediators, which have not been studied in VILI, are significantly up- or downregulated by MS. These results, of course, need to be confirmed by other techniques and need to be examined with animal studies as well as with clinical studies. However, these preliminary data are very encouraging, providing the starting point for further investigation of VILI (Waters et al., 2002).

2.7 Conclusions and Perspectives

Thereby, MS of cellular element of different tissues (cardiomyocytes, fibroblasts, VSMCs, ECs and others) results in activation of various intracellular signaling pathways, regulates gene expression and cell.

Thereby, mechanical stretch of the cellular elements of various tissues (cardiomyocytes, fibroblasts, VSMCs, ECs, and many others) leads to the activation of various intracellular signaling pathways that regulate gene expression and cell function. In the myocardium and in many other tissues intimate intercellular interactions are established under normal conditions; mediator and receptor components of innate immunity play a pivotal role in their regulation. Excessive MR leads to the development of pathological processes, and their reversibility also depends on the parameters of innate immunity. As a result of excessive MS major cellular elements (e.g., cardiomyocytes) and satellite cells (e.g. fibroblasts) may form pathological lesions, the nature of which depends on the pre-emptive destruction of interacting cells.

Searching for drugs with targeted action working at the extracellular, membrane and intracellular levels and which will improve the consequences of cxcessive MS is of undoubted interest and is actual for the treatment of many human pathologies.

Acknowledgements This work was supported by the Russian Foundation for Basic Research (grant No. 09-04-01277-a). AK's Department was supported by Ministry of Education and Science of the Russian Federation. The Order of Ministry of Education and Science of the Russian Federation No. 743 from 01 July 2010, Supplement, Event 4.4, the Period of Financing 2010–2019.

References

Agarwal S, Deschner J, Long P, Verma A, Hofman C, Evans CH, Piesco N (2004) Role of NF-kappaB transcription factors in antiinflammatory and proinflammatory actions of mechanical signals. Arthritis Rheum 50:3541–3548

Albinsson S, Hellstrand P (2007) Integration of signal pathways for stretch-dependent growth and differentiation in vascular smooth muscle cell. Am J Physiol 293:C772–C782

Ali MH, Schumacker PT (2002) Endothelial responses to mechanical stress Where is the mechanosensor? Crit Care Med 30:S198–S206

Al-Lamki RS, Brookes AP, Wang J, Reid MJ, Parameshwar J, Goddard MJ, Tellides G, Wan T, Min W, Pober JS, Bradley JR (2009) TNF receptors differentially signal and are differentially expressed and regulated in the human heart. Am J Transplant 9:2679–2696

Andersson M, Karisson L, Svensson PA, Ulfhammer E, Ekman M, Jernås M, Carlsson LM, Jern S (2005) Differential global gene expression response patterns of human endothelium exposed to shear stress and intraluminal pressure. J Vasc Res 42:441–452

Armant MA, Fenton MJ (2002) Toll-like receptors family of pattern-recognition receptors in mammals. Genome Biol 3(8):3011.1–3011.6

Awolesi MA, Sessa WC, Sumpio BE (1995) Cyclic strain upregulates nitric oxide synthase in cultured bovine aortic endothelial cells. J Clin Invest 96:1449–1454

Banerjee I, Fuseler JW, Intwala AR, Baudino TA (2009) IL-6 loss causes ventricular dysfunction, fibrosis, reduced capillary density and dramatically alters the cell populations of the developing and adult heart. Am J Physiol Heart Circ Physiol 296:H1694–H1704

Banes AJ, Tsuzaki M, Hu P, Brigman B, Brown T, Almekinders L, Lawrence WT, Fischer T (1995) PDGF-BB, IGF-I and mechanical load stimulate DNA synthesis in avian tendon fibroblasts in vitro. J Biomech 28(12):1505–1513

Birukov KG (2005) Regulation of vascular endothelial cell signal transduction and phenotype by mechanical factors. In: Shepro D (ed) Microvascular research: biology and pathology. Academic, San Diego, CA, pp 209–215

Birukov KG, Jacobson JR, Flores AA, Ye SQ, Birukova AA, Verin AD, Garcia JG (2003) Magnitude-dependent regulation of pulmonary endothelial cell barrier function by cyclic stretch. Am J Physiol 285:L785–L797

Bishop JE, Butt R, Dawes K, Laurent G (1998) Mechanical load enhances the stimulatory effect of PDGF on pulmonary artery fibroblast procollagen synthesis. Chest 114(Suppl 1):25S

Boyd JH, Mathur S, Wang Y, Bateman RM, Walley KR (2006) Toll-like receptor stimulation in cardiomyoctes decreases contractility and initiates an NF-κB dependent inflammatory response. Cardiovasc Res 72:384–393

Butt RP, Bishop JE (1997) Mechanical load enhances the stimulatory effect of serum growth factors on cardiac fibroblast procollagen synthesis. J Mol Cell Cardiol 29(4):1141–1151

Chang H, Shyu KG, Wang BW, Kuan P (2003) Regulation of hypoxia-inducible factor-1a by cyclical mechanical stretch in rat vascular smooth muscle cells. Clin Sci 105:447–456

Chen K, Huang J, Gong W, Iribarren P, Dunlop NM, Wang JM (2007) Toll-like receptors in inflammation, infection and cancer. Int Immunopharmacol 7(10):1271–1285

Cheng J, Du J (2007) Mechanical stretch stimulates proliferation of venous smooth muscle cells throgh activation of the insulin-like growth factor-1 receptor. Arterioscler Thromb Vasc Res 27:1744–1751

Cheng JJ, Wung BS, Chao YJ, Wang DL (1996) Cyclic strain enhances adhesion of monocytes to endothelial cells by increasing intercellular adhesion molecule-1 expression. Hypertension 28(3):386–391

Cheng WP, Hung HF, Wang BW, Shyu KG (2008) The molecular regulation of GADD153 in apoptosis of cultured vascular smooth muscle cells by cyclic mechanical stretch. Cardiovasc Res 77:551–559

Chien S (2007) Mechanotransduction and endothelial cell homeostasis: the wisdom of the cell. Am J Physiol 292:H1209–H1224

Cohen CR, Mills I, Du W, Kamal K, Sumpio BE (1997) Activation of the adenylyl cyclase/cyclic AMP/protein kinase A pathway in endothelial cells exposed to cyclic strain. Exp Cell Res 231(1):184–189

Csiszar A, Ahmad M, Smith KE, Labinskyy N, Gao Q, Kaley G, Edwards JG, Wolin MS, Ungvari Z (2006) Bone morphogenetic protein-2 induces proinflammatory endothelial phenotype. Am J Pathol 168:629–638

Csiszar A, Labinskyy N, Smith KE et al (2007) Downregulation of bone morphogenetic protein 4 expression in coronary arterial endothelial cells: role of shear stress and the cAMP/protein kinase A pathway. Arterioscler Thromb Vasc Biol 27:776–782

Cummins PM, Sweeney NO, Killeen MT, Birney YA, Redmond EM, Cahill PA (2007) Cyclic strain-mediated matrix metalloproteinase regulation within the vascular endothelium: a force to be reckoned with. Am J Physiol 292:H28–H42

Dai G, Vaughn S, Zhang Y, Wang ET, Garcia-Cardena G, Gimbrone MA (2007) Biomechanical forces in atherosclerosis-resistant vascular regions regulate endothelial redox balance via phosphoinositiol 3-kinase/Akt-dependnet activation of Nrf2. Circ Res 101:723–733

Das P, Schurman DJ, Smith RL (1997) Nitric oxide and G proteins mediate the response of bovine articular chondrocytes to fluid-induced shear. J Orth Res 15:87–93

Davies PF (1995) Flow-mediated endothelial mechanotransduction. Physiol Rev 75:519–560

Deschner J, Rath-Deschner B, Agarwal S (2006) Regulation of matrix metalloproteinase expression by dynamic tensile strain in rat fibrochondrocytes. Osteoarthritis Cartilage 14:264–272

Dos Santos CC, Slutsky AS (2000) Invited review: mechanisms of ventilator-induced lung injury: a perspective. J Appl Physiol 89:1645–2084

Dreyfuss D, Saumon G (1998) Ventilator-induced lung injury: lessons from experimental studies. Am J Respir Crit Care Med 157:294–323

Fermor B, Weinberg JB, Pisetsky DS, Misukonis MA, Fink C, Guilak F (2002) Induction of cyclooxygenase-2 by mechanical stress through a nitric oxide-regulated pathway. Osteoarthritis Cartilage 10:792–798

Finkel T (1998) Oxygen radicals and signaling. Curr Opin Cell Biol 10:248–253

Frank D, Kuhn C, Brors B, Hanselmann C, Lüdde M, Katus HA, Frey N (2008) Gene expression pattern in biomechanically stretched cardiomyocytes: evidence for a stretch-specific gene program. Hypertension 51:309–318

Frantz S, Ertl G, Bauersachs J (2007) Mechanisms of disease: Toll-like receptors in cardiovascular disease. Nat Clin Pract Cardiovasc Med 4(8):444–454

Gassner R, Buckley MJ, Georgescu H, Studer R, Stefanovich-Racic M, Piesco NP, Evans CH, Agarwal S (1999) Cyclic tensile stress exerts antiinflammatory actions on chondrocytes by inhibiting inducible nitric oxide synthase. J Immunol 163:2187–2192

Guillot L, Balloy V, McCormack FX, Golenbock DT, Chignard M, Si-Tahar M et al (2002) Cutting edge: the immunostimulatory activity of the lung surfactant protein-A involves Toll-like receptor 4. J Immunol 168:5989–5992

Haga JH, Li YS, Chien S (2007) Molecular basis of the effects of mechanical stretch on vascular smooth muscle cells. J Biomech 40:947–960

Harrison DG, Widder J, Grumbach I, Chen W, Webber M, Searles C (2006) Endothelial mechanotransduction, nitric oxide and vascular inflammation. J Int Med 259:351–363

Hirota H, Yoshida K, Kishimoto T, Taga T (1995) Continuous activation of gp130, a signal-transducing receptor component for IL-6-related cytokines, causes myocardial hypertrophy in mice. Proc Natl Acad Sci USA 92:4862–4866

Hsieh HJ, Li NQ, Frangos JA (1991) Shear stress increases endothelial platelet-derived growth factor mRNA levels. Am J Physiol 260(2 Pt 2):H642–H646

Huddleson JP, Ahmad N, Lingrel JB (2006) Up-regulation of the KLF2 transcription factor by fluid shear stress requires nucleolin. J Biol Chem 281(22):15121–15128

Imai Y, Kawano T, Iwamoto S, Nakagawa S, Takata M, Miyasaka K (1999) Intratracheal anti-tumor necrosis factor-α antibody attenuates ventilator-induced lung injury in rabbits. J Appl Physiol 87:510–515

Ingber DE (1997) Tensegrity: The architectural basis of cellular mechanotransduction. Annu Rev Physiol 59:575–599

Joki N, Kaname S, Hirakata M, Hori Y, Yamaguchi T, Fujita T, Katoh T, Kurokawa K (2000) Tyrosine-kinase dependent TGF-beta and extracellular matrix expression by mechanical stretch in vascular smooth muscle cells. Hypertens Res – Clin Exp 23(2):91–99

Joos H, Albrecht W, Laufer S, Reichel H, Brenner RE (2008) IL-1beta regulates FHL2 and other cytoskeleton-related genes in human chondrocytes. Mol Med 14:150–159

Juliano RL, Haskill S (1993) Signal transduction from the extracellular matrix. J Cell Biol 120(3):577–585

Kakisis JD, Liapis CD, Sumpio BE (2004) Effects of cyclic strain on vascular cells. Endothelium 11:17–28

Kaser S, Kaser A, Sandhofer A, Ebenbichler CF, Tilg H, Patsch JR (2003) Resistin messenger-RNA expression is increased by proinflammatory cytokine in vitro. Biochem Biophys Res Commun 309:286–290

Katsumi A, Orr AW, Tzima E, Schwartz MA (2004) Integrins in mechanotransduction. J Biol Chem 279:12001–12004

Khachigian LM, Anderson KR, Halnon NJ, Gimbrone MA Jr, Resnick N, Collins T (1997) Egr-1 is activated in endothelial cells exposed to fluid shear stress and interacts with a novel shear-stress-response element in the PDGF A-chain promoter. Arterioscler Thromb Vasc Biol 17(10):2280–2286

Kinderlerer AR, Ali F, Johns M, Lidington EA, Leung V, Boyle JJ, Hamdulay SS, Evans PC, Haskard DO, Mason JC (2008) KLF-dependent, shear stress-induced expression of CD59: a novel cytoprotective mechanism against complement-mediated injury in the vasculature. J Biol Chem 283:14636–14644

Korenaga R, Yamamoto K, Ohura N, Sokabe T, Kamiya A, Ando J (2001) Sp1-mediated down-regulation of P2X4 receptor gene transcription in endothelial cells exposed to shear stress. Am J Physiol 280(5):H2214–H2221

Kosaki K, Ando J, Korenaga R, Kurokawa T, Kamiya A (1998) Fluid shear stress increases the production of granulocyte-macrophage colony-stimulating factor by endothelial cells via mRNA stabilization. Circ Res 82(7):794–802

Kouri FM, Eickelberg O (2006) Transforming growth factor-alpha, a novel mediator of strain-induced vascular remodeling. Circ Res 99(4):348–350

Koval'chuk LV, Khoreva MV, Varivoda AS, Konstantinova EV, Iudin AA, Mastakova II (2008) Role of innate immunity receptors in development of acute myocardial infarction. Zh Mikrobiol Epidemiol Immunobiol 4:64–68

Laurindo FR, Pedro Mde A, Barbeiro HV, Pileggi F, Carvalho MH, Augusto O, da Luz PL (1994) Vascular free radical release. Ex vivo and in vivo evidence for a flow-dependent endothelial mechanism. Circ Res 74(4):700–709

Lee AA, Delhaas T, McCulloch AD, Villarreal FJ (1999) Differential responses of adult cardiac fibroblasts to in vitro biaxial strain patterns. J Mol Cell Cardiol 31(10):1833–1843

Lehoux S (2006) Redox signaling in vascular responses to shear and stretch. Cardiovasc Res 71:269–279

Lehoux S, Castier Y, Tedgui A (2006) Molecular mechanism of the vascular response to haemodynamic forces. J Int Med 259:381–392

Lemarie CA, Tharaux P, Esposito B, Tedgui A, Lehoux S (2006) Transforming growth factor-*a* mediates nuclear factor *kb* activation in strained arteries. Circ Res 99:434–441

Li H, Sun B (2007) Toll-like receptor 4 in atherosclerosis. J Cell Mol Med 11(1):88–95

Li C, Xu Q (2007) Mechanical stress-initiated signal transduction in vascular smooth muscle cells in vitro and in vivo. Cell Signal 19:881–891

Li Q, Muragaki Y, Hatamura I, Ueno H, Ooshima A (1998) Stretchinduced collagen synthesis in cultured smoothmuscle cells from rabbit aortic media and a possible involvement of angiotensin II and transforming growth factor-beta. J Vasc Res 35(2):93–103

Li C, Hu Y, Sturm G, Wick G, Xu Q (2000) Ras/Rac-Dependent activation of p38 mitogen-activated protein kinases in smooth muscle cells stimulated by cyclic strain stress. Arterioscler Thromb Vasc Biol 20(3):E1–9

Liang F, Gardner DG (1999) Mechanical strain activates BNP gene transcription through a p38/NF-KB-dependent mechanism. J Clin Invest 104:1603–1612

Lin MC, Almus-Jacobs F, Chen HH, Parry GC, Mackman N, Shyy JY, Chien S (1997) Shear stress induction of the tissue factor gene. J Clin Invest 99:737–744

Liu M, Slutsky AS (1997) Anti-inflammatory therapies: application of molecular biology techniques in intensive care medicine. Intensive Care Med 23:718–731

Liu M, Tanswell AK, Post M (1999) Mechanical force-induced signal transduction in lung cells. Am J Physiol 277:L667–L683

Luster AD (1998) Chemokines–chemotactic cytokines that mediate inflammation. N Engl J Med 338:436–445

MacKena D, Summerour SR, Villarreal FJ (2000) Role of mechanical factors in modulating fibroblast function and extracellular matrix synthesis. Card Res 46:257–263

Malek AM, Gibbons GH, Dzau VJ, Izumo S (1993) Fluid shear stress differentially modulates expression of genes encoding basic fibroblast growth factor and platelet-derived growth factor B chain in vascular endothelium. J Clin Invest 92(4):2013–2021

Maniotis AJ, Chen CS, Ingber DE (1997) Demonstration of mechanical connections between integrins, cytoskeletal filaments, and nucleoplasm that stabilize nuclear structure. Proc Natl Acad Sci USA 94(3):849–854

Mathy-Hartert MBS, Sanchez C, Lambert C, Henrotin Y (2008) L'application de forces cycliques d'étirement diminue la production de médiateurs de l'inflammation par les chondrocytes arthrosiques. Rev Rhum 75:999–1000

McNicholas-Bevensee CM, DeAndrade KB, Bradley WE, Dell'Italia LJ, Lucchesi PA, Bevensee MO (2006) Activation of gadolinium-sensitive ion channels in cardiomyocytes in early adaptive stages of volume overload-induced heart failure. Cardiovasc Res 72:262–270

Misch EA, Hawn TR (2008) Toll-like receptor polymorphism and susceptibility to human disease. Clin Sci 114:347–360

Mitsumata M, Fishel RS, Nerem RM, Alexander RW, Berk BC (1993) Fluid shear stress stimulates platelet-derived growth factor expression in endothelial cells. Am J Physiol 265(1 Pt 2): H3–H8

Mohtai M, Gupta MK, Donlon B, Ellison B, Cooke J, Gibbons G, Schurman DJ, Smith RL (1996) Expression of interleukin-6 in osteoarthritic chondrocytes and effects of fluid-induced shear on this expression in normal human chondrocytes in vitro. J Orth Res 14:67–73

Morita T, Yoshizumi M, Kurihara H, Maemura K, Nagai R, Yazaki Y (1993) Shear stress increases heparin-binding epidermal growth factor-like growth factor mRNA levels in human vascular endothelial cells. Biochem Biophys Res Commun 197(1):256–262

Mourgeon E, Isowa N, Keshavjee S, Zhang X, Slutsky AS, Liu M (2000) Mechanical stretch stimulates macrophage inflammatory protein-2 secretion from fetal rat lung cells. Am J Physiol 279:L699–L706

Mowbray AL, Kang DH, Rhee SG, Kang SW, Jo H (2008) Laminar shear stress up-regulates peroxiredoxins (PRX) in endothelial cells: PRX 1 as a mechanosensitive antioxidant. J Biol Chem 283(3):1622–1627

Naka T, Sakoda T, Doi T, Akagami T, Tsujino T, Masuyama T, Ohyanagi M (2008) Mechanical stretch induced interleukin-18 (IL-18) expression through angiotensin subtype 1 receptor (AT1R) and endothelin 1 in cardiomyocytes. Prep Biochem Biotechnol 38:201–212

Narimanbekov IO, Rozycki HJ (1995) Effect of IL-1 blockade on inflammatory manifestations of acute ventilator-induced lung injury in a rabbit model. Exp Lung Res 21:239–254

Naruse K, Sokabe M (1993) Involvement of stretch-activated ion channels in Ca2+ mobilization to mechanical stretch in endothelial cells. Am J Physiol 264(4 Pt 1):C1037–C1044

Nishimura Y, Inoue T, Morooka T, Node K (2008) Mechanical stretch and angiotensin II increase interleukin-13 production and interleukin-13 receptor 2 expression in rat neonatal cardiomyocytes. Circulation 72:647–653

Ohno M, Cooke JP, Dzau VJ, Gibbons GH (1995) Fluid shear stress induces endothelial transforming growth factor beta-1 transcription and production. Modulation by potassium channel blockade. J Clin Invest 95(3):1363–1369

Ohura N, Yamamoto K, Ichioka S, Sokabe T, Nakatsuka H, Baba A, Shibata M, Nakatsuka T, Harii K, Wada Y, Kohro T, Kodama T, Ando J (2003) Global analysis of shear stress-responsive genes in vascular endothelial cells. J Atheroscler Thromb 10:304–313

Orr AW, Helmke BP, Blackman BR, Schwartz MA (2006) Mechanisms of mechanotransduction. Dev Cell 10:11–20

Otani H, Matsuhisa S, Akita Y, Kyoi S, Enoki C, Tatsumi K, Fujiwara H, Hattori R, Imamura H, Iwasaka T (2006) Role of mechanical stress in the form of cardiomyocyte death during the early phase of reperfusion. Circulation 70:1344–1355

Pan J, Singh US, Takahashi T, Oka Y, Palm-Leis A, Herbelin BS, Baker KM (2005) PKC medicates cyclic stretch-induced cardiac hypertrophy through Rho family GTPases and mitogen-activated protein kinases in cardiomyocytes. J Cell Physiol 202:536–553

Paradis P, Dali-Youcef N, Paradis FW, Thibault G, Nermer M (2000) Overexpression of angiotensin II type 1 receptor in cardiomyocytes induces cardiac hypertrophy and remodeling. Proc Natl Acad Sci USA 97:931–936

Paravicini TM, Touyz RM (2006) Redox signaling in hypertension. Cardiovasc Res 71:247–258

Pickvance EA, Oegema TR Jr, Thompson RC Jr (1993) Immunolocalization of selected cytokines and proteases in canine articular cartilage after transarticular loading. J Orth Res 11:313–323

Pimenteal RC, Yamada KA, Kleber AG, Saffitz JE (2002) Autocrine regulation of myocyte Cx43 expression by VEGF. Circ Res 90:671–677

Pimentel DR, Amin JK, Xiao L, Miller T, Viereck J, Oliver-Krasinski J, Baliga R, Wang J, Siwik DA, Singh K, Pagano P, Colucci WS, Sawyer DB (2001) Reactive oxygen species mediate amplitude-dependent hypertrophic and apoptotic responses to mechanical stretch in cardiac myocytes. Circ Res 89:453–460

Pritchard S, Guilak F (2006) Effects of interleukin-1 on calcium signaling and the increase of filamentous actin in isolated and in situ articular chondrocytes. Arthritis Rheum 54: 2164–2174

Pugin J, Dunn I, Jolliet P, Tassaux D, Magnenat JL, Nicod LP, Chevrolet JC (1998) Activation of human macrophages by mechanical ventilation in vitro. Am J Physiol 275:L1040–L1050

Ricard JD, Dreyfuss D, Saumon G (2001) Production of inflammatory cytokines in ventilator-induced lung injury: a reappraisal. Am J Respir Crit Care Med 163:1176–1180

Rosales OR, Sumpio BE (1992) Changes in cyclic strain increase inositol trisphosphate and diacylglycerol in endothelial cells. Am J Physiol 262(4 Pt 1):C956–C962

Ruiz C, Pérez E, García-Martínez O, Díaz-Rodríguez L, Arroyo-Morales M, Reyes-Botella C (2007) Expression of cytokines IL-4, IL-12, IL-15, IL-18, and IFNgamma and modulation by different growth factors in cultured human osteoblast-like cells. J Bone Miner Metab 25(5):286–292

Ruwhof C, van Wamel AET, Egas JM, van der Laarse A (2000) Cyclic stretch induces the release of growth promoting factors from cultured neonatal cardiomyocytes and cardiac fibroblasts. Mol Cell Biochem 208:89–98

Sadoshima J, Izumo S (1993) Mechanical stretch rapidly activates multiple signal transduction pathways in cardiac myocytes: potential involvement of an autocrine/paracrine mechanism. EMBO J. 12:168–192

Sadoshima J, Xu Y, Slayter HS, Izumo S (1993) Autocrine release of angiotensin II mediates stretch-induced hypertrophy of cardiac myocytes in vitro. Cell 75:977–984

Salter DM, Millward-Sadler SJ, Nuki G, Wright MO (2002) Differential responses of chondrocytes from normal and osteoarthritic human articular cartilage to mechanical stimulation. Biorheology 39:97–108

Sanchez Ch, Mathy-Hartert M, Henrotin Y (2011) The mechanosensitivity of cells in joint tissues: role in the pathogenesis of joint diseases. In: Kamkin A, Kiseleva I (eds) Mechanosensitivity in cells and tissues 4. Mechanosensitivity and Mechanotransduction. Springer, New York, NY, pp 297–313

Sandor F, Buc M (2005) Toll-like receptors. I. Structure, function and their ligands. Folia Biologica (Praha) 51:148–156

Saygili E, Rana OR, Saygili E, Reuter H, Frank K, Schwinger RH, Müller-Ehmsen J, Zobel C (2007) Losartan prevents stretch-induced electrical remodeling in cultured atrial neonatal myocytes. Am J Physiol 292:H2898–H2905

Shyu KG (2009) Cellular and molecular effects of mechanical stretch on vascular cells and cardiac myocytes. Clin Sci 116:377–389

Shyu KG, Chen JJ, Shih NL, Wang DL, Chang H, Lien WP, Liew CC (1995) Regulation of human cardiac myosin heavy chain genes by cyclical mechanical stretch in cultured cardiocytes. Biochem Biophys Res Commun 210:567–573

Shyu KG, Ko WH, Yang WS, Wang BW, Kuan P (2005a) Insulin-like growth factor-1 mediates stretch-induced upregulation of myostatin expression in neonatal rat cardiomyocytes. Cardiovasc Res 68:405–414

Shyu KG, Chao YM, Wang BW, Kuan P (2005b) Regulation of discoidin domain receptor 2 by cyclic mechanical stretch in cultured rat vascular smooth muscle cells. Hypertension 46(3):614–621

Shyu KG, Wang BW, Lin CM, Chang H (2010) Cyclic stretch enhances the expression of Toll-like receptor 4 gene in cultured cardiomyocytes via p38 MAP kinase and NF-κB pathway. J Biomed Sci 17:15

Shyy JY, Chien S (2002) Role of integrins in endothelial mechanosensing of shear stress. Circ Res 91:769–775

Shyy JY, Lin MC, Han J, Lu Y, Petrime M, Chien S (1995) The cis-acting phorbol ester "12-O-tetradecanoylphorbol 13-acetate"-responsive element is involved in shear stress-induced monocyte chemotactic protein 1 gene expression. Proc Natl Acad Sci USA 92: 8069–8073

Simon BA (2001) Message in a model. Am J Respir Crit Care Med 163:1043–1044

Sitrin RG, Pan PM, Srikanth S, Todd RF 3rd (1998) Fibrinogen activates NF-κB transcription factors in mononuclear phagocytes. J Immunol 161:1462–1470

Skutek M, van Griensven M, Zeichen J, Brauer N, Bosch U (2001) Cyclic mechanical stretching modulates secretion pattern of growth factors in human tendon fibroblasts. Eur J Appl Physiol 86(1):48–52

Sokabe T, Yamamoto K, Ohura N, Nakatsuka H, Qin K, Obi S, Kamiya A, Ando J (2004) Differential regulation of urokinase-type plasminogen activator expression by fluid shear stress in human coronary artery endothelial cells. Am J Physiol 287(5):H2027–H2034

Standley PR, Obards TJ, Martina CI (1999) Cyclic stretch regulates autocrine IGF-1 in vascular smooth muscle cells: implications in vascular hyperplasia. Am J Physiol 39:E697–E705

Sterpetti AV, Cucina A, Morena AR, Di Donna S, D'Angelo LS, Cavalarro A, Stipa S (1993) Shear stress increases the release of interleukin-1 and interleukin-6 by aortic endothelial cells. Surgery 114:911–914

Szabo-Fresnais N, Lefebvre F, Germain A, Fischmeister R, Pomérance M (2010) A new regulation of IL-6 production in adult cardiomyocytes by beta-adrenergic and IL-1 beta receptors and induction of cellular hypertrophy by IL-6 trans-signalling. Cell Signal 22(7):1143–1152

Takata M, Abe J, Tanaka H, Kitano Y, Doi S, Kohsaka T, Miyasaka K (1997) Intraalveolar expression of tumor necrosis factor- gene during conventional and high-frequency ventilation. Am J Respir Crit Care Med 156:272–279

Takeda K, Akira S (2005) Toll-like receptors in innate immunity. International Immunol 17(1): 1–14

Takeda K, Kaisho T, Akira S (2003) Toll-like receptors. Annu Rev Immunol 21:335–376

Tanabe Y, Saito M, Ueno A, Nakamura M, Takeishi K, Nakayama K (2000) Mechanical stretch augments PDGF receptor beta expression and protein tyrosine phosphorylation in pulmonary artery tissue and smooth muscle cells. Mol Cell Biochem 215(1–2):103–113

van Thienen JV, Fledderus JO, Dekker RJ, Rohlena J, van Ijzendoorn GA, Kootstra NA, Pannekoek H, Horrevoets AJ (2006) Shear stress sustains atheroprotective endothelial KLF2 expression more potently than statins through mRNA stabilization. Cardiovasc Res 72:231–240

Torsoni AS, Marin TM, Velloso LA, Franchini KG (2005) RhoA/ROCK signaling is critical to FAK activation by cyclic stretch in cardiac myocytes. Am J Physiol 289:H1488–H1496

Traub O, Berk BC (1998) Laminar shear stress: mechanisms by which endothelial cells transduce an atheroprotective force. Arterioscler Thromb Vasc Biol 18:677–685

Tremblay LN, Slutsky AS (1998) Ventilator-induced injury: from barotrauma to biotrauma. Proc Assoc Am Physicians 110:482–488

Tremblay L, Valenza F, Ribeiro SP, Li J, Slutsky AS (1997) Injurious ventilatory strategies increase cytokines and c-fos m-RNA expression in an isolated rat lung model. J Clin Invest 99: 944–952

Tsan MF, Gao B (2004) Heat shock protein and innate immunity. Cell Mol Immunol 1(4):274–279

Tsuda A, Stringer BK, Mijailovich SM, Rogers RA, Hamada K, Gray ML (1999) Alveolar cell stretching in the presence of fibrous particles induces interleukin-8 responses. Am J Respir Cell Mol Biol 21:455–462

Uhlig S (2002) Ventilation-induced lung injury and mechanotransduction. Stretching it too far? Am J Physiol 282:L892–L896

Vlahakis NE, Schroeder MA, Limper AH, Hubmayr RD (1999) Stretch induces cytokine release by alveolar epithelial cells in vitro. Am J Physiol 277:L167–L173

Wamel JET van, Ruwhof C, van der Valk-Kokshoorn, Schrier PI, van der Laarse A (2002) Stretch-induced paracrine hypertrophic stimuli increase TGF-β1 expression in cardiomyocytes. Mol Cell Biochem 236:147–153

Wang JH-C, Thampatty BP (2006) An introductory review of cell mechanobiology. Biomechan Mod Mechanobiol 5:1–16

Wang DL, Wung BS, Peng YC, Wang JJ (1995) Mechanical strain increases endothelin-1 gene expression via protein kinase C pathway in human endothelial cells. J Cell Physiol 163(2): 400–406

Wang BW, Chang H, Lin S, Kuan P, Shyu KG (2003) Induction of matrix metalloproteinase-14 and -2 by cyclical mechanical stretch is mediated by tumor necrosis factor-a in cultured human umbilical vein endothelial cell. Cardiovasc Res 59:460–469

Wang BW, Hung HF, Chang H, Kuan P, Shyu KG (2007a) Mechanical stretch enhances the expression of resistin gene in cultured cardiomyocytes via tumor necrosis factor-a. Am J Physiol 293:H2305–H2312

Wang XL, Fu A, Raghavakaimal S, Lee HC (2007b) Proteomic analysis of vascular endothelial cells in response to laminar shear stress. Proteomics 7:588–596

Wasserman SM, Mehraban F, Komuves LG, Yang RB, Tomlinson JE, Zhang Y, Spriggs F, Topper JN (2002) Gene expression profile of human endothelial cells exposed to sustained fluid shear stress. Physiol Genomics 12:13–23

Waters CM (2004) Reactive oxygen species in mechanotransduction. Am J Physiol 287: L484–L485

Waters CM, Sporn PHS, Liu M, Fredberg JJ (2002) Cellular biomechanics in the lung. Am J Physiol Lung Cell Mol Physiol 283:L503–L509

Wernig F, Mayr M, Xu Q (2003) Mechanical stretch-induced apoptosis in smooth muscle cells is mediated by beta1-integrin signaling pathways. Hypertension 41(4):903–911

Wilson E, Mai Q, Sudhir K, Weiss RH, Ives HE (1993) Mechanical strain induces growth of vascular smooth muscle cells via autocrine action of PDGF. J Cell Biol 123:741–747

Woo CH, Shishido T, McClain C, Lim JH, Li JD, Yang J, Yan C, Abe J (2008) Extracellular signal-regulated kinase 5 SUMOylation antagonizes shear stress-induced antiinflammatory response and endothelial nitric oxide synthase expression in endothelial cells. Circ Res 102(5): 538–545

Wung BS, Cheng JJ, Shyue SK, Wang DL (2001) NO modulates monocyte chemotactic protein-1 expression in endo- thelial cells under cyclic strain. Arterioscler Thromb Vasc Biol 21:1941–1947

Yamazaki T, Komuro I, Kudoh S, Zou Y, Shiojima I, Hiroi Y, Mizuno T, Maemura K, Kurihara H, Aikawa R, Takano H, Yazaki Y (1996) Endothelin-1 is involved in mechanical stress-induced cardiomyocyte hypertrophy. J Biol Chem 271:3221–3228

Zampetaki A, Zhang Z, Hu Y, Xu Q (2005) Biomechanical stress induces IL-6 expression in smooth muscle cells via Ras/Rac1-p38 MAPK-NF-kappaB signaling pathways. Am J Physiol 288(6):H2946–H2954

Zeidan A, Broman J, Hellstrand P, Sward K (2003) Cholesterol dependence of vascular ERK1/2 activation and growth in response to stretch: role of endothelin-1. Arterioscler Thromb Vasc Res 23:1528–1534

Zhang Y, Lee TS, Kolb EM, Sun K, Lu X, Sladek FM, Kassab GS, Garland T Jr, Shyy JY (2006) AMP-activated protein kinase is involved in endothelial NO synthase activation in response to shear stress. Arterioscler Thromb Vasc Biol 26:1281–1287

Zheng W, Seftor EA, Meininger CJ, Hendrix MJ, Tomanek RJ (2001) Mechanisms of coronary angiogenesis in response to stretch: role of VEGF and TGF-beta. Am J Physiol 280:H909–H917

Zou Y, Akazawa H, Qin Y, Sano M, Takano H, Minamino T, Makita N, Iwanaga K, Zhu W, Kudoh S, Toko H, Tamura K, Kihara M, Nagai T, Fukamizu A, Umemura S, Iiri T, Fujita T, Komuro I (2004) Mechanical stress activates angiotensin II type 1 receptor without the involvement of angiotensin II. Nat Cell Biol 6:499–506

Chapter 3
Mechanosensitive Pro-inflammatory Gene Expression in Vascular Cells

Marco Cattaruzza, Andreas H. Wagner, and Markus Hecker

Abstract A delicate balance between circumferential wall tension (CWT) or stretch and unidirectional fluid shear stress (FSS), the two principle haemodynamic forces to which the resident cells of the vessel wall are exposed to, governs their phenotype. FSS by virtue of its stimulatory effect on endothelial nitric oxide (NO) synthase activity and expression has been designated as an anti-inflammatory and homeostatic force. In contrast, CWT has been marked as a potentially detrimental pro-inflammatory force causing, e.g. formation of reactive oxygen species, stimulation of stress-activated protein kinases and a prolonged rise in intracellular free calcium. Moreover, with zyxin localised to focal adhesions, a mechanosensitive protein has been characterised that specially transduces an increase in CWT to the nucleus of both endothelial and smooth muscle cells where it orchestrates a complex and partially pro-inflammatory change in gene expression. Tilting the balance between FSS and CWT towards CWT as the result of an inadequately low FSS, e.g. at arterial bifurcations, or volume and/or pressure overload as, e.g. in hypertension, is generally thought to be responsible for both adaptive and maladaptive vascular remodelling processes including arteriogenesis, atherosclerosis, restenosis following angioplasty, and hypertension-induced arterial remodelling. Starting with a summary of the molecular mechanisms governing CWT and FSS-mediated signal transduction in vascular cells, the differential and variable impact of haemodynamically induced pro-inflammatory gene expression on these remodelling processes is discussed herein.

Keywords Endothelial cells · Vascular smooth muscle cells · Stretch · Shear stress · Pro-inflammatory gene expression · Transcriptional regulation · Remodelling

3.1 Introduction

Volume or pressure overload resulting in a supra-physiological increase in (circumferential) wall tension and hence stretching of the cells of the arterial or venous vessel wall is considered to be a pivotal haemodynamic factor for

---

M. Hecker (✉)
Division of Cardiovascular Physiology, Institute of Physiology and Pathophysiology, University of Heidelberg, 69120 Heidelberg, Germany
e-mail: hecker@physiologie.uni-hd.de

A. Kamkin, I. Kiseleva (eds.), *Mechanical Stretch and Cytokines*, Mechanosensitivity in Cells and Tissues 5, DOI 10.1007/978-94-007-2004-6_3,

both adaptive and maladaptive remodelling processes in the cardiovascular system (Lehoux et al., 2006). These include but are not restricted to arteriogenesis, atherosclerosis, vein graft disease, hypertension-induced arterial remodelling and restenosis following angioplasty and/or stenting. In this context, several transcriptional activator and/or repressor molecules have been characterised which share their sensitivity towards tensile stress and control stretch-induced changes in pro-inflammatory gene expression in vascular cells. In addition, supra-physiological stretching of vascular cells augments the level of oxidative stress these cells normally are exposed to (Wang et al., 2001) and, as a consequence, may cause de-differentiation towards a pro-inflammatory phenotype by way of oxidative protein modification.

As outlined below, the aforementioned remodelling processes, except atherosclerosis, do not normally comprise a pronounced inflammatory component; often this is either transitory or even absent. In contrast to the bidirectional stretching of endothelial and smooth muscle cells, unidirectional fluid shear stress, i.e. the viscous drag exerted by the flowing blood, which mainly affects the endothelial cell monolayer of small resistance-sized arteries and arterioles (Helmke and Davies, 2002), is generally considered to limit rather than exacerbate pro-inflammatory gene expression in vascular cells. However, at bifurcations or curvatures, especially of large conduit arteries, unidirectional shear stress is largely decreased and may in fact become oscillatory because of the quasi-turbulent blood flow at these sites (Glagov et al., 1988; Sharma et al., 2010). As a consequence, endothelial cell nitric oxide (NO) formation is markedly diminished since both endothelial NO synthase (NOS-3) activity and expression strongly depend on a certain level of laminar blood flow hence unidirectional shear stress (Hecker et al., 1993; Förstermann and Kleinert, 1995).

Conversely, the quasi-turbulent blood flow at these atherosclerosis predilection sites (Fig. 3.1) results in an additional volume-dependent rhythmic deformation of the vessel wall that affects endothelial cells and smooth muscle cells alike. It is still unclear whether the resulting changes in pro-inflammatory gene expression in these cells are solely due to their bidirectional stretching or to the loss of the unidirectional shear stress hence endothelial cell NO biosynthesis as a critical anti-inflammatory haemodynamic force. This review summarises what is currently known about the mechanisms underlying mechanosensitive changes in pro-inflammatory gene expression and phenotype regulation in vascular cells, focussing on the aforementioned transcriptional activator and/or repressor molecules.

3.2 Significance of Fluid Shear Stress as an Anti-atherosclerotic Haemodynamic Force

As mentioned above, unidirectional fluid shear stress is critical to up-regulating NOS-3 expression and activity in endothelial cells and thus maintaining their anti-inflammatory phenotype (Uematsu et al., 1995; Boon and Horrevoets, 2009; Pan, 2009). This is best exemplified by the finding that upon isolation and culturing of human primary endothelial cells under static conditions not only NOS-3 expression

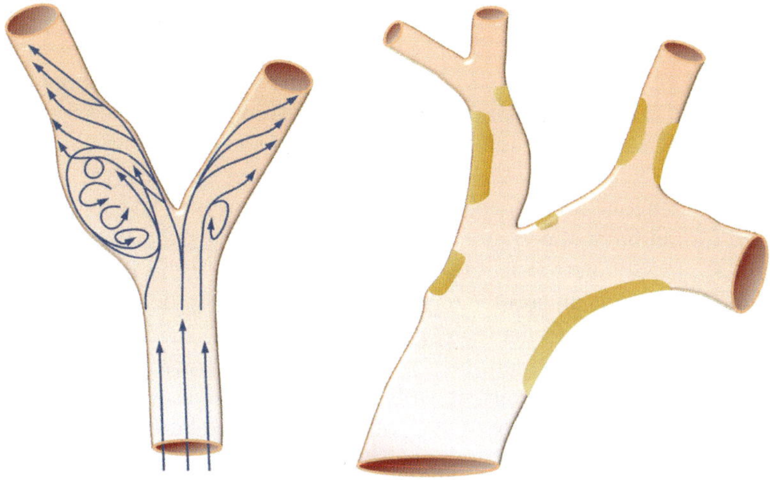

Fig. 3.1 Endothelial dysfunction at sites of haemodynamic stress. Scheme of the typical flow profile at arterial bifurcations (*left panel*) and anatomic scheme of the aortic arch depicting such sites where fluid shear stress and circumferential wall tension (CWT) are in imbalance (*right panel*). Here unidirectional fluid shear stress is reduced while CWT is enhanced, resulting in endothelial dysfunction and, consecutively, atherosclerotic lesion formation

and activity is markedly down-regulated within 6–12 h but also the expression of chemokines and adhesion molecules is strongly enhanced. Similar, albeit less pronounced effects on chemokine and adhesion molecule expression can be observed in human endothelial cells under static conditions following blockade of NOS-3 activity (Zeiher et al., 1995). Moreover, there is a relative reduction of NOS-3 expression in atherosclerosis-prone regions of the mouse aorta (Fig. 3.1; Won et al., 2007) while chronic intermediate exercise and the associated flow-induced increase in shear stress up-regulate expression of the enzyme (Sessa et al., 1994). Although there are conflicting data on the role of oscillatory shear stress in NOS-3 expression (Ziegler et al., 1998; Silacci et al., 2000), it appears that it is the lack of unidirectional shear stress, and thus the decrease in endothelial cell NO biosynthesis, which is responsible for the increased abundance of adhesion molecules and the accumulation of chemokines at atherosclerosis predilection sites which augment the recruitment of pro-inflammatory leukocytes (Busse and Fleming, 1996). This influx of leukocytes, mainly monocyte/macrophages and T cells, is aided by the fact that close to the endothelial cell lining of these arteries, blood flow decreases to nil and shortly becomes even negative. As a consequence the leukocytes are given the chance to roll over the endothelial cell surface and to eventually become activated for transmigration into the sub-endothelial space (Hsiai et al., 2003) where they set off and/or exacerbate atherosclerotic lesion formation. In terms of flow conditions, arterial bifurcations or curvatures thus mimic the situation in post-capillary venules where the velocity of flow normally is lowest and where most leukocytes undergo diapedesis into the inflamed tissue (Du et al., 1995).

In addition to the loss of laminar blood flow hence unidirectional, shear stress at arterial bifurcations or curvatures, rhythmic distensions of the vessel wall ensue because of augmented reflections of the pulse wave. As a result of these rhythmic and augmented changes in circumferential wall tension both endothelial and smooth muscle cells are stretched longitudinally in a cyclic and bidirectional manner. This in turn up-regulates both NADPH oxidase expression and activity, e.g. through enhanced protein kinase C-dependent assembly of the enzyme, and hence leads to an increased formation of superoxide anions in both types of cells (Wung et al., 1997; Cheng et al., 1998; Guest et al., 2006). One consequence of this increased superoxide anion formation is an accelerated neutralisation of endothelial cell-derived NO by way of peroxynitrite formation (Tedgui and Mallat, 2001). While peroxynitrite, depending on its concentration, may per se elicit both protective and deleterious effects on the endothelial cells, it would appear that it is primarily the further reduction in the bioavailability of NO which lifts the brake on pro-inflammatory gene expression (Klotz et al., 2002) resulting in what is collectively referred to as endothelial dysfunction. This dysfunction, primarily reflecting a shift of the endothelial cell phenotype from the anti-inflammatory and anti-thrombotic towards the pro-inflammatory and pro-thrombotic state is widely considered as the decisive factor for both the initiation and propagation of atherosclerosis.

The special haemodynamic situation at arterial bifurcations or curvatures thus promotes this chronic and maladaptive remodelling process of the arterial vessel wall by causing alterations of three pivotal parameters: (1) A decline in unidirectional shear stress that reduces endothelial cell NO formation; (2) an increase in CWT and thus cyclic stretching of both endothelial and smooth muscle cells resulting in an enhanced formation of superoxide anions and hence neutralisation of NO; and (3) the marked deceleration of blood flow velocity facilitating pro-inflammatory monocyte and T cell diapedesis into the vessel wall (Fig. 3.2).

3.2.1 The T-786C SNP of the nos-3 *Gene as a Genetic Determinant of Endothelial Dysfunction*

There have been many attempts recently, most notably by way of genome-wide association studies, to associate genetic variations, namely single nucleotide polymorphisms (SNP), with an increased risk for coronary heart disease (CHD) and thus myocardial infarction (Musunuru and Kathiresan, 2010). However, even when combined, most of the variations identified so far, only modestly increase the risk for CHD and can only explain approximately 2.5% of the inheritability of the disease. There is one perhaps less well recognised genetic variation that could be directly linked to endothelial dysfunction, and that is the T-786C SNP of the endothelial NO synthase gene (*nos-3*).

According to several small to intermediate scale association studies, the homozygous variant of this SNP, the CC genotype, occurs in 12–15% of all Caucasians (Cattaruzza et al., 2004a; Rossi et al., 2003). Independently of other primary risk

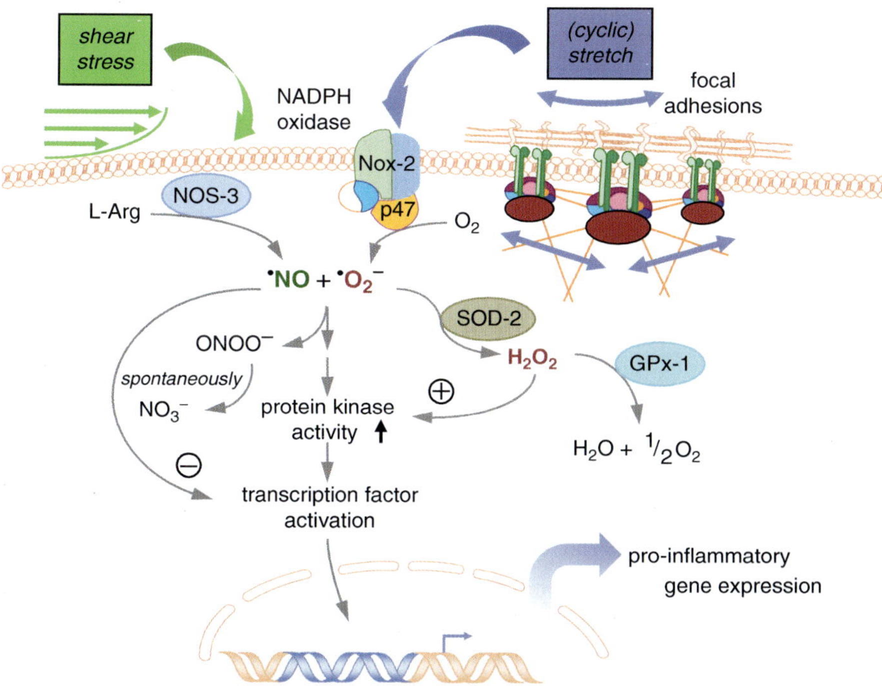

Fig. 3.2 Mechanosensitive redox-signalling in endothelial cells. Fluid shear stress and CWT regulate redox-signalling by adjusting the balance between nitric oxide (NO) and superoxide anions (O_2^-). These two radicals are important anti-inflammatory and pro-inflammatory second messengers, respectively, which control a plethora of signalling pathways in a mutually antagonistic manner. The degradation of NO and O_2^- adds another level of complexity to the signalling as this gives rise to an additional potential second messenger, peroxynitrite ($\cdot ONOO^-$), and affects the abundance of another, i.e. hydrogen peroxide (H_2O_2) derived from the dismutation of O_2^- by, e.g. manganese-containing superoxide-dismutase (SOD-2). Glutathione peroxidase-1 (GPx-1) is required to limit the abundance of H_2O_2 which may also give rise to the formation of potentially harmful hydroxyl radicals

factors for atherosclerosis, it increases the relative risk to contract CHD between 1.7 and 2.5-fold. Remarkably, it also confers a similar relative risk to contract rheumatoid arthritis (Melchers et al., 2006) which appears to be an even more potent risk factor for CHD than diabetes (van Halm et al., 2009). The T-786C SNP, which is situated in the promoter region of the *nos-3* gene, in fact decreases basal and especially stimulus-dependent expression of the gene and thus the NO-synthesizing capacity of the endothelial cells. The two stimuli to which the C-type promoter is much less sensitive are unidirectional fluid shear stress and the anti-Th1 cell cytokine interleukin-10 (IL-10). IL-10 via activation of the transcription factor signal transducer and activator of transcription 3 (Stat3; Cattaruzza et al., 2003) elicits a several fold increase in NOS-3 expression in endothelial cells derived from TT genotype (i.e., genetically normal) individuals which is absent in endothelial cells derived from CC genotype individuals. The most likely reason for this insensitivity to IL-10 is the steric hindrance of Stat3 binding to the *nos-3* promoter at a position approximately 60 base pairs upstream of the SNP by an as yet unidentified DNA-binding protein. One functional consequence of the C-type *nos-3* gene variant is a greatly diminished blockade by IL-10 of pro-inflammatory gene expression elicited by pro-inflammatory cytokines such as tumour necrosis factor α (TNFα), interferon-γ (IFNγ) and/or interleukin-1β (IL-1β) in endothelial cells derived from individuals with the CC genotype. Weakening of endothelial cell NO synthesis, i.e. endothelial dysfunction, may thus explain the increased relative risk of affected individuals to contract rheumatoid arthritis and CHD alike.

Normally re-exposure of endothelial cells cultured under static conditions to unidirectional shear stress raises the level of NOS-3 expression back to that of freshly isolated cells, suggesting that this haemodynamic force maintains abundance of the enzyme in native endothelial cells at a physiologically meaningful level. The transcription factor(s) responsible for this maintenance of NOS-3 expression have not been unambiguously identified thus far. While some investigators have favoured different members of the *krueppel*-like family of transcription factors (SenBanerjee et al., 2004; Hamik et al., 2007), others have identified a so-called shear stress-responsive element in the promoter of the human *nos-3* gene that facilitates interaction with the generic pro-inflammatory transcription factor NF-κB (Davis et al., 2004; Hay et al., 2003). Basal transcription of the *nos-3* gene, on the other hand, appears to be controlled by prototypic maintenance factors such as Sp1 or GATA2 (Laumonnier et al., 2000). In our hands, exposure to fluid shear stress results in the binding of an as yet unidentified transcription factor to the aforementioned Stat family cis-regulatory element. As in the case of the insensitivity to IL-10, a plausible explanation for the shear stress insensitivity of the C-type promoter would be steric hindrance of the yet to be identified shear stress-dependent transcription factor by a protein binding with high affinity to the sequence motif around position -786 of the *nos-3* gene.

Given the functional importance of an adequate endothelial cell NO synthesis capacity to control pro-inflammatory gene expression in the vessel wall, it is difficult to reconcile why the T-786C SNP of the *nos*-3 gene does not feature more prominently in the genome-wide association studies for CHD (Musunuru and Kathiresan, 2010) or fails to accelerate atherogenesis in affected individuals (Cattaruzza et al., 2004a; Rossi et al., 2003; Colombo et al., 2003). One obvious reason is the existence of compensatory mechanisms such as an increased expression of manganese-containing superoxide dismutase-2 (SOD-2) in CC genotype endothelial cells in response to fluid shear stress (Asif et al., 2009a; b). As a consequence, the greatly reduced amounts of NO generated by these cells are much better protected against neutralisation by superoxide anions, thus retaining a critical anti-inflammatory activity. Another compensatory mechanism appears to be the shear stress-induced enhanced release of anti-inflammatory prostaglandins from the endothelial cells which may be responsible for the impeded transmigration and activation (according to the expression of pro-inflammatory gene products) of monocytes through a monolayer of CC genotype as compared to TT genotype endothelial cells.

3.2.2 Alterations in Endothelial Cell Antioxidant Enzyme Activity or Expression as a Protective Mechanism in Atherogenesis

As described before, SOD-2 may provide an important protective effect on the bioavailability of NO in the vessel wall. In terms of the anti-inflammatory effects of NO, especially on the expression of pro-inflammatory gene products by the endothelial cells themselves, protection against the development of atherosclerosis strongly hinges on tilting the balance slightly in favour of NO over superoxide anions. This rather delicate balance is clearly shifted towards superoxide anion formation at atherosclerosis predilection sites with the decline in unidirectional shear stress, and thus NO formation, and increased cyclic stretching of the endothelial cells which enhance NADPH oxidase activity (Silacci et al., 2001). A systematic screening for changes in pro- or antioxidant enzyme expression in human cultured endothelial cells revealed that short-term exposure to fluid shear stress up-regulates NOS-3, the NADPH oxidase subunit p67phox and, namely in CC genotype endothelial cells, SOD-2 (Asif et al., 2009b). In response to cyclic stretch, on the other hand, only expression of GSH peroxidase-1 (GPx-1), which is responsible for detoxifying hydrogen peroxide derived from the superoxide dismutase reaction, and of heme oxygenase-1 were distinctly elevated. In contrast, catalase expression was in fact down-regulated upon prolonged exposure to cyclic stretch (Wagner et al., 2009).

Knockdown of GPx-1 expression reinforced both basal and stretch-induced expression of pro-inflammatory gene products such as vascular cell adhesion molecule-1 (VCAM-1), monocyte chemoattractant protein-1 (MCP-1 or CCL2) and CD40 as well as the adhesion of monocytes to the endothelial cells under physiological flow conditions. Expression of these pro-inflammatory gene products was clearly dependent on an enhanced formation of reactive oxygen species (ROS), namely superoxide anions dismutating to hydrogen peroxide (Wagner et al., 2009). Hydrogen peroxide itself or a reactive metabolite like the hydroxyl radical can activate stress-sensitive protein kinases, such as c-Jun N-terminal kinase (JNK; Hojo et al., 2002), p38 mitogen-activated protein (MAP) kinase and Rho kinase (Fig. 3.3). This leads, for example, to activation of the transcription factors activator protein-1 (AP-1) and CCAAT/enhancer-binding protein (C/EBP) which in turn up-regulate expression of the aforementioned pro-inflammatory gene products (Collins et al., 1995). Since the affinity of catalase for hydrogen peroxide in vascular cells is rather low, it seems that GPx-1 is the critical enzyme controlling stretch-induced ROS-mediated changes in gene expression in both endothelial and smooth muscle cells at near physiological levels of hydrogen peroxide. Overexpression of superoxide dismutases does not seem to constitute a meaningful therapeutic intervention in this context as it may lead to a supra-physiological accumulation of hydrogen peroxide and eventually to an excessive formation of deleterious hydroxyl radicals.

A pivotal role of GPx-1 in balancing stretch-induced oxidative stress-mediated changes in pro-inflammatory gene expression may also be inferred from the finding that the most prevalent oxidative modification of proteins in (vascular) cells, i.e. hydroxyl radical-mediated carbonylation (Cattaruzza and Hecker, 2008) of the

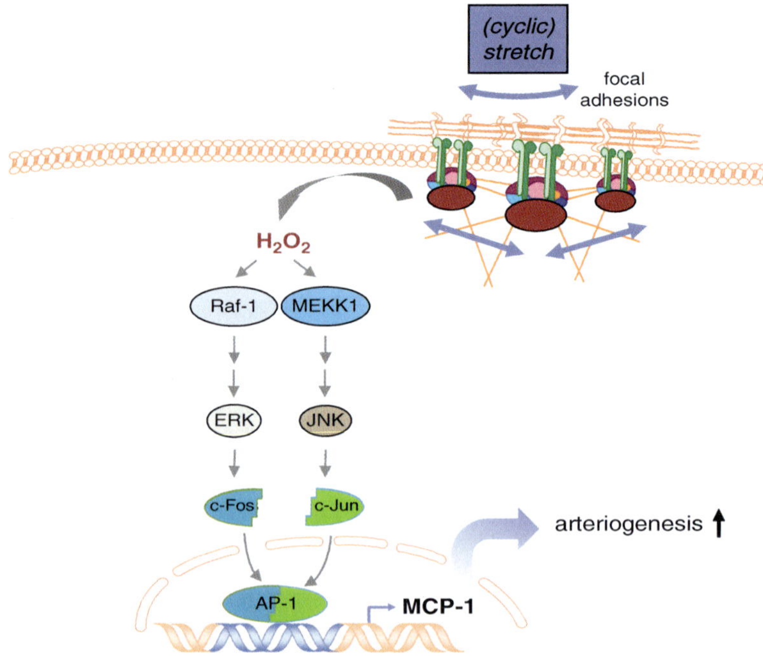

Fig. 3.3 CWT-induced MCP-1 expression during arteriogenesis. The onset of arteriogenesis is associated with a CWT-induced increase in monocyte chemoattractant protein-1 (MCP-1/CCL2) expression by the arteriolar smooth muscle cells. It is mediated by the stretch-induced activation of the transcription factor AP-1, presumably via generation of excess hydrogen peroxide (H_2O_2) that affects the downstream effector kinases ERK1/2 and JNK. This results in the activation of c-Jun and an increased expression of the immediate early gene c-Fos which heterodimerise to the active transcription factor that interacts with the promoter of the *ccl2* gene

enzyme, does not result in its aggregation and proteasomal degradation but rather appears to reinforce its activity. Stretch-induced oxidative stress thus results in an increase in both GPx-1 expression and activity. This, in turn, may play a crucial role in maintaining the anti-inflammatory phenotype of vascular cells.

3.2.3 Peroxynitrite Protection Against Stretch-Induced Pro-inflammatory Gene Expression

Peroxynitrite-catalysed tyrosine nitration, the second most prevalent oxidative or rather nitrosative modification of proteins, has long been considered as a marker of inflamed vascular tissue, such as in the atherosclerotic vessel wall. There are, however, only few examples where tyrosine nitration results in an irreversible alteration of protein function and subsequent degradation by the proteasome (Souza et al., 2000). One membrane-bound protein in human endothelial cells that is both stretch-sensitive and prone to tyrosine nitration is CD40. Ligation of this co-stimulatory receptor, either by platelet or T cell-derived CD154, constitutes one of the most potent pro-inflammatory stimuli for endothelial cells. It therefore plays a pivotal role in the initiation and progression of atherosclerosis (Phipps et al., 2001; Lutgens et al., 1999). Cyclic stretching of human cultured endothelial cells results in a transient up-regulation of both CD40 mRNA and protein that is mediated by the ROS-dependent activation of AP-1 followed by a rapid decline in CD40 protein to approximately 25% of baseline. This loss of CD40, which coincides with an enhanced formation of peroxynitrite in the stretched endothelial cells, is caused by proteasomal degradation of the membrane-bound protein following nitration of one of its two N-terminal tyrosine residues which appears to be readily accessible by peroxynitrite. Before internalisation and degradation, endothelial cell CD40 is functionally inactivated through this oxidative protein modification by way of dissociation of certain TRAF molecules from the receptor that signal into the cell and the nucleus (Hildebrandt et al., 2009).

This novel mechanism may explain why CD40 is absent in straight arteries where shear stress-dependent NO synthesis and stretch-induced superoxide anion formation give rise to significant amounts of peroxynitrite. In veins, shear stress-dependent NO synthesis is rather low while at arterial bifurcations and curvatures stretch-dependent superoxide anion formation clearly exceeds that of NO so that peroxynitrite levels remain comparatively low. It is precisely in these blood vessels and at these arteriosclerosis predilection sites where CD40 expression is readily detected (Korff et al., 2007), and where it presumably aids in the transmigration (and stimulation) of pro-inflammatory monocytes and T effector cells into the vessel wall and further into the inflamed tissue in both a physiologically meaningful (adaptive immune response) and pathophysiological context (atherosclerosis). Interestingly, CD154 stimulation of CD40-expressing cells, namely antigen-presenting cells including endothelial cells, may per se up-regulate peroxynitrite levels in these cells (Davis and Zou, 2005), implying the existence of a self-limiting mechanism by which these cells are capable of preventing an exaggerated expression of this Janus-faced co-stimulatory molecule.

3.3 Stretch-Induced Remodelling Processes Comprising a Comparatively Limited Pro-inflammatory Response

In atherosclerosis an imbalance between shear stress-dependent NO biosynthesis and stretch-induced superoxide anion formation in conjunction with the greatly decelerated blood flow at the vessel wall plays a decisive role in altering the phenotype of both endothelial and smooth muscle cells along with the development of a chronic recurrent inflammation in the vessel wall. In contrast, all other clinically relevant remodelling processes seem to preferentially hinge on an increase in CWT, hence stretch, accompanied by a less pronounced pro-inflammatory response.

3.3.1 Arteriogenesis

Arteriogenesis, for instance, i.e. the adaptive enlargement of collateral arterioles on occlusion of the main feeding artery, although initially comprising many features of the early phase of atherosclerosis, does not develop as a result of an imbalance between increased (cyclic) stretch and reduced unidirectional shear stress. In lieu thereof, this remodelling process appears to be initiated by a flow, hence volume-dependent shift, in CWT (Demicheva et al., 2008) or a flow-dependent rise in unidirectional shear stress (for a comprehensive review see Schaper, 2009), or both (Chen et al., 2010). On the other hand, like atherosclerosis arteriogenesis strongly depends on an initial influx of mononuclear leukocytes, namely monocytes, into the vessel wall. In contrast to atherosclerosis, though, these monocytes do not transmigrate through the arteriolar endothelial cells but extravasate into the perivascular space, presumably via the post-capillary venules (Behm et al., 2008).

A crucial factor for this influx of monocytes and their differentiation into macrophages is a mechanosensitive up-regulation of MCP-1 (CCL2) expression and release in the arteriolar vessel wall. Unlike atherosclerosis where the dysfunctional endothelial cells appear to be the main source of this chemokine, MCP-1 seems to be predominantly derived from the arteriolar smooth muscle cells in arteriogenesis (Demicheva et al., 2008). Vascular smooth muscle cells are not capable of sensing alterations in shear stress directly but only indirectly through enhanced NO biosynthesis by the arteriolar endothelial cells, which would in fact inhibit rather than augment smooth muscle cell MCP-1 expression and/or release (Zeiher et al., 1995). It is therefore unlikely that this haemodynamic force accounts for this pivotal step in arteriogenesis. Moreover, when calculating changes of the two haemodynamic forces in the collateral arterioles following experimental occlusion of, e.g. the femoral artery of a mouse (hind limb ischemia model), fluid shear stress is only elevated by about 1.2-fold. In spite of the steep pressure drop and thus rise in flow following ligation of the femoral artery, the resulting increase in shear stress, which is inversely related to the third power of the inner radius, is almost offset by the concomitant volume-dependent distension of the arteriole and ensuing increase of the inner radius. Contrariwise CWT, the magnitude of which (according to Laplace's law) equals the level of transmural pressure multiplied by the quotient of inner radius and wall thickness, rises more prominently by about 2.3-fold (Demicheva et al., 2008). It is of note, however, that in this case, as a result of the increase in flow driven by the elevated longitudinal pressure gradient, the rise in CWT is due to the volume-dependent distension of the arteriolar vessel wall rather than an increase in transmural pressure per se. Accordingly, stretch-dependent MCP-1 formation and/or release predominantly derive from the arteriolar smooth muscle cells, and this effect can be mimicked experimentally with both mouse and human cultured smooth muscle cells exposed to cyclic stretch. In this context, it is worth mentioning that MCP-1 not only is a potent chemokine for mononuclear cells but may also act as a growth factor (Schepers et al., 2006) to expand the arteriolar smooth muscle cell mass in arteriogenesis (Fig. 3.4).

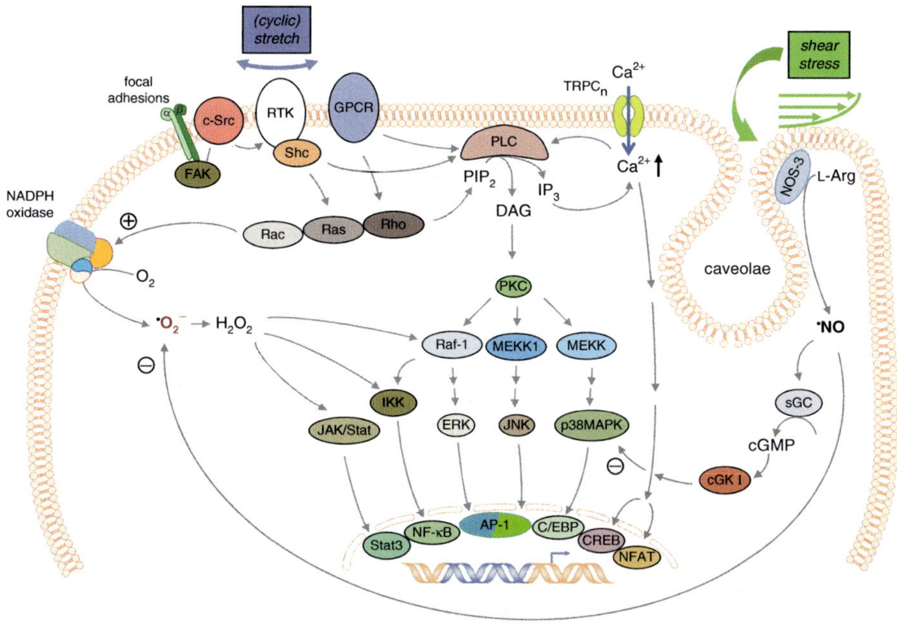

Fig. 3.4 Mechanosensitive pro-atherosclerotic signalling pathways in endothelial and vascular smooth muscle cells. The figure schematically depicts a selection of pro-atherosclerotic signalling pathways known to be mechanosensitive. It integrates endothelial and smooth muscle cell pathways in one scheme; however, most of the pathways shown are not specific for a given cell type apart from unidirectional fluid shear stress. Reduced shear stress in endothelial cells results in a decline of bioavailable NO which, in turn, decreases the intracellular cGMP level and thus the activity of cGMP-dependent protein kinase I (cGKI) which can no longer effectively block G-protein-coupled receptor (GPCR) signalling (Schlossmann and Desch, 2009). Likewise, the control by NO of mitogen activated protein kinase (MAPK) signalling via cGKI activation of MAPK phosphatase-3 is diminished (Sharma et al., 2002). The simultaneous rise in CWT, on the other hand, leads to the activation of NADPH oxidase and, in turn, to that of several MAPKs which control transcription factors such as Stat1/3, NF-κB, AP-1 or C/EBP that are pivotal for pro-inflammatory gene expression in these cells. This can also be stimulated by the secondary rise in intracellular calcium, e.g. via stretch-sensitive ion channels, leading to nuclear translocation of the transcription factors CREB and NFAT. c-Src, tyrosine kinase Src; DAG, dicacylglycerol; ERK, extracellular signal-related kinase(s); FAK, focal adhesion kinase; IKK, IκB kinase; IP_3: inositol trisphosphate; JAK/Stat: janus kinase/signal transducer and activator of transcription pathway; MEKK, mitogen-activated kinase kinase; PIP_2: phosphatidylinositol 4,5-bisphosphate; Raf-1, rapid accelerated fibrosarcoma kinase-1; RTK, receptor tyrosine kinase; sGC, soluble isoform of guanylyl cyclase

3.3.2 Remodelling Processes Due to a Prolonged Rise in Transmural Pressure

Venous bypass vasculopathy and hypertension-induced arterial remodelling have in common that the level of (hydrostatic) pressure the vessel wall is exposed to, and thus CWT, is virtually perpetually increased well beyond the physiological level for either type of vessel. Initially the arterial vessel wall attempts to offload this overt mechanical strain by active vasoconstriction through the formation and release, namely from the endothelial cells, of potent vasoconstrictors such as endothelin-1 (ET-1). Because of the lack of sufficient numbers of contractile smooth muscle cells, pressurised veins typically start to remodel by increasing the mass of mostly synthetic smooth muscle cells by way of hyperplasia and the expression of matrix metalloproteinases and matrix proteins, which ultimately results in a thickened and much stiffer vessel wall. Arteries and arterioles remodel in a similar manner when active vasoconstriction is not sufficient to lower CWT to physiological levels, and especially when the rise in transmural pressure is maintained for prolonged periods. The difference between arteries and arterioles in this respect is that arterial smooth muscle cells preferentially undergo hypertrophy while arteriolar smooth muscle cells increase their mass like veins by true hyperplasia. In this context, the mitogen-like activity of vasoconstrictor compounds such as ET-1 becomes much more prominent. The increase in smooth muscle cell mass and stiffness of the extracellular matrix, in turn, will lead to a perpetuation of arterial or venous resistance at a supra-physiological level, and thus of the rise in transmural pressure, which as a noxious stimulus can no longer be abated. In contrast to atherosclerosis or arteriogenesis, these remodelling processes do not seem to have a significant pro-inflammatory component.

3.3.3 Restenosis

Moreover, a rather brief pressure trauma of the arterial vessel wall during balloon angioplasty and/or stenting also leads to a remodelling process that, albeit somewhat accelerated, very much resembles the secondary reaction of the arteriolar smooth muscle cells to a prolonged increase in transmural pressure as in arterial or pulmonary hypertension. The vessel wall thickens by forming a neointima with mostly synthetic smooth muscle cells that have rebuilt the extracellular matrix. If this remodelling process is not halted, restenosis will inevitably ensue. If atherosclerotic arteries are distended, which is the indication for such interventional procedures in the patient, a strong pro-inflammatory component must be considered. Mechanistically, we do not know much about the pro-inflammatory response in restenosis, as there is essentially no adequate animal experimental model available to study this. Typically either healthy blood vessels are subjected to balloon injury, or those derived from small (mouse) or large (pig) experimental animals that are hypercholesterolemic for genetic reasons or because of a high cholesterol diet. This generally results in a marked pro-inflammatory response mostly driven by mononuclear leukocytes infiltrating the artificial lesion and de-differentiating to foam cells (Kelkenberg et al., 2002). In healthy arteries undergoing balloon angioplasty or stenting, it is smooth muscle cell hyperplasia as well as the remodelling of the extracellular matrix without any appreciable inflammatory component that appear to be responsible for the ensuing restenosis (Schwartz et al., 2004).

Therefore, to prevent restenosis in the clinic, typically cytostatic agents are applied by way of drug-eluting stents or coated balloons. The resulting enhanced risk of in-stent thrombosis due to the lack of a cellular overcoat on the stent struts is usually controlled by long-term treatment with at least two anti-platelet drugs.

3.4 Unspecific Stretch-Sensitive Transcription Factors Controlling Pro-inflammatory Gene Expression

As outlined above, localisation of atherosclerotic plaques seems to correlate with areas of reduced laminar and/or increased oscillatory shear stress, with cyclic stretch being enhanced as well (Zhao et al., 1995). Cyclic deformation appears to promote atherosclerosis through an increased formation of ROS, namely superoxide, associated with an impaired bioavailabiliy of NO (Guzik et al., 2002). Therefore, cyclic stretch is a likely candidate for promoting local plaque formation (Howard et al., 1997; Wagner et al., 2001).

Superoxide anions are generated by a variety of enzymes including xanthine oxidase, cyclooxygenase, and NADPH oxidase (Molavi and Mehta, 2004). Among them, the latter enzyme appears to dominate in vascular cells (Kawashima and Yokoyama, 2004). In the early stages of atherosclerosis, superoxide anions seem to be produced by the NOX-2 or NOX-4 containing NADPH oxidase, which is mainly localised in the endothelium (Goettsch et al., 2009). In advanced lesions, NOX-1 or NOX-4 containing vascular smooth muscle cell NADPH oxidase serve as an additional source of superoxide anions (Kawashima and Yokoyama, 2004; Deliri and McNamara, 2007). The resulting increase in ROS within the cell, mainly reflecting secondary hydrogen peroxide formation, serves as a signal for the initiation and perpetuation of pro-inflammatory gene expression (Wagner et al., 2009), including co-stimulatory and adhesion molecules as well as chemokines and cytokines. This change in pro-inflammatory gene expression is mainly brought about by transcription factors such as AP-1, C/EBP or nuclear factor-κB (NF-κB) which are either directly or indirectly sensitive to what is collectively referred to as oxidative stress (Sen and Packer, 1996; Kumar et al., 2003).

3.4.1 Redox-Sensitive Transcription Factors

AP-1 which in fact is not a single transcription factor but typically a heterodimer composed of Jun, Fos or ATF (activating transcription factor) subunits that bind to a common cis-regulatory element (5'-TGACTCA-3') in the promoter of their target genes (Karin et al., 1997), is not a specific mechanosensitive transcription factor but rather reacts to cellular stress including haemodynamic and hence oxidative stress. Thus, either an upstream protein kinase or phosphatase or protein kinases such as JNK capable of directly phosphorylating members of the AP-1 family, thus making them competent to translocate to the nucleus, appear to be redox-sensitive (Hojo et al., 2002; McCubrey et al., 2006). In addition, oxidation of certain sulfhydryl groups may enhance the DNA-binding activity of, e.g. c-Jun/Fos heterodimers (Abate et al., 1990). In contrast, stretch-induced activation of p38-MAPK and subsequently that of certain members (β and δ) of the C/EBP family of transcription factors also seems to involve a redox-sensitive step. It is unclear at present, though, at which level this apparently hydrogen peroxide-mediated effect occurs (Wagner et al., 2009). As for AP-1, the existence of a regulatory thiol redox-sensitive signalling cascade leading to the activation of the negative regulator C/EBPς has been verified (CHOP/Gadd153; Scott and Loo, 2007).

Interestingly, both types of transcription factors can exert different, sometimes opposing effects on the expression of pro-inflammatory or remodelling-associated gene products, depending on the type of cell and vascular bed involved. Thus, expression of GPx-1 in response to an increase in CWT in both native and cultured endothelial cells seems to be mediated by a p38-MAPK-dependent activation of C/EBPβ or NF-IL6 (human equivalent), while JNK-dependent activation of AP-1 blocks rather than stimulates the stretch-induced expression of this protective antioxidant enzyme (Wagner et al., 2009). In human venous as well as in porcine aortic cultured endothelial cells or in rabbit native venous endothelial cells, stretch-dependent activation of AP-1 is critical for the up-regulation of prepro-ET-1 gene expression, the precursor of ET-1 (Lauth et al., 2000). In rabbit native arterial endothelial cells, on the other hand, prepro-ET-1 gene expression is driven by C/EBPβ. Similarly, in rat vascular smooth muscle cells stretch-dependent expression of the endothelin B-type receptor which primarily accounts for the (maladaptive) remodelling effects of ET-1, e.g. in vein graft disease, restenosis following angioplasty/stenting or arterial or pulmonary hypertension is mediated by C/EBPβ rather than AP-1 (Lauth et al., 2000). However, the biological effects of ET-1 in this context (smooth muscle cell apoptosis and/or hypertrophy/hyperplasia, altered matrix protein expression) strongly depend on the composition of the extracellular matrix, which seems to be altered by the smooth muscle cells themselves (Cattaruzza et al., 2002a).

ROS formation induced by uniaxial cyclic stretch leads to NF-κB activation in human fibroblasts via the signalling cascade stretch-activated channel activation – increase in intracellular calcium – formation of ROS – phosphorylation and subsequent degradation of the regulatory IκB subunit (Amma et al., 2005). There are further results demonstrating that exposure of human cultured arterial and venous endothelial cells to cyclic stretch leads to a protein kinase C (PKC)-dependent activation of AP-1, NF-κB and cAMP response element-binding protein (CREB), which interestingly could not be observed in bovine arterial endothelial cells (Du et al., 1995). Moreover, there was a temporal disparity in the stretch-induced activation of AP-1 and NF-κB peaking at 4 hours as compared to that of CREB which increased in a biphasic manner at 15 minutes and 24 hours.

PKC isoforms contain unique structural features that are susceptible to oxidative modification (Gopalakrishna and Jaken, 2000). The N-terminal regulatory domain contains zinc-binding, cysteine-rich motifs that are readily oxidised by peroxide. When oxidised, the auto-inhibitory function of the regulatory domain is compromised and, consequently, cellular PKC activity is disinhibited or enhanced. The sequential activation of such PKC isoforms (α and ε) contributes to activation of the Raf-1-ERK1/2 axis, and thus to an increased transcriptional activity of Elk1, a prototypic ERK1/2 substrate (Cheng et al., 2001).

Apparently a large variety of pro-inflammatory gene products can be controlled by the transcription factors mentioned above at the level of transcription. These include, e.g. co-stimulatory molecules such as CD40 and potentially ephrinB2 (Korff et al., 2008), adhesion molecules like ICAM-1, VCAM-1 or E-selectin, and chemokines such as MCP-1 or IL-8, which collectively play an important role in the direct or indirect remodelling response of the vessel wall to an inadequate haemodynamic situation (Kinlay et al., 2001; Libby, 2007). In both arteriogenesis and restenosis following angioplasty/stenting with their subacute rather than chronic inflammatory component, activation of AP-1 and expression of AP-1-dependent gene products such as prepro-ET-1, and in particular MCP-1, seem to play a pivotal role (Schaper and Scholz, 2003). In arteriogenesis, it is the stretch-induced expression of MCP-1 by the arteriolar smooth muscle cells that is mediated by AP-1 (Demicheva et al., 2008). Consequently, monocyte infiltration and differentiation into macrophages in the remodelling arterioles as well as arteriogenesis itself is virtually abolished following in vivo administration (ear artery ligation model, topical formulation) of a specific AP-1 decoy oligonucleotide (Demicheva et al., 2008). Also, in mice lacking the AP-1 family member JunB, arteriogenesis is strongly impaired. Conversely, a prolonged rise in mean arterial blood pressure, and hence CWT, clearly augments collateral arteriolar arteriogenesis in hypertensive BPH as compared to normotensive BPL mice (hind limb ischemia model), corroborating the pivotal role of the stretch-dependent rise in smooth muscle cell MCP-1 expression.

3.4.2 Calcium-Sensitive Transcription Factors

Another family of transcription factors that can be activated by stretch and that stimulates the expression of pro-inflammatory gene products in vascular and cardiac myocytes comprises the nuclear factors of activated T cells (NFAT). Like CREB, which is directly phosphorylated by the calcium/calmodulin-dependent protein kinase II (CaMKII), NFATc1 through c4 are typically activated through an increase in intracellular calcium (via different pathways including stretch-sensitive ion channels such as the transient receptor potential (TRP) channels or receptor tyrosine kinases and G-protein-coupled receptors that activate phospholipase Cγ and δ, respectively) leading to a CaMKII-dependent phosphorylation, hence activation, of the phosphatase calcineurin and to the subsequent nuclear translocation of the dephosphorylated NFAT protein (Hill-Eubanks et al., 2003). As a consequence, a certain subset of pro-inflammatory mediators including cytokines such as IL-6 is up-regulated (Nilsson et al., 2006) that may contribute to stretch-induced arterial remodelling and reorganisation of the extracellular matrix.

3.5 Zyxin, a Specific Mechanosensitive Transcription Factor in Endothelial and Smooth Muscle Cells

In hypertension, increased pressure causes a chronic increase in CWT. As discussed in detail before, this increase in CWT causes a compensatory hypertrophic (mainly conduit arteries) or hyperplastic (mainly resistance-sized arteries) increase in wall thickness to normalise CWT. The downside of this ostensibly meaningful adaptive response, however, is an increase in total peripheral resistance leading to a supra-physiological cardiac afterload. Although much is known about clinical consequences and molecular events, e.g. the involvement of chemokines as trophic factors for smooth muscle cells (Ishibashi et al., 2004), in late pressure-induced arterial remodelling in hypertension, the signalling events specifically induced by a rise in CWT and their consequences for the onset of this process are largely elusive.

How should a signalling protein look like that specifically transduces changes in CWT in vascular cells? When we first thought about such a hypothetical protein, we postulated that this molecule must be associated with the actual sensor for CWT, in our opinion integrin-containing structures like focal adhesions (for review see Frangos et al., 2001). Moreover, this 'mechanotransducer' had to be activatable and capable of transducing the signal, i.e. an increase in CWT, to the nucleus where it would orchestrate the first steps in adapting the cellular phenotype to the altered haemodynamic situation. Such a protein, however, was not known at the time.

While analysing the effects of pro-inflammatory cytokines on gene expression in vascular smooth muscle cells in an unrelated study, we became interested in the molecular architecture of zyxin, one of the few gene products which were suppressed in this setting (Cattaruzza et al., 2002b). Zyxin is firmly associated to focal adhesion by α-actinin-binding domains and actA domains that interact with other focal adhesion-associated proteins such as VASP (Reinhard et al., 1995; 1999) but additionally contains three C-terminal LIM domains, a type of zinc finger motif known to play a role in signal transduction and potentially transcriptional regulation (Wang and Gilmore, 2003; Beckerle, 1997). Zyxin thus fulfilled at least two of our three criteria for a true mechanotransducer (Fig. 3.5 upper panel) and, consequently, we started to explore the biology of zyxin.

Zyxin is the first member of a family of highly homologous proteins containing 3 LIM-domains (Bach, 2000) that has been described in mammalian cells (Crawford and Beckerle, 1991; Crawford et al., 1992). In various cell types, the protein binds to and is involved in the organisation of the actin cytoskeleton especially at cell-matrix contacts, namely focal adhesions and cell-cell contacts (Vasioukhin et al., 2000; Hansen and Beckerle, 2006), by stabilising actin (stress) fibres and thereby inhibiting cell motility (Hoffman et al., 2006). Interestingly, this activity is reinforced under conditions of high mechanical strain in order to stabilise the cellular structure (Yoshigi et al., 2005; Hirata et al., 2008a, b; Colombelli et al., 2009; Ngu et al., 2010).

However, when considering the rather complex structure of the molecule, a dual role of zyxin as a structural protein and mechanotransducer is highly suggestive. In fact, Ajuba and the potential proto-oncogene lipoma-preferred protein (LPP), the

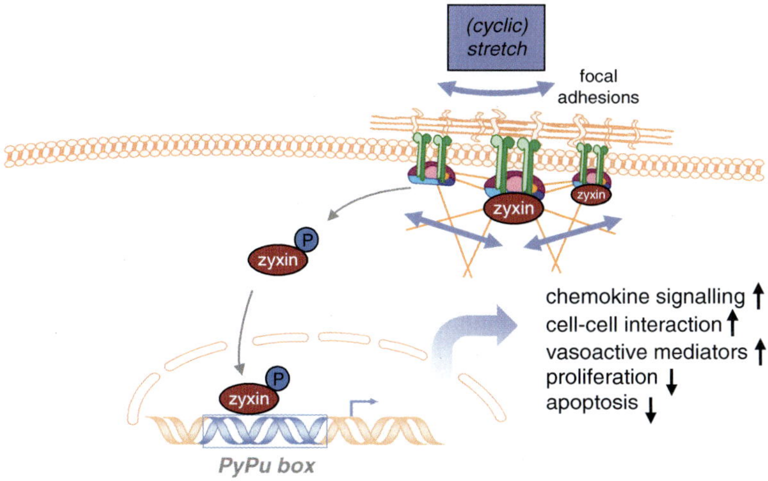

Fig. 3.5 CWT-induced zyxin signalling in vascular cells. Increased CWT causes dissociation of the cytoskeletal protein zyxin from the focal adhesions, presumably through phosphorylation of the protein. Zyxin then translocates to the nucleus and, as a transcription factor binding to a prototypic poly-pyrimidine motif (PyPu box), orchestrates a complex change in gene expression leading to the onset of hypertension-induced arterial remodelling. Although there is a prominent pro-inflammatory component, other mechanisms and mediators such as vasoactive autacoids or direct cell cycle control may play a dominant role in this presumably adaptive remodelling process

two other members of this LIM domain family of proteins, have been implicated in signal transduction (Kanungo et al., 2000; Petit et al., 2000). Moreover, selectively expressed domains of zyxin or LPP are capable of entering the nucleus (Nix et al., 2001) and, as a fusion protein, also of transactivating a reporter gene in a yeast model (Petit et al., 2000). In cardiomyocytes, zyxin as whole protein is capable of translocating into the nucleus (Kato et al., 2005).

3.5.1 Zyxin Is a Mechanotransducer in Vascular Cells

Therefore, we started to analyse the localisation and function of zyxin in vascular smooth muscle cells as well as endothelial cells in more detail. The protein was in fact localised exclusively to focal adhesions and stress fibres in both types of cells at baseline. However, upon exposure to cyclic stretch, a surrogate stimulus for CWT, the protein associated more prominently with the stress fibres (Cattaruzza et al., 2004b; Yoshigi et al., 2005) and, most intriguingly, rapidly translocated to the nucleus (Cattaruzza et al., 2004b; Fig. 3.5 lower panel). Subsequently, this CWT-induced nuclear translocation of zyxin was related to distinct changes in gene expression in the vascular smooth muscle cells, corroborating our assumption that zyxin in fact transduces changes in CWT to the nucleus. Other stimuli such as exposure to pro-inflammatory cytokines or osmotic stress, or, in endothelial cells, to laminar shear stress, failed to elicit an appreciable translocation of zyxin to the nucleus, implying CWT or stretch to be a specific and perhaps exclusive stimulus for zyxin activation and nuclear translocation (Cattaruzza et al., 2004b).

By comparing human cultured endothelial cells with or without siRNA-mediated knockdown of zyxin, stretch-induced changes in gene expression mediated by this protein were further analysed in a comprehensive manner by employing a genome-wide DNA microarray approach. It revealed that about 67% of all stretch-sensitive gene products are controlled by zyxin. Moreover, zyxin exerts this effect by binding to a specific motif in the promoter of these genes, a stretch of pyrimidine bases termed pyrimidine-purine box. It thus acts as a true transcription factor (Wójtowicz et al., 2010). The genetic program that is orchestrated by zyxin in endothelial cells exposed to strong alterations in CWT is highly suggestive. On the one hand, reminiscent of the changes in gene expression in arteriogenesis, cell adhesion and chemokine expression as well as various intracellular signalling pathways are highly up-regulated. This can be exemplified by the fact that among the 20 most profoundly induced gene products, 3 cell adhesion molecules (VCAM-1, ICAM-1 and E-selectin) and 5 chemokines (CXCL8/IL-8, CCL2/MCP-1, CX3CL1/fractalkine, CXCL5 and CXCL11; Wójtowicz et al., 2010) can be found. On the other hand, proliferation as well as apoptosis of the vascular smooth muscle cells is halted and the cells are sensitised to vasoactive mediators such as endothelin-1 or norephinephrine (Wójtowicz et al., 2010). These findings implicate zyxin in some kind of cellular assessment procedure to test whether the early changes in gene expression in response to an increased CWT (such as, e.g. that of the prepro-ET-1 gene; Macarthur et al., 1994) are sufficient to restore normal cell, and hence vessel function, or whether additional measures such as a more fundamental change in phenotype as in pressure-induced arterial remodelling are required. To this end, the long-term effects of a zyxin deficiency on gene expression in cells of the vessel wall, and its functional implications are underway.

With zyxin, therefore, a specific CWT-transducing protein has been characterised in vascular cells. Although the systematic analysis of zyxin-dependent alterations in gene expression in endothelial and smooth muscle cells revealed comparable and rather suggestive changes of their transcriptome, its functional consequences still need to be explored further.

3.6 Circumferential Wall Tension and Maintenance of the Vascular Smooth Muscle Cell Phenotype

Vascular smooth muscle cells, albeit highly differentiated, possess a distinct plasticity, ranging from a synthetic phenotype characterized by fast proliferation, matrix synthesis and (often) pro-inflammatory gene expression to a strictly contractile and non-proliferating phenotype under stable conditions. Moreover, as discussed in an exemplary way in the preceding paragraphs, under certain conditions contractile smooth muscle cells undergo a synthetic and proliferative reprogramming that can be induced by a plethora of exogenous factors such as ET-1 (Kida et al., 2010), chemokines such as, e.g. IL-8 or MCP-1 (Yue et al., 1994; Porreca et al., 1997), or platelet-derived growth factor (Kida et al., 2010), to name just a few.

Therefore, it is understandable that increases in CWT may readily lead to a synthetic smooth muscle cell phenotype mediated by one of these factors as discussed above for, e.g. MCP-1 in arteriogenesis. However, this phenotypic reprogramming is rapidly reversible when the exposure or sensitivity to these synthetic stimuli decreases (Kawai-Kowase and Owens, 2007). Moreover, pathways stabilising the contractile smooth muscle phenotype have been characterised. Important and well established in this regard is the myocardin-serum response factor (SRF) system which, in turn, is modulated by numerous co-transcription factors (Du et al., 2003; Yoshida et al., 2003; Wang et al., 2004). Less characterised but potentially also highly significant may be the transcriptional repressor Zfm1 that does not act at the single gene level but rather at the level of the basal transcription machinery, namely by interacting with certain proliferative and pro-apoptotic co-factors of the TATA-box binding protein (Zhang et al., 1998a; Zhang and Childs, 1998b; Cattaruzza et al., 2002b). Both myocardin and Zfm1 stabilise the expression of gene products prototypic for the contractile phenotype but inhibit synthetic gene expression as well as proliferation.

Therefore, as the exposure to chemokines or ET-1 is not sufficient to explain the long-lasting phenotype switch observed in smooth muscle cells exposed to chronically elevated CWT, the impact of CWT and/or stretch-induced changes in pro-inflammatory gene expression on myocardin and Zfm1 must be understood. However, to date only limited information is available in this regard, except for two recent reports suggesting that the mechanical load of smooth muscle cells may in fact modulate myocardin-induced gene expression. Thus, stretching of venous smooth muscle cells seems to induce a myocardin-mediated hypertrophic gene programme (Hellstrand and Albinsson, 2005), and cardiomyocytes reveal a similar response through activation of MRTF-A, a co-transcription factor highly homologous to myocardin (Kuwahara et al., 2010).

Interestingly, also unidirectional fluid shear stress, which, as discussed above, may be regarded as an antagonistic haemodynamic force to CWT, seems capable of interfering with the myocardin system. The main effector kinase activated by endothelial cell-derived NO in smooth muscle cells, the cGMP-dependent protein kinase I, apparently causes the degradation of Elk-1, which is one of the inhibitory co-factors competing with myocardin for SRF binding, and in this way stabilises the contractile smooth muscle cell phenotype (Choi et al., 2010). Since the consequences of an imbalance between CWT and fluid shear stress for this level of smooth muscle cell phenotype control are as yet poorly defined, a thorough analysis of stretch-induced changes in myocardin and Zfm1 function seems warranted and may in fact be quite insightful.

3.7 Conclusions and Perspectives

In summary, an altered balance between unidirectional fluid shear stress as the cardinal anti-inflammatory haemodynamic force and circumferential wall tension or stretch basically promoting pro-inflammatory gene expression ultimately determines the degree of inflammation and, consequently, the type of remodelling process the affected vessel is subjected to. Decreased unidirectional fluid shear stress in the face of enhanced pulsatile stretch causes endothelial dysfunction, and hence atherosclerosis, namely at arterial bifurcations or curvatures. In contrast to this maladaptive remodelling process, an increase in the longitudinal pressure gradient preferentially results in a volume-dependent distension, hence stretching of the arteriolar vessel wall, and sets off arteriogenesis which by forming a natural bypass constitutes an adaptive remodelling process. Even though arteriogenesis like atherosclerosis requires the recruitment of mononuclear leukocytes and thus a pro-inflammatory component, it is self-limiting, whereas atherosclerosis typically deteriorates into a chronic recurrent inflammation of the arterial vessel wall. A chronic rise in transmural pressure elicits the characteristic hypertrophic or hyperplastic, ultimately maladaptive, remodelling response in arteries or arterioles in hypertension. In contrast to atherosclerosis or arteriogenesis, hypertension-induced arterial remodelling does not comprise a noteworthy pro-inflammatory response, but strongly depends on a phenotypic shift of the vascular smooth muscle cells from the physiological contractile to the pathophysiological synthetic state. Pro-inflammatory mediators that are formed by both endothelial and smooth muscle cells during the aforementioned remodelling processes comprise a multitude of chemokines and cytokines as well as various receptor-ligand dyads required for cell-cell but also for cell-matrix interactions. Their expression is largely controlled by stress-responsive rather than mechanosensitive transcription factors and may in fact be the result of an imbalance between antioxidants such as NO and pro-oxidants such as superoxide anions, and in particular hydrogen peroxide, which affects several pivotal protein kinase pathways in vascular cells. The cytoskeletal protein zyxin, on the other hand, seems to mark the first stretch-specific mechanotransducer in both endothelial and smooth muscle cells that inter alia controls the expression of various chemokines and cell-cell interaction molecules (Fig. 3.6).

What is missing? Naturally, this review cannot cover all aspects of mechanosensitive pro-inflammatory gene expression and is therefore focussed on the signalling pathways leading to an alteration in the transcriptome of vascular cells. Posttranscriptional or epigenetic mechanisms have not been addressed as such data are still quite scarce. Despite the fact that microRNAs are likely to affect gene expression, and hence the phenotype, of these cells, apparently no comprehensive attempt has been made as yet to determine the microRNA signature of endothelial or vascular smooth muscle cells exposed to fluid shear stress and/or (cyclic) stretch. MicroRNAs implicated in vascular remodelling so far encompass miR-143, miR-145 and miR-26a, which are up-regulated in vascular smooth muscle cells in response to a pressure trauma of the vessel wall such as in restenosis (Xin et al., 2009) or following mechanical stretch (Mohamed et al., 2010). In

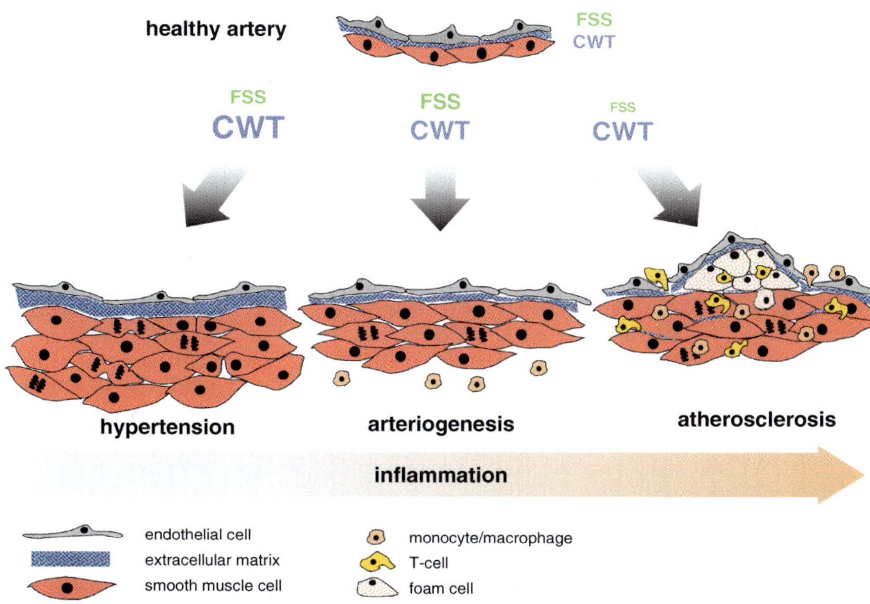

Fig. 3.6 Haemodynamic forces inducing different types of vascular remodelling processes. An alteration of the balance of forces (usually a decrease of the ratio of fluid shear stress to CWT) determines the degree of inflammation and, consequently, the type of remodelling process the affected vessel is subjected to. While a chronic rise in transmural pressure elicits the hypertrophic or hyperplastic remodelling response in hypertension, an increase in the longitudinal pressure gradient may favour arteriogenesis instead. Decreased unidirectional fluid shear stress in the face of enhanced pulsatile stretch causes endothelial dysfunction, and hence atherosclerosis. Common to all these remodelling processes is a varying degree of pro-inflammatory gene expression and smooth muscle cell dedifferentiation. Inflammation is highest in atherosclerosis followed by arteriogenesis and hypertension-induced arterial remodelling where it is virtually absent

endothelial cells, miR-21 may contribute to the shear stress-dependent stabilisation of the anti-inflammatory phenotype (Weber et al., 2010) The target transcripts of these microRNAs, though, remain to be elucidated in order to judge the relative significance of this posttranscriptional mechanism for adaptive or maladaptive vascular remodelling as compared to the transcription factors described herein.

Another interesting aspect that could not be covered here is the site-specific heterogeneity of vascular cells (Hastings et al., 2007; Davies et al., 2010) and the role of the microenvironment with regard to their response to haemodynamic stimuli. In this context, it is well known that the stiffness of extracellular matrix by way of outside-in-signalling via extracellular matrix-integrin interactions greatly influences the phenotype of these cells which, in turn, can actively alter the composition of the extracellular matrix, and hence its stiffness. How this mechanistically relates to the expression of pro-inflammatory gene products in these cells, however, remains to be worked out.

Acknowledgements The authors work is supported by the German Research Foundation (DFG), the Federal Ministry of Research and Education (BMBF), the European Commission and the German Cardiac Society (DGK). We are indebted to Dr. Gerd König for the making of the figures and critically reading the manuscript.

References

Abate C, Patel L, Rauscher FJ 3rd, Curran T (1990) Redox regulation of fos and jun DNA-binding activity in vitro. Science 249:1157–1161

Amma H, Naruse K, Ishiguro N, Sokabe M (2005) Involvement of reactive oxygen species in cyclic stretch-induced NF-κB activation in human fibroblast cells. Br J Pharmacol 145:364–373

Asif AR, Hecker M, Cattaruzza M (2009a) Disinhibition of SOD-2 expression to compensate for a genetically determined NO deficit in endothelial cells – brief report. Arterioscler Thromb Vasc Biol 29:1890–1893

Asif AR, Oellerich M, Armstrong VW, Hecker M, Cattaruzza M (2009b) T-786C polymorphism of the NOS-3 gene and the endothelial cell response to fluid shear stress-a proteome analysis. J Proteome Res 8:3161–3168

Bach I (2000) The LIM domain: regulation by association. Mech Dev 91:5–17

Beckerle MC (1997) Zyxin: zinc fingers at sites of cell adhesion. Bioessays 19:949–957

Behm CZ, Kaufmann BA, Carr C, Lankford M, Sanders JM, Rose CE, Kaul S, Lindner JR (2008) Molecular imaging of endothelial vascular cell adhesion molecule-1 expression and inflammatory cell recruitment during vasculogenesis and ischemia-mediated arteriogenesis. Circulation 117:2902–2911

Boon RA, Horrevoets AJ (2009) Key transcriptional regulators of the vasoprotective effects of shear stress. Hamostaseologie 29:39–43

Busse R, Fleming I (1996) Endothelial dysfunction in atherosclerosis. J Vasc Res 33:181–194

Cattaruzza M, Hecker M (2008) Protein carbonylation and decarboylation: a new twist to the complex response of vascular cells to oxidative stress. Circ Res 102:273–274

Cattaruzza M, Berger MM, Ochs M, Fayyazi A, Füzesi L, Richter J, Hecker M (2002a) Deformation-induced endothelin B receptor-mediated smooth muscle cell apoptosis is matrix-dependent. Cell Death Differ 9:219–226

Cattaruzza M, Schäfer K, Hecker M (2002b) Cytokine-induced down-regulation of zfm1/splicing factor-1 promotes smooth muscle cell proliferation. J Biol Chem 277:6582–6589

Cattaruzza M, Słodowski W, Stojakovic M, Krzesz R, Hecker M (2003) Interleukin-10 induction of nitric-oxide synthase expression attenuates CD40-mediated interleukin-12 synthesis in human endothelial cells. J Biol Chem 278:37874–37880

Cattaruzza M, Guzik TJ, Słodowski W, Pelvan A, Becker J, Halle M, Buchwald AB, Channon KM, Hecker M (2004a) Shear stress insensitivity of endothelial nitric oxide synthase expression as a genetic risk factor for coronary heart disease. Circ Res 95:841–847

Cattaruzza M, Lattrich C, Hecker M (2004b) Focal adhesion protein zyxin is a mechanosensitive modulator of gene expression in vascular smooth muscle cells. Hypertension 43: 726–730

Chen Z, Rubin J, Tzima E (2010) Role of PECAM-1 in arteriogenesis and specification of preexisting collaterals. Circ Res 107:1355–1363

Cheng JJ, Wung BS, Chao YJ, Wang DL (1998) Cyclic strain-induced reactive oxygen species involved in ICAM-1 gene induction in endothelial cells. Hypertension 31:125–130

Cheng JJ, Wung BS, Chao YJ, Wang DL (2001) Sequential activation of protein kinase C (PKC)-α and PKC- ϵ contributes to sustained Raf/ERK1/2 activation in endothelial cells under mechanical strain. J Biol Chem 276:31368–31375

Choi C, Sellak H, Brown FM, Lincoln TM 2010 cGMP-dependent protein kinase and the regulation of vascular smooth muscle cell gene expression: possible involvement of Elk-1 sumoylation. Am J Physiol Heart Circ Physiol 299:H1660–H1670

Collins T, Read MA, Neish AS, Whitley MZ, Thanos D, Maniatis T (1995) Transcriptional regulation of endothelial cell adhesion molecules: NF-κB and cytokine-inducible enhancers. FASEB J 9:899–909

Colombelli J, Besser A, Kress H, Reynaud EG, Girard P, Caussinus E, Haselmann U, Small JV, Schwarz US, Stelzer EH (2009) Mechanosensing in actin stress fibers revealed by a close correlation between force and protein localization. J Cell Sci 122:1665–1679

Colombo MG, Paradossi U, Andreassi MG, Botto N, Manfredi S, Masetti S, Biagini A, Clerico A (2003) Endothelial nitric oxide synthase gene polymorphisms and risk of coronary artery disease. Clin Chem 49:389–395

Crawford AW, Beckerle MC (1991) Purification and characterization of zyxin, an 82,000-dalton component of adherens junctions. J Biol Chem 266:5847–5853

Crawford AW, Michelsen JW, Beckerle MC (1992) An interaction between zyxin and α-actinin. J Cell Biol 116:1381–1393

Davies PF, Civelek M, Fang Y, Guerraty MA, Passerini AG (2010) Endothelial heterogeneity associated with regional athero-susceptibility and adaptation to disturbed blood flow in vivo. Semin Thromb Hemost 36:265–275

Davis B, Zou MH (2005) CD40 ligand-dependent tyrosine nitration of prostacyclin synthase in vivo. Circulation 112:2184–2192

Davis ME, Grumbach IM, Fukai T, Cutchins A, Harrison DG (2004) Shear stress regulates endothelial nitric-oxide synthase promoter activity through nuclear factor κB binding. J Biol Chem 279:163–168

Deliri H, McNamara CA (2007) Nox 4 regulation of vascular smooth muscle cell differentiation marker gene expression. Arterioscler Thromb Vasc Biol 27:12–14

Demicheva E, Hecker M, Korff T (2008) Stretch-induced activation of the transcription factor activator protein-1 controls monocyte chemoattractant protein-1 expression during arteriogenesis. Circ Res 103:477–484

Du W, Mills I, Sumpio BE (1995) Cyclic strain causes heterogeneous induction of transcription factors, AP-1, CRE binding protein and NF-κB, in endothelial cells: species and vascular bed diversity. J Biomech 28:1485–1491

Du KL, Ip HS, Li J, Chen M, Dandre F, Yu W, Lu MM, Owens GK, Parmacek MS (2003) Myocardin is a critical serum response factor cofactor in the transcriptional program regulating smooth muscle cell differentiation. Mol Cell Biol 23:2425–2437

Förstermann U, Kleinert H (1995) Nitric oxide synthase: expression and expressional control of the three isoforms. Naunyn Schmiedebergs Arch Pharmacol 352:351–364

Frangos SG, Knox R, Yano Y, Chen E, Di Luozzo G, Chen AH, Sumpio BE (2001) The integrin-mediated cyclic strain-induced signaling pathway in vascular endothelial cells. Endothelium 8:1–10

Glagov S, Zarins C, Giddens DP, Ku DN (1988) Hemodynamics and atherosclerosis. Insights and perspectives gained from studies of human arteries. Arch Pathol Lab Med 112:1018–1031

Goettsch C, Goettsch W, Muller G, Seebach J, Schnittler HJ, Morawietz H (2009) Nox4 overexpression activates reactive oxygen species and p38 MAPK in human endothelial cells. Biochem Biophys Res Commun 380:355–360

Gopalakrishna R, Jaken S (2000) Protein kinase C signaling and oxidative stress. Free Radic Biol Med 28:1349–1361

Guest TM, Vlastos G, Alameddine FM, Taylor WR (2006) Mechanoregulation of monocyte chemoattractant protein-1 expression in rat vascular smooth muscle cells. Antioxid Redox Signal 8:1461–1471

Guzik TJ, Mussa S, Gastaldi D, Sadowski J, Ratnatunga C, Pillari R, Channon KM (2002) Mechanisms of increased vascular superoxide production in human diabetes mellitus: role of NAD(P)H oxidase and endothelial nitric oxide synthase. Circulation 105:1656–1662

Hamik A, Lin Z, Kumar A, Balcells M, Sinha S, Katz J, Feinberg MW, Gerzsten RE, Edelman ER, Jain MK (2007) Kruppel-like factor 4 regulates endothelial inflammation. J Biol Chem 282:13769–13779

Hansen MD, Beckerle MC (2006) Opposing roles of zyxin/LPP ACTA repeats and the LIM domain region in cell-cell adhesion. J Biol Chem 281:16178–16188
Hastings NE, Simmers MB, McDonald OG, Wamhoff BR, Blackman BR (2007) Atherosclerosis-prone hemodynamics differentially regulates endothelial and smooth muscle cell phenotypes and promotes pro-inflammatory priming. Am J Physiol Cell Physiol 293:C1824–C1833
Hay DC, Beers C, Cameron V, Thomson L, Flitney FW, Hay RT (2003) Activation of NF-κB nuclear transcription factor by flow in human endothelial cells. Biochim Biophys Acta 1642:33–44
Hecker M, Mülsch A, Bassenge E, Busse R (1993) Vasoconstriction and increased flow: two principal mechanisms of shear stress-dependent endothelial autacoid release. Am J Physiol 265:H828–H833
Hellstrand P, Albinsson S (2005) Stretch-dependent growth and differentiation in vascular smooth muscle: role of the actin cytoskeleton. Can J Physiol Pharmacol 83:869–875
Helmke BP, Davies PF (2002) The cytoskeleton under external fluid mechanical forces: hemodynamic forces acting on the endothelium. Ann Biomed Eng 30:284–296
Hildebrandt A, Hecker M, Wagner AH (2009) Cyclic stretch-mediated tyrosine nitration of CD40 impairs CD40 ligand-stimulated IL12p40 expression in human endothelial cells. Acta Physiol 195:O529
Hill-Eubanks DC, Gomez MF, Stevenson AS, Nelson MT (2003) NFAT regulation in smooth muscle. Trends Cardiovasc Med 13:56–62
Hirata H, Tatsumi H, Sokabe M (2008a) Mechanical forces facilitate actin polymerization at focal adhesions in a zyxin-dependent manner. J Cell Sci 121:2795–2804
Hirata H, Tatsumi H, Sokabe M (2008b) Zyxin emerges as a key player in the mechanotransduction at cell adhesive structures. Commun Integr Biol 1:192–195
Hoffman LM, Jensen CC, Kloeker S, Wang CL, Yoshigi M, Beckerle MC (2006) Genetic ablation of zyxin causes Mena/VASP mislocalization, increased motility, and deficits in actin remodeling. J Cell Biol 172:771–782
Hojo Y, Saito Y, Tanimoto T, Hoefen RJ, Baines CP, Yamamoto K, Haendeler J, Asmis R, Berk BC (2002) Fluid shear stress attenuates hydrogen peroxide-induced c-Jun NH2-terminal kinase activation via a glutathione reductase-mediated mechanism. Circ Res 91:712–718
Howard AB, Alexander RW, Nerem RM, Griendling KK, Taylor WR (1997) Cyclic strain induces an oxidative stress in endothelial cells. Am J Physiol 272:C421–C427
Hsiai TK, Cho SK, Wong PK, Ing M, Salazar A, Sevanian A, Navab M, Demer LL, Ho CM (2003) Monocyte recruitment to endothelial cells in response to oscillatory shear stress. FASEB J 17:1648–1657
Ishibashi M, Hiasa K, Zhao Q, Inoue S, Ohtani K, Kitamoto S, Tsuchihashi M, Sugaya T, Charo IF, Kura S, Tsuzuki T, Ishibashi T, Takeshita A, Egashira K (2004) Critical role of monocyte chemoattractant protein-1 receptor CCR2 on monocytes in hypertension-induced vascular inflammation and remodeling. Circ Res 94:1203–1210
Kanungo J, Pratt SJ, Marie H, Longmore GD (2000) Ajuba, a cytosolic LIM protein, shuttles into the nucleus and affects embryonal cell proliferation and fate decisions. Mol Biol Cell 11:3299–3313
Karin M, Liu Z, Zandi E (1997) AP-1 function and regulation. Curr Opin Cell Biol 9:240–246
Kato T, Muraski J, Chen Y, Tsujita Y, Wall J, Glembotski CC, Schaefer E, Beckerle M, Sussman MA (2005) Atrial natriuretic peptide promotes cardiomyocyte survival by cGMP-dependent nuclear accumulation of zyxin and Akt. J Clin Invest 115:2716–2730
Kawai-Kowase K, Owens GK (2007) Multiple repressor pathways contribute to phenotypic switching of vascular smooth muscle cells. Am J Physiol Cell Physiol 292:C59–C69
Kawashima S, Yokoyama M (2004) Dysfunction of endothelial nitric oxide synthase and atherosclerosis. Arterioscler Thromb Vasc Biol 24:998–1005
Kelkenberg U, Wagner AH, Sarhaddar J, Hecker M, von der Leyen HE (2002) CCAAT/enhancer-binding protein decoy oligodeoxynucleotide inhibition of macrophage-rich vascular lesion formation in hypercholesterolemic rabbits. Arterioscler Thromb Vasc Biol 22:949–954

Kida T, Chuma H, Murata T, Yamawaki H, Matsumoto S, Hori M, Ozaki H (2010) Chronic treatment with PDGF-BB and endothelin-1 synergistically induces vascular hyperplasia and loss of contractility in organ-cultured rat tail artery. Atherosclerosis. [PMID: 21129745, Epub ahead of print]

Kinlay S, Libby P, Ganz P (2001) Endothelial function and coronary artery disease. Curr Opin Lipidol 12:383–389

Klotz LO, Schroeder P, Sies H (2002) Peroxynitrite signaling: receptor tyrosine kinases and activation of stress-responsive pathways. Free Radic Biol Med 33:737–743

Korff T, Aufgebauer K, Hecker M (2007) Cyclic stretch controls the expression of CD40 in endothelial cells by changing their transforming growth factor-β1 response. Circulation 116:2288–2297

Korff T, Braun J, Pfaff D, Augustin HG, Hecker M (2008) Role of ephrinB2 expression in endothelial cells during arteriogenesis: Impact on smooth muscle cell migration and monocyte recruitment. Blood 112:73–81

Kumar A, Knox AJ, Boriek AM (2003) CCAAT/enhancer-binding protein and activator protein-1 transcription factors regulate the expression of interleukin-8 through the mitogen-activated protein kinase pathways in response to mechanical stretch of human airway smooth muscle cells. J Biol Chem 278:18868–18876

Kuwahara K, Kinoshita H, Kuwabara Y, Nakagawa Y, Usami S, Minami T, Yamada Y, Fujiwara M, Nakao K (2010) Myocardin-related transcription factor A is a common mediator of mechanical stress- and neurohumoral stimulation-induced cardiac hypertrophic signaling leading to activation of brain natriuretic peptide gene expression. Mol Cell Biol. 30:4134–4148

Laumonnier Y, Nadaud S, Agrapart M, Soubrier F (2000) Characterization of an upstream enhancer region in the promoter of the human endothelial nitric-oxide synthase gene. J Biol Chem 275:40732–40741

Lauth M, Wagner AH, Cattaruzza M, Orzechowski HD, Paul M, Hecker M (2000) Transcriptional control of deformation-induced preproendothelin-1 gene expression in endothelial cells. J Mol Med 78:441–450

Lehoux S, Castier Y, Tedgui A (2006) Molecular mechanisms of the vascular responses to haemodynamic forces. J Intern Med 259:381–392

Libby P (2007) Inflammatory mechanisms: the molecular basis of inflammation and disease. Nutr Rev 65:S140–S146

Lutgens E, Gorelik L, Daemen MJ, de Muinck ED, Grewal IS, Koteliansky VE, Flavell RA (1999) Requirement for CD154 in the progression of atherosclerosis. Nat Med 5:1313–1316

Macarthur H, Warner TD, Wood EG, Corder R, Vane JR (1994) Endothelin-1 release from endothelial cells in culture is elevated both acutely and chronically by short periods of mechanical stretch. Biochem Biophys Res Commun 200:395–400

McCubrey JA, Lahair MM, Franklin RA (2006) Reactive oxygen species-induced activation of the MAP kinase signaling pathways. Antioxid Redox Signal 8:1775–1789

Melchers I, Blaschke S, Hecker M, Cattaruzza M (2006) The -786C/T single-nucleotide polymorphism in the promoter of the gene for endothelial nitric oxide synthase: insensitivity to physiologic stimuli as a risk factor for rheumatoid arthritis. Arthritis Rheum 54:3144–3151

Mohamed JS, Lopez MA, Boriek AM (2010) Mechanical stretch up-regulates microRNA-26a and induces human airway smooth muscle hypertrophy by suppressing glycogen synthase kinase-3β. J Biol Chem 285:29336–29347

Molavi B, Mehta JL (2004) Oxidative stress in cardiovascular disease: molecular basis of its deleterious effects, its detection, and therapeutic considerations. Curr Opin Cardiol 19:488–493

Musunuru K, Kathiresan S (2010) Genetics of coronary artery disease. Annu Rev Genomics Hum Genet 11:91–108

Ngu H, Feng Y, Lu L, Oswald SJ, Longmore GD, Yin FC (2010) Effect of focal adhesion proteins on endothelial cell adhesion, motility and orientation response to cyclic strain. Ann Biomed Eng 38:208–222

Nilsson LM, Sun ZW, Nilsson J, Nordström I, Chen YW, Molkentin JD, Wide-Swensson D, Hellstrand P, Lydrup ML, Gomez MF (2006) Novel blocker of NFAT activation inhibits IL-6 production in human myometrial arteries and reduces vascular smooth muscle cell proliferation. Am J Physiol Cell Physiol 292:C1167–C1178

Nix DA, Fradelizi J, Bockholt S, Menichi B, Louvard D, Friederich E, Beckerle MC (2001) Targeting of zyxin to sites of actin membrane interaction and to the nucleus. J Biol Chem 276:34759–34767

Pan S (2009) Molecular mechanisms responsible for the atheroprotective effects of laminar shear stress. Antioxid Redox Signal 11:1669–1682

Petit MM, Fradelizi J, Golsteyn RM, Ayoubi TA, Menichi B, Louvard D, Van de Ven WJ, Friederich E (2000) LPP, an actin cytoskeleton protein related to zyxin, harbors a nuclear export signal and transcriptional activation capacity. Mol Biol Cell 11:117–129

Phipps RP, Koumas L, Leung E, Reddy SY, Blieden T, Kaufman J (2001) The CD40-CD40 ligand system: a potential therapeutic target in atherosclerosis. Curr Opin Investig Drugs 2:773–777

Porreca E, Di Febbo C, Reale M, Castellani ML, Baccante G, Barbacane R, Conti P, Cuccurullo F, Poggi A (1997) Monocyte chemotactic protein 1 (MCP-1) is a mitogen for cultured rat vascular smooth muscle cells. J Vasc Res 34:58–65

Reinhard M, Jouvenal K, Tripier D, Walter U (1995) Identification, purification, and characterization of a zyxin-related protein that binds the focal adhesion and microfilament protein VASP (vasodilator-stimulated phosphoprotein. Proc Natl Acad Sci USA 92:7956–7960

Reinhard M, Zumbrunn J, Jaquemar D, Kuhn M, Walter U, Trueb B (1999) An α-actinin binding site of zyxin is essential for subcellular zyxin localization and α-actinin recruitment. J Biol Chem 274:13410–13418

Rossi GP, Taddei S, Virdis A, Cavallin M, Ghiadoni L, Favilla S, Versari D, Sudano I, Pessina AC, Salvetti A (2003) The T-786C and Glu298Asp polymorphisms of the endothelial nitric oxide gene affect the forearm blood flow responses of Caucasian hypertensive patients. J Am Coll Cardiol 41:938–945

Schaper W (2009) Collateral circulation: past and present. Basic Res Cardiol 104:5–21

Schaper W, Scholz D (2003) Factors regulating arteriogenesis. Arterioscler Thromb Vasc Biol 23:1143–1151

Schepers A, Eefting D, Bonta PI, Grimbergen JM, de Vries MR, van Weel V, de Vries CJ, Egashira K, van Bockel JH, Quax PH (2006) Anti-MCP-1 gene therapy inhibits vascular smooth muscle cells proliferation and attenuates vein graft thickening both in vitro and in vivo. Arterioscler Thromb Vasc Biol 26:2063–2069

Schlossmann J, Desch M (2009) cGK substrates. Handb Exp Pharmacol 191:163–193

Schwartz RS, Chronos NA, Virmani R (2004) Preclinical restenosis models and drug-eluting stents: still important, still much to learn. J Am Coll Cardiol 44:1373–1385

Scott DW, Loo G (2007) Curcumin-induced GADD153 upregulation: modulation by glutathione. J Cell Biochem 101:307–320

Sen CK, Packer L (1996) Antioxidant and redox regulation of gene transcription. FASEB J 10:709–720

SenBanerjee S, Lin Z, Atkins GB, Greif DM, Rao RM, Kumar A, Feinberg MW, Chen Z, Simon DI, Luscinskas FW, Michel TM, Gimbrone MA Jr, García-Cardeña G, Jain MK (2004) KLF2 Is a novel transcriptional regulator of endothelial proinflammatory activation. J Exp Med 199:1305–1315

Sessa WC, Pritchard K, Seyedi N, Wang J, Hintze TH (1994) Chronic exercise in dogs increases coronary vascular nitric oxide production and endothelial cell nitric oxide synthase gene expression. Circ Res 74:349–353

Sharma GD, Nguyen HT, Antonov AS, Gerrity RG, von Geldern T, Pandey KN (2002) Expression of atrial natriuretic peptide receptor-A antagonizes the mitogen-activated protein kinases (Erk2 and P38MAPK) in cultured human vascular smooth muscle cells. Mol Cell Biochem 233:165–173

Sharma SK, Sweeny J, Kini AS (2010) Coronary bifurcation lesions: a current update. Cardiol Clin 28:55–70

Silacci P, Formentin K, Bouzourène K, Daniel F, Brunner HR, Hayoz D (2000) Unidirectional and oscillatory shear stress differentially modulate NOS III gene expression. Nitric Oxide 4:47–56

Silacci P, Desgeorges A, Mazzolai L, Chambaz C, Hayoz D (2001) Flow pulsatility is a critical determinant of oxidative stress in endothelial cells. Hypertension 38:1162–1166

Souza JM, Choi I, Chen Q, Weisse M, Daikhin E, Yudkoff M, Obin M, Ara J, Horwitz J, Ischiropoulos H (2000) Proteolytic degradation of tyrosine nitrated proteins. Arch Biochem Biophys 380:360–366

Tedgui A, Mallat Z (2001) Anti-inflammatory mechanisms in the vascular wall. Circ Res 88: 877–887

Uematsu M, Ohara Y, Navas JP, Nishida K, Murphy TJ, Alexander RW, Nerem RM, Harrison DG (1995) Regulation of endothelial cell nitric oxide synthase mRNA expression by shear stress. Am J Physiol 269:C1371–C1378

van Halm VP, Peters MJ, Voskuyl AE, Boers M, Lems WF, Visser M, Stehouwer CD, Spijkerman AM, Dekker JM, Nijpels G, Heine RJ, Bouter LM, Smulders YM, Dijkmans BA, Nurmohamed MT (2009) Rheumatoid arthritis versus diabetes as a risk factor for cardiovascular disease: a cross-sectional study, the CARRE Investigation. Ann Rheum Dis 68:1395–1400

Vasioukhin V, Bauer C, Yin M, Fuchs E (2000) Directed actin polymerization is the driving force for epithelial cell-cell adhesion. Cell 100:209–219

Wagner AH, Schroeter MR, Hecker M (2001) 17β-Estradiol inhibition of NADPH oxidase expression in human endothelial cells. FASEB J 15:2121–2130

Wagner AH, Kautz O, Fricke K, Zerr-Fouineau M, Demicheva E, Güldenzoph B, Bermejo JL, Korff T, Hecker M (2009) Upregulation of glutathione peroxidase offsets stretch-induced proatherogenic gene expression in human endothelial cells. Arterioscler Thromb Vasc Biol 29:1894–1901

Wang DS, Proffit D, Tsao PS (2001) Mechanotransduction of endothelial oxidative stress induced by cyclic strain. Endothelium 8:283–291

Wang Y, Gilmore TD (2003) Zyxin and paxillin proteins: focal adhesion plaque LIM domain proteins go nuclear. Biochim Biophys Acta 1593:115–120

Wang Z, Wang DZ, Hockemeyer D, McAnally J, Nordheim A, Olson EN (2004) Myocardin and ternary complex factors compete for SRF to control smooth muscle gene expression. Nature 428:185–189

Weber M, Baker MB, Moore JP, Searles CD (2010) MiR-21 is induced in endothelial cells by shear stress and modulates apoptosis and eNOS activity. Biochem Biophys Res Commun 393:643–648

Wójtowicz A, Babu SS, Li L, Gretz N, Hecker M, Cattaruzza M (2010) Zyxin mediation of stretch-induced gene expression in human endothelial cells. Circ Res 107:898–902

Won D, Zhu SN, Chen M, Teichert AM, Fish JE, Matouk CC, Bonert M, Ojha M, Marsden PA, Cybulsky MI (2007) Relative reduction of endothelial nitric-oxide synthase expression and transcription in atherosclerosis-prone regions of the mouse aorta and in an in vitro model of disturbed flow. Am J Pathol 171:1691–1704

Wung BS, Cheng JJ, Hsieh HJ, Shyy YJ, Wang DL (1997) Cyclic strain-induced monocyte chemotactic protein-1 gene expression in endothelial cells involves reactive oxygen species activation of activator protein 1. Circ Res 81:1–7

Xin M, Small EM, Sutherland LB, Qi X, McAnally J, Plato CF, Richardson JA, Bassel-Duby R, Olson EN (2009) MicroRNAs miR-143 and miR-145 modulate cytoskeletal dynamics and responsiveness of smooth muscle cells to injury. Genes Dev 23:2166–2178

Yoshida T, Sinha S, Dandré F, Wamhoff BR, Hoofnagle MH, Kremer BE, Wang DZ, Olson EN, Owens GK (2003) Myocardin is a key regulator of CArG-dependent transcription of multiple smooth muscle marker genes. Circ Res 92:856–864

Yoshigi M, Hoffman LM, Jensen CC, Yost HJ, Beckerle MC (2005) Mechanical force mobilizes zyxin from focal adhesions to actin filaments and regulates cytoskeletal reinforcement. J Cell Biol 171:209–215

Yue TL, Wang X, Sung CP, Olson B, McKenna PJ, Gu JL, Feuerstein GZ (1994) Interleukin-8. A mitogen and chemoattractant for vascular smooth muscle cells. Circ Res 75:1–7
Zeiher AM, Fisslthaler B, Schray-Utz B, Busse R (1995) Nitric oxide modulates the expression of monocyte chemoattractant protein 1 in cultured human endothelial cells. Circ Res 76:980–986
Zhang D, Childs G (1998b) Human ZFM1 protein is a transcriptional repressor that interacts with the transcription activation domain of stage-specific activator protein. J Biol Chem 273: 6868–6877
Zhang D, Paley AJ, Childs G (1998a) The transcriptional repressor ZFM1 interacts with and modulates the ability of EWS to activate transcription. J Biol Chem 273:18086–18091
Zhao S, Suciu A, Ziegler T, Moore JEJR, Burki E, Meister JJ, Brunner HR (1995) Synergistic effects of fluid shear stress and cyclic circumferential stretch on vascular endothelial cell morphology and cytoskeleton. Arterioscler Thromb Vasc Biol 15:1781–1786
Ziegler T, Silacci P, Harrison VJ, Hayoz D (1998) Nitric oxide synthase expression in endothelial cells exposed to mechanical forces. Hypertension 32:351–355

Chapter 4
Cytokines, Heart and Calcium Current in Sepsis

Milan Stengl, Miroslav Prucha, and Martin Matejovic

Abstract Sepsis has been defined as the systemic host response to infection with an overwhelming systemic production of cytokines leading to generalized endothelial and epithelial damages, to the changes in immune and neuroendocrine systems and consequently to (multiple) organ dysfunction. The myocardial contractile performance is significantly impaired in severe sepsis and septic shock. In this review the major cytokines involved in the development of sepsis are characterized with special emphasis on cardiac function and on regulation of cardiac calcium current.

Keywords Calcium current · Cytokines · Heart · Sepsis

4.1 Introduction

The major cause of health concern that claims a large number of lives worldwide every year are the infectious diseases and related sepsis. Severe sepsis, defined as sepsis associated with acute organ dysfunction, results from a generalized inflammatory response to infection (Martin et al., 2003; Angus et al., 2001). Annually, there are 900,000 and 750,000 diagnosed sepsis cases in the Europe and USA resulting in hundreds of thousands of deaths (Martin et al., 2003; Angus et al., 2001). Sepsis is the tenth most common cause of death in the United States, and the situation is also very similar in European Union countries (Martin et al., 2003; Angus et al., 2001). The mortality associated with severe sepsis is high, 20–40% of the patients die. Major epidemiological studies have found that due to its progressively increasing incidence, sepsis is a fundamental medical problem of the new millennium and has wide-reaching socio-economic consequences. The consequences are reflected in most disciplines of medicine where sepsis negatively affects the results of treatment. In addition, the resources required to treat septic patients create an enormous burden on the health care system. There is thus a critical need to improve our understanding of the sepsis pathophysiology and to develop innovative and efficacious therapies for the treatment of this deadly disease. The pathophysiology of sepsis,

---

M. Stengl (✉)
Department of Physiology, Faculty of Medicine in Plzen, Charles University in Prague, 301 66 Plzen, Czech Republic
e-mail: milan.stengl@lfp.cuni.cz

A. Kamkin, I. Kiseleva (eds.), *Mechanical Stretch and Cytokines*, Mechanosensitivity in Cells and Tissues 5, DOI 10.1007/978-94-007-2004-6_4,

however, represents an extremely complicated subject with a complex cross-talk of a number of organs, tissues, regulatory and signalling pathways. Despite intensive research there still remains a vast number of obscure issues that need to be clarified in more detail. Within these limits, this review summarizes the current knowledge about several aspects of the sepsis: from an overview of inflammatory mediators produced in sepsis to organ impairment, i.e. myocardial depression, and further to the molecular and cellular mechanisms of this impairment and the role of calcium current in it.

4.2 Cytokines in Sepsis

Sepsis is a systemic inflammatory response to infection. The ethiological agents involved in septic states, include: bacteria, viruses, fungi and parasites. In Gram negative sepsis endotoxin represents the underlying pathogenic structure (Parker and Watkins, 2001). Immune activation by Gram-positive bacteria is far less well understood, but lipoteichoic acids (LTA) is a key Gram-positive immunostimulatory component (Draing et al., 2008). As a result of their effects on the leukocytes and cells of parenchymatous organs, there is a marked production of pro- and anti-inflammatory mediators (Calandra et al., 1990; Cassatella et al., 1993; Damas et al., 1989; Emery and Salmon, 1991). Two main cytokines are important for the development of an inflammatory cascade induced by endotoxin: tumor necrosis factor-α (TNF-α) and interleukin-1β (IL-1β). These cytokines stimulate the immune response, including, among others, generation of arachidonic acid, expression of integrins, complement and production of nitric oxide. In addition, cytokines activate neutrophils and increase procoagulant activity of endothelial cells, resulting in tissue damage characteristic of sepsis.

Early sepsis is associated with profound elevation of inflammatory mediators in blood, such as TNF-α, IL-1, IL-8, IL-12, gamma interferon (IFN-γ) and IL-6. They play a crucial role in initiating effective immune-inflammatory process against infection. The current concept of inflammatory response in sepsis represents the dogma of the SIRS (systemic inflammatory response syndrome) to CARS (compensatory anti-inflammatory response syndrome) transition. CARS is characterized by the production of anti-inflammatory mediators such as transforming growth factor β (TGF-β), IL-4, IL-10, IL-13. These cytokines may be released in an effort to counteract ongoing inflammation. Interestingly, some studies showed that anti-inflammatory response is present already in the early stages of sepsis and hyper-inflammatory response is present also in the late phase of sepsis (Tschaikowsky et al., 2002; Osuchowski et al., 2007). The cascade of cytokines production is mediated by JAK/STAT proteins with the subsequent effect on Th1/Th2 balance (Scott et al., 2002).

The release of inflammatory mediators, i.e. leukotrienes, complement components, cytokines and production of antigen-antibody complexes cause accumulation of neutrophils and monocytes in the site of damage as well as in remote organs. In addition, T cells contribute to the protective immune responses in sepsis via their ability to produce crucial immune mediators (Reim et al., 2009). Activated CD4+lymphocytes are divided into two subgroups with antagonistic cytokine profiles; type TH1 with TNF-α, interferon–γ and interleukin-2 (proinflammatory) and type TH2 with IL-4 and IL-10 production (antiinflammatory). While IFN-γ and TNF-α increase the ability of phagocytes to kill infectious agents, IL-4 and IL-10 support mainly humoral responses to extracellular pathogens. Factors responsible for the differentiation of cells are not fully known, but in the case of infection important factors include the type of pathogen, the amount of infectious inoculum and the localization of infection (Abbas et al., 1996).

IL-10 has been identified as being specifically responsible for the deactivation of monocytes. A time connection between the dynamics of IL-10 and TNF-α has been established (de Waal Malefyt et al., 1991; Durez et al., 1993). IL-10 acts as a class II MHC molecules suppressor, inhibiting antigen-specific proliferation of T lymphocytes. Moreover, it inhibits the INF-γ production by lymphocytes – an effect independent of monocytes, and inhibits the production of pro-inflammatory cytokines and chemokines by activated macrophages.

IL-12 family of cytokines is an important determinant of T cell response. IL-12, IL-27 and IL-23 are cytokines that share homology at the receptor and subunit. IL-12 is known to play an important role in defence against bacterial infection via the induction of IFN-gamma production by T and NK cells (Gee et al., 2009). In addition, IL-10 and IL-12 are crucial cytokines in Thl/Th2 polarization (O'Garra and Murphy, 2009). Interleukin-17 is a recently described pro-inflammatory cytokine, which is important in auto-immunity and infection. Th17 cells-produced IL-17 promotes chemotaxis and neutrophils maturation. Despite the protective role of Th17 cell in experimental studies (Ye et al., 2001; Iwakura et al., 2008), its role in human sepsis needs further elucidation (White et al., 2010). High mobility group B protein-1 (HMGB1) is a critical late mediator of experimental sepsis. HMGB is a constitutive intracellular protein and is released from activated monocytes and macrophages into extracellular space. HMGB1 facilitates recognition of microbial products, stimulates cell migration, inhibits phagocytic elimination of apoptotic neutrophils, and activates immune cells. Administration of anti-HMGB1 antibodies in experimental models protects animals from lethal experimental sepsis (Huang et al., 2010). IL-33 is a recently identified member of the IL-1 family. IL-33 reduces the systemic pro-inflammatory response without affecting a shift from Thl to Th2 (Alves-Filho et al., 2010). In this context it seems to be a promising therapeutic agent in sepsis. The dominant type of mediator production is important for the type of pro- or anti-inflammatory response, while its intensity defines the degree of compartmentalization of the immune response.

Cytokines produced in septic patients activate coagulation system and endothelium (Schouten et al., 2008; Stief et al., 2007). Tissue factor (TF) activated by TNF-α, IL-2, and IL-6 has a key role in the coagulation cascade. By contrast, cytokines cause a reduction of fibrinolysis, and a decrease in protein C and antithrombin III. This pro-coagulation state accentuates an inflammatory response in sepsis. Furthermore, TNF-α, IL-1 and other mediators lead to substantial endothelial dysfunction, which significantly contribute to microvascular failure in sepsis (Levi, 2010; Ait-Oufella et al., 2010).

It is well recognized in clinical practice that the outcome of septic patients is dependent on the site of infection as well as on the causative agent(s). Different causative agents are associated with different production of cytokines (Opal et al., 2003; Damas et al., 1989). An example is the influence of causative agents on TNF-α production and the role of TNF-α in the defense against infection. As already pointed out, TNF-α is a principal mediator, which has a negative role in the pathogenesis of sepsis. The binding of TNF-α to its receptors – p55 and p75 – drives the activation of inflammatory and coagulation cascades with its adverse effects. But TNF-α has also a fundamental positive influence on the successful management of infection by the microorganism. A study by Echtenacher et al. (2003) demonstrated a relationship between the mortality and TNF-α production. After cecal ligation and puncture (CLP) in mice, a state of immunoparalysis developed during which super-infection led to increased lethality. Injection of recombinant human TNF-α before or at the time of super-infection conferred protection against Salmonella infection. TNF-α substitution during this state of immunoparalysis can be beneficial or deleterious, depending on the location of TNF-α activity in the animal, timing of TNF-α administration or the type of superinfection (Echtenacher et al., 2003). The importance of TNF-α in the endogenous response to infection is demonstrated by the use of inhibitors TNF for the treatment of a number of chronic inflammatory conditions such as Crohn's disease or rheumatoid arthritis. Here, the inhibition of TNF-α resulted in reactivation of latent tuberculosis or new sepsis. The causative agents were primarily commensal or commonly non-pathogenic for humans. On the contrary, no such dependence has been shown for IL-6. The German study using a similar experimental model demonstrated a link between mortality and level of IL-6, but the lack of IL-6 did not influence the mortality rate (Remick et al., 2005).

Hence, the important characteristic of sepsis is the interaction between two subjects, the macro- and the microorganism. The concept of a hyperinflammatory syndrome that dominated the past two decades has been challenged and at present, sepsis is seen more as a dynamic syndrome characterized by many often antagonistic pathways. The inflammation does not act as a primary physiological compensatory mechanism and rather oscillates between the phase of hyperinflammatory response and anergy or immunoparalysis. We must not forget the important role of the microorganism. The ability of different pathogens to drive pro- and anti-inflammatory cytokines production, their different tricks to escape the surveillance of the immune system, their diverse range of toxin production and their different sensitivity to treatments are equally important in the pathogenesis of sepsis.

4.3 Heart in Sepsis

Myocardial depression is a well-recognized manifestation of sepsis and septic shock (Court et al., 2002; Rudiger and Singer, 2007; Merx and Weber, 2007). In general, cardiovascular dysfunction in sepsis was shown to be associated with an increased mortality (Parrillo et al., 1990). Left ventricular systolic functions, stroke volume and ejection fraction, are significantly reduced in septic patients (Parker et al., 1984; Jardin et al., 1999). Also left ventricular diastolic functions are impaired, slower ventricular filling and aberrant ventricular relaxation were demonstrated using echocardiography (Jafri et al., 1990; Poelaert et al., 1997; Munt et al., 1998). Interestingly, survivors of septic shock showed lower ejection fraction and higher end-diastolic volumes compared to nonsurvivors, perhaps suggesting some protective effect of myocardial depression in sepsis (Parker et al., 1984). Also right ventricular dysfunction is common in sepsis and the reported functional changes resemble those in the left ventricle. Decreased right ventricular ejection fraction and right ventricular dilation were demonstrated and a close temporal parallel between right and left ventricular dysfunction was found (Parker et al., 1990).

A large number of studies, both clinical and preclinical, were performed in recent decades and various mechanisms contributing to the development of cardiac dysfunction were identified. Cardiac dysfunction in sepsis is obviously multifactorial. The relative contribution of individual mechanisms that will allow identification of preferential therapeutic targets, however, remains to be determined.

4.3.1 Circulatory Changes

According to the original theory the myocardial depression in sepsis was supposed to result from global myocardial ischemia from hypoperfusion. However, the coronary blood flow in septic patients was found to be normal or even elevated (Cunnion et al., 1986). Despite the unchanged (or increased) coronary blood flow some local hypoperfusion cannot be excluded. In endotoxemic canine hearts heterogeneous cardiac blood flow was documented and possible focal ischemia suggested (Groeneveld et al., 1991). Nevertheless, using a hypoxic marker [18F]fluoromisonidazole, no evidence of cellular hypoxia was detected in hearts of septic rats (Hotchkiss et al., 1991). Similarly, in a canine model of sepsis with documented myocardial depression the intracellular free energy levels were maintained, suggesting that the high-energy synthetic capabilities do not limit cardiac function and arguing again against myocardial ischemia (Solomon et al., 1994).

4.3.2 Circulating Myocardial Depressant Substances

In a seminal study, Parrillo et al. (1985) demonstrated a negative inotropic effect in rat cardiac myocytes of serum taken from acute phase septic patients with reduced left ventricular ejection fraction. Serum from nonseptic patients, either critically ill or with structural heart disease as cause of depressed ejection fraction, or from healthy volunteers did not affect the shortening of rat cardiac myocytes. A strong correlation between the sepsis-induced reduction of left ventricular ejection fraction in vivo and negative inotropic effect in rat cardiac myocytes induced by serum from the same septic patients was demonstrated. In extension of these findings the ultrafiltration experiments revealed that the depressant serum activity did not pass through 10,000 or 30,000 Da filters. The depressant activity was only found in supernatants containing molecules above 10,000 and 30,000 Da (Reilly et al., 1989). In another ultrafiltration study, cardiotoxic effects in vitro of ultrafiltrates from septic patients with reduced left ventricular contractility work index were associated with increased concentrations of interleukin-1, interleukin-8, and C3a (Hoffmann et al., 1999). All these result strongly suggest that the circulating substances rather than myocardial ischemia are involved in pathogenesis of septic myocardial dysfunction. The list of potential candidates for myocardial depressant substance/s is huge. A recent proteomic study identified 30 protein forms representing 22 unique proteins whose plasma levels were increased and 6 forms of 5 unique proteins with significantly decreased plasma levels during sepsis (Thongboonkerd et al., 2009). Most of these altered proteins are related to inflammatory responses, some of them are involved in oxidative and nitrosative stress. From all potential mediators tumor necrosis factor- α (TNF-α), interleukin-1β (IL-1) and interleukin-6 (IL-6) seem to play a central role and attract higher attention.

4.3.3 Cytokines (TNF-α, IL-1, IL-6)

TNF-α has significant hemodynamic effects that mimic those of septic shock: hypotension, decreased systemic vascular resistance, reduced ejection fraction and myocardial depression (Tracey et al., 1986; Eichenholz et al., 1992; Finkel et al., 1992). In a pilot study, application of anti-TNF-α antibodies improved left ventricular function (Vincent et al., 1992). Larger studies with TNF-α monoclonal antibodies or TNF-α receptor fusion protein, however, failed to improve overall survival (Abraham et al., 1995, 1997, 1998; Fisher et al., 1996). Although activated macrophages are the predominant source of TNF-α, it is synthesized also by cardiac myocytes (Kapadia et al., 1995) suggesting an autocrine regulatory function of TNF-α in the heart. Early and delayed negative inotropic effects of TNF-α can be distinguished (Meldrum, 1998). The early negative inotropic effect of TNF-α may involve inhibition of L-type calcium current (Krown et al., 1995) and accumulation of sphingosine (Oral et al., 1997) and seems to ne nitric oxide independent (Yokoyama et al., 1993). The delayed effect appears to be associated with inducible nitric oxide synthase induction and subsequent nitric oxide-induced myofilament desensitization for calcium (Goldhaber et al., 1996; Finkel et al., 1992; Murray and Freeman, 1996; Oral et al., 1997).

IL-1 is synthesized by monocytes, macrophages and neutrophils in response to TNF-α and its production is increased in sepsis (Hesse et al., 1988). IL-1 also induces depression of cardiac contractility (Hosenpud et al., 1989) and this effect is associated with induction of nitric oxide synthase (Stein et al., 1996). Similar to TNF-α, application of recombinant human IL-1 receptor antagonist in randomized clinical trial failed to demonstrate a statistically significant reduction in mortality (Fisher et al., 1994; Opal et al., 1997). An interesting phenomenon is the apparent synergistic action of TNF-α and IL-1. In combination, TNF-alpha and IL-1beta induced depression of rat myocardial cell contractility at substantially lower concentrations than when applied either of them alone (Kumar et al., 1996). Furthermore, the serum myocardial depressant activity was eliminated only by removal of both TNF-alpha and IL-1beta, removal of either of them alone was not sufficient. The synergistic depressant action of TNF-α and IL-1 was confirmed in human atrial trabeculae (Cain et al., 1999).

When IL-1, TNF-α and IL-6 were monitored in critically ill surgical patients with documented infections, only IL-6 was always increased in serum of acutely ill patients. Furthermore, IL-6 correlated well with APACHE II score, and the mortality rate increased significantly in the group of patients with high IL-6 serum levels (Damas et al., 1992). Increased plasma levels of IL-6 were also documented in clinically relevant animal model of septic shock (Sykora et al., 2009). Therefore, IL-6 appears to be a good marker of severity during bacterial infection. IL-6 was also associated with depression of myocardial contractility. Analysis of depressant effects of human septic serum in rat cardiac myocytes revealed an association of negative inotropic effect with increased levels of IL-6, IL-8, IL-10 cytokines (Joulin et al., 2007). Interestingly, in the same study IL-1 and TNF-α were not increased in septic serum.

4.3.4 Cardiac Cellular Changes

4.3.4.1 Mitochondria

Mitochondria represent a central structure in cellular metabolism and a complex mitochondrial dysfunction in sepsis was demonstrated by a number of studies (Watts et al., 2004; Crouser, 2004; Carré and Singer, 2008; Harrois et al., 2009). Both cardiac mitochondrial ultrastructure and function were reported to be impaired. Destruction of mitochondria and ultrastructural mitochondrial changes including heterogeneity with patchy disruption of inner and outer membranes and variable swelling were documented by electron microscopy (Hersch et al., 1990; Watts et al., 2004; Suliman et al., 2004). The activities of mitochondrial respiratory chain enzymes are inhibited in sepsis. In septic baboon heart mitochondria decreased activities of Complex I and Complex II were found (Gellerich et al., 1999). Also oxidation of cytochrome c by myocardial cytochrome c oxidase, the terminal oxidase in the electron transport chain, is inhibited in sepsis (Levy et al., 2004). The important role of cytochrome c in mitochondrial as well as myocardial dysfunction was confirmed by experiments, in which application of exogenous cytochrome c was able to restore cytochrome oxidase activity and to improve myocardial function and survival (Piel et al., 2007, 2008). Mitochondrial uncoupling could also contribute to the decreased cardiac mechanical efficiency. Uncoupling proteins (UCP) 2 and 3 were detected in human heart (Murray et al., 2004). UCP2 and UCP3 are able to uncouple ATP production from mitochondrial respiration, thereby decreasing energy metabolism efficiency; UCP2 has been also suggested to affect the production of reactive oxygen species (Schrauwen and Hesselink, 2002). In septic rat heart, an increased UCP2 mRNA expression was found, however, there was no detectable UCP2 protein in mitochondria isolated from either control or septic hearts (Roshon et al., 2003). The absence of UCP-2 protein expression argues against the hypothesis that UCP-2 contributes to decreased cardiac mechanical efficiency in septic shock. Finally, pharmacological inhibition of mitochondrial permeability transition pore was able to reduce myocardial dysfunction and mortality rate suggesting a critical role for mitochondrial permeability transition in septic organ dysfunction (Larche et al., 2006).

4.3.4.2 Contractility

In the heart, transsarcolemmal calcium entry during action potential triggers massive release of calcium from the sarcoplasmic reticulum through ryanodine receptor release channels and that in turn activates the actin/myosin interaction (Bers, 2002). Therefore, the processes involved in cellular calcium handling represent obvious candidates to be affected by sepsis and consequently to underlie the myocardial depression and as such they were intensively scrutinized by a number of studies. Sarcolemmal L-type calcium current was reported to be depressed in sepsis (Zhong et al., 1997; Stengl et al., 2010). The calcium-induced calcium-release from sarcoplasmic reticulum in endotoxemic dogs was reduced (Liu and Wu, 1991), probably due to decreased number of ryanodine receptor release channels (Wu and Liu, 1992; Dong et al., 2001). In another study (Cohen et al., 2006), decreased myocardial contraction, calcium release and ryanodine binding were found, however the calcium release channel protein content did not change, suggesting a posttranslational functional modification of the channels. The calcium uptake by sarcoplasmic reticulum calcium ATPase was also found to be impaired (Wu et al., 2001a), an effect related to decreased phosphorylation of phospholamban (Wu et al., 2002; Hassoun et al., 2008). In conclusion, the myocardial cellular calcium handling is clearly altered in sepsis and these multiple alterations contribute to the myocardial depression. Interestingly, the sarcoplasmic reticulum calcium handling dysfunction was shown to coincide with mitochondrial calcium accumulation and consequent impaired respiratory capacity (Hassoun et al., 2008), thus linking the contractile and mitochondrial dysfunction.

Finally, significant changes were described on the level of the contractile apparatus proteins. A number of structural cellular changes including scattered foci of actin/myosin contractile apparatus disruption were observed in long-term human septic myocardium (Rossi et al., 2007). Changes in mRNA and protein expression patterns of myocardial contractile proteins were reported in endotoxemic rats (Patten et al., 2001). Myofilament calcium sensitivity was decreased in skinned myocardial fibers from endotoxemic rabbits (Tavernier et al., 1998, 2001). Reduced myofibrillar calcium sensitivity was confirmed in isolated endotoxemic rabbit hearts (Takeuchi et al., 1999) and in septic rats (Wu et al., 2001b). In rats with chronic peritoneal sepsis an increased maximal and minimal Mg-ATPase activity, together with decreased sensitivity to calcium in myofilament proteins were found (Powers et al., 1998).

Multiple changes in adrenergic signaling were also suggested to contribute to myocardial dysfunction in sepsis. On cellular level, downregulation of β-adrenergic receptors (Shepherd et al., 1987; Tang and Liu, 1996), downregulation of G_S-protein (Matsuda et al., 2000) or upregulation of G_i-protein (Böhm et al., 1995) were reported. These cardiac cellular changes may be consequent to high circulating plasma catecholamine levels (Bocking et al., 1979; Benedict and Rose, 1992; Hahn et al., 1995) and/or apoptosis in cardiovascular autonomic centers leading to autonomic failure (Sharshar et al., 2003). From other signaling cascades possibly involved in pathogenesis of septic shock the caspases receive increasing attention. Recently, an important relationship between endotoxin-induced caspase activation and reduced contractile reserve and sarcomere disarray was shown at the level of single left ventricular cardiac myocytes (Lancel et al., 2005).

4.4 Cardiac L-Type Calcium Current (I_{CaL})

I_{CaL} represents one of crucial ionic currents of cardiac myocytes. By triggering calcium release from the sarcoplasmic reticulum I_{CaL} couples sarcolemmal excitation with contraction of the myocyte (Fabiato and Fabiato, 1975; Bers, 2002). Furthermore, I_{CaL} substantially contributes to the long cardiac action potential plateau phase (Fig. 4.1) and may be involved in arrhythmogenesis (for review see McDonald et al., 1994; Hofmann et al., 1999; Benitah et al., 2010).

4.4.1 Structure

I_{CaL} channel consists of four subunits: the largest main subunit α1C (Cav1.2, encoded by CACNA1C gene) and auxiliary subunits β (Cavβ, encoded by CACNB gene), α2 and δ (α2δ subunit complex encoded by a single gene CACNA2D) (Bodi et al., 2005). The main subunit that incorporates the conduction pore, voltage sensor, gating apparatus and binding sites for regulation, is sufficient to express channel activity. The function of α subunit is modulated by β subunit. Co-expression of α and

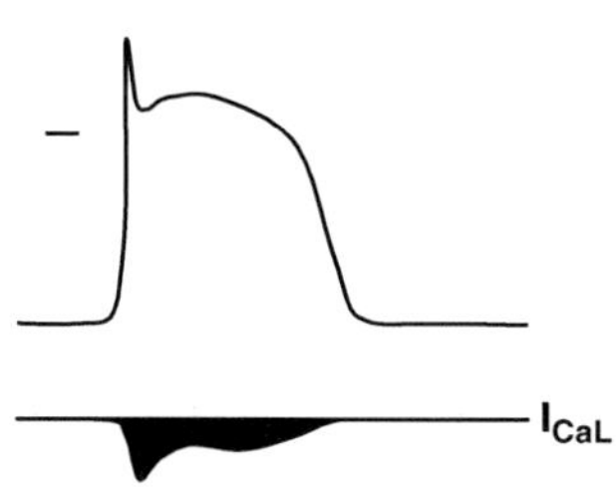

Fig. 4.1 Time course of cardiac ventricular action potential and of I_{CaL}. *Upper panel*, cardiac ventricular action potential. The *horizontal bar* marks 0 mV. The spike and dome configuration of action potential is typical for epicardial and midmyocardial cells. *Bottom panel*, I_{CaL}. After activation the current slowly inactivates according to the voltage profile of action potential

β subunits increases the peak current (Josephson and Varadi, 1996) and the presence of β subunit seems necessary for channel facilitation (Kamp et al., 2000).

4.4.2 Function

The channel is voltage-dependent, activated by depolarizations. The activation threshold is around −30 mV, the peak current occurs at 0 to +10 mV and the current-voltage relationship is bell-shaped with reversal at +60 to +70 mV (McDonald et al., 1994). The voltage dependences of activation and inactivation overlap producing window current (Hirano et al., 1992). The inactivation of I_{CaL} is both voltage– and calcium-dependent, the calcium-dependent inactivation being the faster process (McDonald et al., 1994). The calcium–dependent inactivation represents a negative feedback mechanism when a large increase in intracellular calcium due to release from sarcoplasmic reticulum accelerates the inactivation, thus limiting further calcium influx into the cell (Sipido et al., 1995). The I_{CaL} channel is 500–1000 times more permeable for divalent cations than for monovalent ones (Tsien et al., 1987). The conductance of the channel increases with the concentration (maximal conductance of 5 pS for calcium); at a physiological calcium concentration half-maximal conductance was found (Guia et al., 2001). I_{CaL} is the main inward current during the plateau phase of the cardiac action potential.

The activation of I_{CaL} is slower than that of fast sodium current (I_{Na}), approximately synchronous with inactivation of transient outward current (I_{to}). During the plateau I_{CaL} gradually declines due to voltage- and calcium dependent inactivation (Fig. 4.1) (Linz and Meyer, 2000). Reactivation of I_{CaL} during action potential plateau significantly contributes to the genesis of early afterdepolarization, a cellular proarrhythmic event (Volders et al., 2000).

4.4.3 Modulation

I_{CaL} is markedly enhanced by β-adrenergic agonists. The signaling pathway involves β-adrenergic receptor, G_s-protein, adenylate cyclase, cAMP and protein kinase A (Trautwein et al., 1987). The phosphorylation of the channel leads to an increased channel open probability and channel availability (Tsien et al., 1986; Ochi and Kawashima, 1990). From 3 β-receptor subtypes (β1–β3) in the heart, the effects of β1-adrenergic receptor stimulation on I_{CaL} are the most pronounced (van der Heyden et al., 2005). From α-adrenergic receptors the α1A-subtype is the most abundant in the heart (van der Heyden et al., 2005). Stimulation of α-adrenergic receptors with activation of protein kinase C as the final step of signaling cascade produces complex effects and both stimulation (Alden et al., 2002) and inhibition (Hu et al., 2000) of I_{CaL} were reported.

Cytoskeletal modulation of cardiac I_{CaL} receives increasing attention (Calaghan et al., 2004). Disruption of microtubules by colchicine increased I_{CaL} density in rat cardiac myocytes (Gomez et al., 2000). In contrast, depolymerization of the subsarcolemmal actin cytoskeleton by cytochalasin D was shown to be associated with reduction of I_{CaL} (Rueckschloss and Isenberg, 2001). The link between the actin-based cytoskeleton and T-tubular I_{CaL} channels may be represented by Ahnak protein (Hohaus et al., 2002). Ahnak protein is abundantly expressed in cardiomyocytes, it is phosphorylated by protein kinase A and associates to the intracellular regulatory β2 subunit of I_{CaL} channels (Haase et al., 1999; Haase, 2007). The ahnak-C1 domain may act as physiological brake on I_{CaL}, when under basal condition I_{CaL} conductance is inhibited by strong interaction between dephosphorylated ahnak and dephosphorylated β2 subunit of I_{CaL} channel. Phosphorylation of ahnak and/or β2 subunit relieves the inhibitory effects and increases I_{CaL} (Haase et al., 2005; Haase, 2007).

4.5 I_{CaL} in Sepsis

In a number of experimental rodent models of endotoxemia and sepsis a shortening of action potential duration with reduction of I_{CaL} was reported (Hung and Lew, 1993; Zhong et al., 1997). In endotoxemic rabbits the number of cardiac dihydropyridine receptors (i.e. I_{CaL} channels) was reduced (Lew et al., 1996). In ventricular myocytes from endotoxemic guinea pigs I_{CaL} density was decreased and that was associated with shortening of action potential duration, decreased cell contraction, and reduced systolic intracellular calcium concentration (Zhong et al., 1997). Shortening of action potential duration and decreased cell contraction was found also in endotoxemic rabbit cardiac myocytes (Hung and Lew, 1993). Controversial results were reported in septic rats. On one hand, a downregulation of cardiac dihydropyridine receptors in late sepsis preceded by upregulation in early sepsis was found (Hsu et al., 2007); on the other hand, no differences in action potential duration and I_{CaL} were reported (Wu et al., 1993; Fig. 4.2).

Recently, the effects of sepsis on I_{CaL} and action potential duration were investigated in a clinically more relevant non-rodent larger-order animal model, porcine model of hyperdynamic septic shock that closely mimics clinical scenario including standard resuscitative measures (Stengl et al., 2010). In this porcine model a significant shortening of cardiac repolarization induced by sepsis was found both in vivo (shortening of heart rate-corrected QT-interval) and in vitro (shortening of action potential duration). I_{CaL} density was also decreased and the analysis of I_{CaL} kinetics suggested that this reduction may be mainly due to a decreased number of I_{CaL} channels.

Action potential-clamp experiments revealed that the calcium influx into the cardiac cells through I_{CaL} channels is reduced in sepsis by a double mechanism: primary reduction of I_{CaL} probably due to a decreased number of functional I_{CaL} channels and decreased action potential duration, i.e. decreased time period spent at voltages necessary for I_{CaL} activation (action potential plateau). Thus, in sepsis I_{CaL} of reduced density flows into the cell for shorter time (Fig. 4.3).

Possible mechanisms leading to I_{CaL} reduction were also addressed. Since TNF-α was reported to inhibit I_{CaL} in rat cardiac myocytes (Krown et al., 1995; Li et al.,

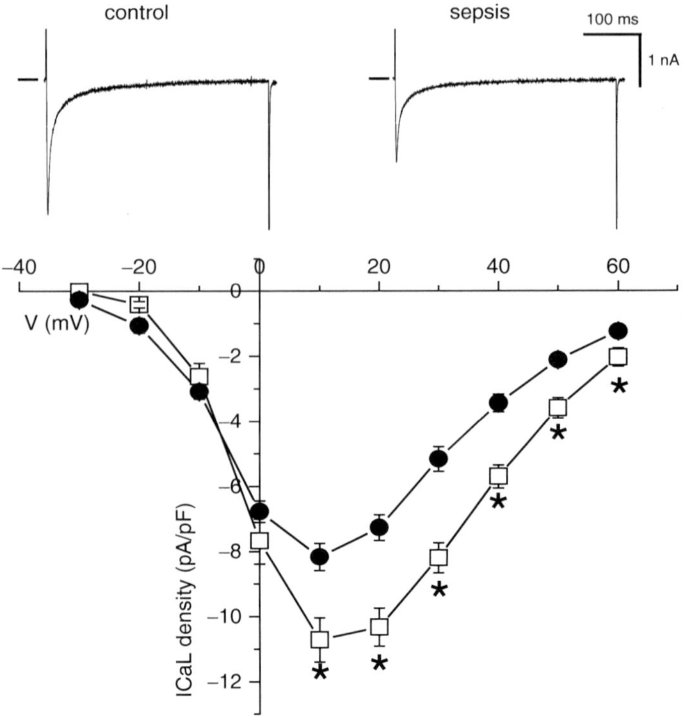

Fig. 4.2 Reduced I_{CaL} density in septic cardiac myocytes. *Upper panel*, representative recordings of I_{CaL} in cardiac myocytes from control and septic pigs at 0 mV. *Bottom panel*, voltage dependence of I_{CaL} measured in control (*open squares*) and septic cardiac myocytes (*filled circles*). $^{*}p < 0.05$

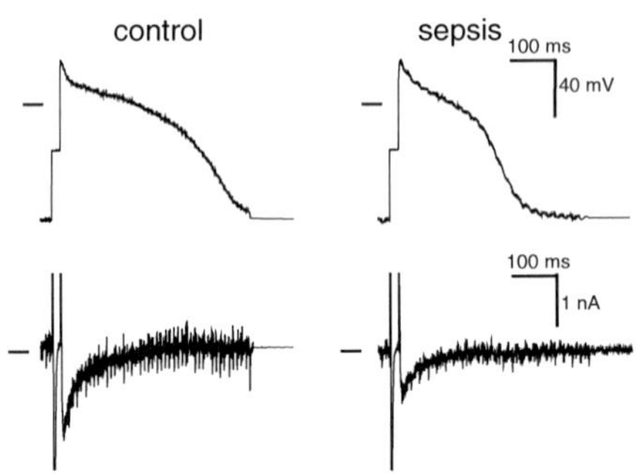

Fig. 4.3 I_{CaL} activated by action potential waveforms (action potential-clamp). I_{CaL} traces (*bottom panels*) activated by control action potential in a control cell (*left panels*) and by septic action potential in a septic cell (*right panels*)

2003), possible involvement of TNF-α was investigated. Neither acute, nor chronic (several hours-lasting incubation) application of TNF-α was able to influence action potential duration or I_{CaL} density. Therefore, TNF-α probably does not contribute to the reduction of I_{CaL} in this experimental model. In general, the consistency between the rodent endotoxemic models and the clinically more relevant porcine model of hyperdynamic septic shock suggests that the reduction of I_{CaL} and shortening of cardiac repolarization in sepsis represent a phenomenon of general nature that may be expected also in patients with severe sepsis and septic shock. Appropriate therapeutic measures should be considered, especially in patients, in whom the cardiac calcium handling is already compromised, e.g. patients with heart failure, patients with calcium-channel blocker therapy.

4.6 Conclusion and Perspectives

In last decades, intensive research accumulated vast information about mechanisms of sepsis. Despite this effort, however, the sepsis continues to claim a large number of lives worldwide and its incidence increases. In general, there is a failure to successfully translate the experimental findings into clinical practice and improved therapy. Obviously there is a number of reasons for this discrepancy between the experimental research and clinical practice. One of the most important is the vast complexity of the disease that is characterized by an overwhelming production of inflammatory mediators. This leads to an extremely complicated network of up- and down-regulated crosstalks on various levels: from the whole organisms to single organs/cells/molecules. To get a full picture of this regulatory dyssynchronization with identification of critical points, at which the therapy should be aimed, will be a very challenging task. Second, there is a lack of really clinically relevant models; most of the data is nowadays obtained in small rodent (transgenic) models that do not correspond well to usual clinical situation and human sepsis development. Therefore, in our opinion the first logical step must be development of a clinically relevant model that would reflect well the clinical situation and would allow a successful translation of the experimental findings to the clinics. Application of new powerful experimental technologies (e.g. genomics, proteomics) to such a model will certainly deliver novel findings that will eventually be translated into therapeutic and diagnostic improvements.

In this review the recent experimental work was summarized with emphasis on cardiac function. We hope that it will help to stimulate further research into the mechanisms of the sepsis and eventually to improve the prognosis of sepsis patients.

Acknowledgement This work was supported by the Research Project MSM 0021620819 (Replacement of and support to some vital organs).

References

Abbas AK, Murphy KM, Sher A (1996) Functional diversity of helper T lymphocytes. Nature 383:787–793

Abraham E, Wunderink R, Silverman H, Perl TM, Nasraway S, Levy H, Bone R, Wenzel RP, Balk R, Allred R et al (1995) Efficacy and safety of monoclonal antibody to human tumor necrosis factor alpha in patients with sepsis syndrome. A randomized, controlled, double-blind, multicenter clinical trial. TNF-alpha MAb Sepsis Study Group. JAMA 273:934–941

Abraham E, Glauser MP, Butler T, Garbino J, Gelmont D, Laterre PF, Kudsk K, Bruining HA, Otto C, Tobin E, Zwingelstein C, Lesslauer W, Leighton A (1997) p55 Tumor necrosis factor receptor fusion protein in the treatment of patients with severe sepsis and septic shock. A randomized controlled multicenter trial. Ro 45-2081 Study Group. JAMA 277:1531–1538

Abraham E, Anzueto A, Gutierrez G, Tessler S, San Pedro G, Wunderink R, Dal Nogare A, Nasraway S, Berman S, Cooney R, Levy H, Baughman R, Rumbak M, Light RB, Poole L, Allred R, Constant J, Pennington J, Porter S (1998) Double-blind randomised controlled trial of monoclonal antibody to human tumour necrosis factor in treatment of septic shock. NORASEPT II Study Group. Lancet 351:929–933

Ait-Oufella H, Maury E, Lehoux S, Guidet B, Offenstadt G (2010) The endothelium: physiological functions and role in microcirculatory failure during severe sepsis. Intensive Care Med 36:1286–1298

Alden KJ, Goldspink PH, Ruch SW, Buttrick PM, García J (2002) Enhancement of L-type Ca^{2+} current from neonatal mouse ventricular myocytes by constitutively active PKC-betaII. Am J Physiol 282:C768–C774

Alves-Filho JC, Sonego F, Souto O, Freitas F, Verri A, Auxiliadora-Martins A, Filho M, Mc Kenzie AB, Xu AN, D Q, Cunha F, Liew FY (2010) Interleukin–33 attenuates sepsis by enhancing neutrophil influx to the site of infection. Nature Med 16:703–712

Angus DC, Linde-Zwirble WT, Lidicker J, Clermont G, Carcillo J, Pinsky MR (2001) Epidemiology of severe sepsis in the United States: analysis of incidence, outcome, and associated costs of care. Crit Care Med 29:1303–1310

Benedict CR, Rose JA (1992) Arterial norepinephrine changes in patients with septic shock. Circ Shock 38:165–172

Benitah JP, Alvarez JL, Gómez AM (2010) L-type Ca^{2+} current in ventricular cardiomyocytes. J Mol Cell Cardiol 48:26–36

Bers DM (2002) Cardiac excitation-contraction coupling. Nature 415:198–205

Bocking JK, Sibbald WJ, Holliday RL, Scott S, Viidik T (1979) Plasma catecholamine levels and pulmonary dysfunction in sepsis. Surg Gynecol Obstet 148:715–719

Bodi I, Mikala G, Koch SE, Akhter SA, Schwartz A (2005) The L-type calcium channel in the heart: the beat goes on. J Clin Invest 115:3306–3317

Böhm M, Kirchmayr R, Gierschik P, Erdmann E (1995) Increase of myocardial inhibitory G-proteins in catecholamine-refractory septic shock or in septic multiorgan failure. Am J Med 98:183–186

Cain BS, Meldrum DR, Dinarello CA, Meng X, Joo KS, Banerjee A, Harken AH (1999) Tumor necrosis factor-alpha and interleukin-1beta synergistically depress human myocardial function. Crit Care Med 27:1309–1318

Calaghan SC, Le Guennec JY, White E (2004) Cytoskeletal modulation of electrical and mechanical activity in cardiac myocytes. Prog Biophys Mol Biol 84:29–59

Calandra T, Baumgartner JD, Grau GE, Wu MM, Lambert PH, Schellekens J, Verhoef J, Glauser MP (1990) Prognostic values of tumor necrosis factor/cachectin, interleukin-1, interferon-alpha, and interferon-gamma in the serum of patients with septic shock. Swiss-Dutch J5 Immunoglobulin Study Group. J Infect Dis 161:982–987

Carré JE, Singer M (2008) Cellular energetic metabolism in sepsis: the need for a systems approach. Biochim Biophys Acta 1777:763–771

Cassatella MA, Meda L, Bonora S, Ceska M, Constantin G (1993) Interleukin 10 (IL-10) inhibits the release of proinflammatory cytokines from human polymorphonuclear leukocytes. Evidence for an autocrine role of tumor necrosis factor and IL-1 beta in mediating the production of IL-8 triggered by lipopolysaccharide. J Exp Med 178:2207–2211

Cohen RI, Wilson D, Liu SF (2006) Nitric oxide modifies the sarcoplasmic reticular calcium release channel in endotoxemia by both guanosine-3',5' (cyclic) phosphate-dependent and independent pathways. Crit Care Med 34:173–181

Court O, Kumar A, Parrillo JE, Kumar A (2002) Clinical review: Myocardial depression in sepsis and septic shock. Critical Care 6:500–508

Crouser ED (2004) Mitochondrial dysfunction in septic shock and multiple organ dysfunction syndrome. Mitochondrion 4:729–741

Cunnion RE, Schaer GL, Parker MM, Natanson C, Parrillo JE (1986) The coronary circulation in human septic shock. Circulation 73:637–644

Damas P, Reuter A, Gysen P, Demonty J, Lamy M, Franchimont P (1989) Tumor necrosis factor and interleukin-1 serum levels during severe sepsis in humans. Crit Care Med 17:975–978

Damas P, Ledoux D, Nys M, Vrindts Y, De Groote D, Franchimont P, Lamy M (1992) Cytokine serum level during severe sepsis in human IL-6 as a marker of severity. Ann Surg 215: 356–362

de Waal Malefyt R, Haanen J, Spits H, Roncarolo MG, te Velde A, Figdor C, Johnson K, Kastelein R, Yssel H, de Vries JE (1991) Interleukin 10 (IL-10) and viral IL-10 strongly reduce antigen-specific human T cell proliferation by diminishing the antigen-presenting capacity of monocytes via downregulation of class II major histocompatibility complex expression. J Exp Med 174:915–924

Dong LW, Wu LL, Ji Y, Liu MS (2001) Impairment of the ryanodine-sensitive calcium release channels in the cardiac sarcoplasmic reticulum and its underlying mechanism during the hypodynamic phase of sepsis. Shock 16:33–39

Draing C, Sigel S, Deininger S, Traub S, Munke R, Mayer C, Hareng L, Hartung T, von Aulock S, Hermann C (2008) Cytokine induction by Gram-positive bacteria. Immunobiology 213:285–296

Durez P, Abramowicz D, Gérard C, Van Mechelen M, Amraoui Z, Dubois C, Leo O, Velu T, Goldman M (1993) In vivo induction of interleukin 10 by anti-CD3 monoclonal antibody or bacterial lipopolysaccharide: differential modulation by cyclosporin A. J Exp Med 177: 551–555

Echtenacher B, Urbaschek R, Weigl K, Freudenberg MA, Männel DN (2003) Treatment of experimental sepsis-induced immunoparalysis with TNF. Immunobiology 208:381–389

Eichenholz PW, Eichacker PQ, Hoffman WD, Banks SM, Parrillo JE, Danner RL, Natanson C (1992) Tumor necrosis factor challenges in canines: patterns of cardiovascular dysfunction. Am J Physiol 263:H668–H675

Emery P, Salmon M (1991) The immune response: systemic mediators of inflammation. Br J Hosp Med 45:164–168

Fabiato A, Fabiato F (1975) Contractions induced by a calcium-triggered release of calcium from the sarcoplasmic reticulum of single skinned cardiac cells. J Physiol 249:469–495

Finkel MS, Oddis CV, Jacob TD, Watkins SC, Hattler BG, Simmons RL (1992) Negative inotropic effects of cytokines on the heart mediated by nitric oxide. Science 257:387–389

Fisher CJ Jr, Dhainaut JF, Opal SM, Pribble JP, Balk RA, Slotman GJ, Iberti TJ, Rackow EC, Shapiro MJ, Greenman RL et al (1994) Recombinant human interleukin 1 receptor antagonist in the treatment of patients with sepsis syndrome. Results from a randomized, double-blind, placebo-controlled trial. Phase III rhIL-1ra Sepsis Syndrome Study Group. JAMA 271: 1836–1843

Fisher CJ Jr, Agosti JM, Opal SM, Lowry SF, Balk RA, Sadoff JC, Abraham E, Schein RM, Benjamin E (1996) Treatment of septic shock with the tumor necrosis factor receptor:Fc fusion protein. The soluble TNF Receptor Sepsis Study Group. N Engl J Med 334:1697–1702

Gee K, Guzzo C, Che Mat NF, Ma W, Kumar A (2009) The IL-12 family of cytokines in infection, inflammation and autoimmune disorders. Inflamm Allergy Drug Targets 8:40–52

Gellerich FN, Trumbeckaite S, Hertel K, Zierz S, Müller-Werdan U, Werdan K, Redl H, Schlag G (1999) Impaired energy metabolism in hearts of septic baboons: diminished activities of Complex I and Complex II of the mitochondrial respiratory chain. Shock 11:336–341

Goldhaber JI, Kim KH, Natterson PD, Lawrence T, Yang P, Weiss JN (1996) Effects of TNF-alpha on $[Ca^{2+}]_i$ and contractility in isolated adult rabbit ventricular myocytes. Am J Physiol 271:H1449–55

Gomez AM, Kerfant BG, Vassort G (2000) Microtubule disruption modulates Ca^{2+} signaling in rat cardiac myocytes. Circ Res 86:30–36

Groeneveld AB, van Lambalgen AA, van den Bos GC, Bronsveld W, Nauta JJ, Thijs LG (1991) Maldistribution of heterogeneous coronary blood flow during canine endotoxin shock. Cardiovasc Res 25:80–88

Guia A, Stern MD, Lakatta EG, Josephson IR (2001) Ion concentration-dependence of rat cardiac unitary L-type calcium channel conductance. Biophys J 80:2742–2750

Haase H (2007) Ahnak, a new player in beta-adrenergic regulation of the cardiac L-type Ca^{2+} channel. Cardiovasc Res 73:19–25

Haase H, Podzuweit T, Lutsch G, Hohaus A, Kostka S, Lindschau C, Kott M, Kraft R, Morano I (1999) Signaling from beta-adrenoceptor to L-type calcium channel: identification of a novel cardiac protein kinase A target possessing similarities to AHNAK. FASEB J 13:2161–2172

Haase H, Alvarez J, Petzhold D, Doller A, Behlke J, Erdmann J, Hetzer R, Regitz-Zagrosek V, Vassort G, Morano I (2005) Ahnak is critical for cardiac Ca(V)1.2 calcium channel function and its beta-adrenergic regulation. FASEB J 19:1969–1977

Hahn PY, Wang P, Tait SM, Ba ZF, Reich SS, Chaudry IH (1995) Sustained elevation in circulating catecholamine levels during polymicrobial sepsis. Shock 4:269–273

Harrois A, Huet O, Duranteau J (2009) Alterations of mitochondrial function in sepsis and critical illness. Curr Opin Anaesthesiol 22:143–149

Hassoun SM, Marechal X, Montaigne D, Bouazza Y, Decoster B, Lancel S, Neviere R (2008) Prevention of endotoxin-induced sarcoplasmic reticulum calcium leak improves mitochondrial and myocardial dysfunction. Crit Care Med 36:2590–2596

Hersch M, Gnidec AA, Bersten AD, Troster M, Rutledge FS, Sibbald WJ (1990) Histologic and ultrastructural changes in nonpulmonary organs during early hyperdynamic sepsis. Surgery 107:397–410

Hesse DG, Tracey KJ, Fong Y, Manogue KR, Palladino MA Jr, Cerami A, Shires GT, Lowry SF (1988) Cytokine appearance in human endotoxemia and primate bacteremia. Surg Gynecol Obstet 166:147–153

Hirano Y, Moscucci A, January CT (1992) Direct measurement of L-type Ca^{2+} window current in heart cells. Circ Res 70:445–455

Hoffmann JN, Werdan K, Hartl WH, Jochum M, Faist E, Inthorn D (1999) Hemofiltrate from patients with severe sepsis and depressed left ventricular contractility contains cardiotoxic compounds. Shock 12:174–180

Hofmann F, Lacinová L, Klugbauer N (1999) Voltage-dependent calcium channels: from structure to function. Rev Physiol Biochem Pharmacol 139:33–87

Hohaus A, Person V, Behlke J, Schaper J, Morano I, Haase H (2002) The carboxyl-terminal region of ahnak provides a link between cardiac L-type Ca^{2+} channels and the actin-based cytoskeleton. FASEB J 16:1205–1216

Hosenpud JD, Campbell SM, Mendelson DJ (1989) Interleukin-1-induced myocardial depression in an isolated beating heart preparation. J Heart Transplant 8:460–464

Hotchkiss RS, Rust RS, Dence CS, Wasserman TH, Song SK, Hwang DR, Karl IE, Welch MJ (1991) Evaluation of the role of cellular hypoxia in sepsis by the hypoxic marker [18F]fluoromisonidazole. Am J Physiol 261:R965–72

Hsu C, Wu G, Yang SL, Hsu HK, Yang RC, Tang C, Liu MS (2007) Intracellular redistribution of dihydropyridine receptor in the rat heart during the progression of sepsis. J Surg Res 141: 146–152

Hu K, Mochly-Rosen D, Boutjdir M (2000) Evidence for functional role of epsilonPKC isozyme in the regulation of cardiac Ca^{2+} channels. Am J Physiol 279:H2658–64

Huang W, Tang Y, Li L (2010) HMGB1, a potent proinflammatory cytokine in sepsis. Cytokine 51:119–126

Hung J, Lew WY (1993) Cellular mechanisms of endotoxin-induced myocardial depression in rabbits. Circ Res 73:125–134

Iwakura Y, Nakae S, Saijo S, Ishigame H (2008) The roles of IL-17A in inflammatory immune responses and host defense against pathogens. Immunol Rev 226:57–79

Jafri SM, Lavine S, Field BE, Bahorozian MT, Carlson RW (1990) Left ventricular diastolic function in sepsis. Crit Care Med 18:709–714

Jardin F, Fourme T, Page B, Loubières Y, Vieillard-Baron A, Beauchet A, Bourdarias JP (1999) Persistent preload defect in severe sepsis despite fluid loading: A longitudinal echocardiographic study in patients with septic shock. Chest 116:1354–1359

Josephson IR, Varadi G (1996) The beta subunit increases Ca^{2+} currents and gating charge movements of human cardiac L-type Ca^{2+} channels. Biophys J 70:1285–1293

Joulin O, Petillot P, Labalette M, Lancel S, Neviere R (2007) Cytokine profile of human septic shock serum inducing cardiomyocyte contractile dysfunction. Physiol Res 56:291–297

Kamp TJ, Hu H, Marban E (2000) Voltage-dependent facilitation of cardiac L-type Ca channels expressed in HEK-293 cells requires beta -subunit. Am J Physiol 278:H126–H255.

Kapadia S, Lee J, Torre-Amione G, Birdsall HH, Ma TS, Mann DL (1995) Tumor necrosis factor-alpha gene and protein expression in adult feline myocardium after endotoxin administration. J Clin Invest 96:1042–1052

Krown KA, Yasui K, Brooker MJ, Dubin AE, Nguyen C, Harris GL, McDonough PM, Glembotski CC, Palade PT, Sabbadini RA (1995) TNF alpha receptor expression in rat cardiac myocytes: TNF alpha inhibition of L-type Ca^{2+} current and Ca^{2+} transients. FEBS Lett 376:24–30

Kumar A, Thota V, Dee L, Olson J, Uretz E, Parrillo JE (1996) Tumor necrosis factor alpha and interleukin 1beta are responsible for in vitro myocardial cell depression induced by human septic shock serum. J Exp Med 183:949–958

Lancel S, Joulin O, Favory R, Goossens JF, Kluza J, Chopin C, Formstecher P, Marchetti P, Neviere R (2005) Ventricular myocyte caspases are directly responsible for endotoxin-induced cardiac dysfunction. Circulation 111:2596–2604

Larche J, Lancel S, Hassoun SM, Favory R, Decoster B, Marchetti P, Chopin C, Neviere R (2006) Inhibition of mitochondrial permeability transition prevents sepsis-induced myocardial dysfunction and mortality. J Am Coll Cardiol 48:377–385

Levi M (2010) The coagulant response in sepsis and inflammation. Hamostaseologie 30:10–12

Levy RJ, Vijayasarathy C, Raj NR, Avadhani NG, Deutschman CS (2004) Competitive and noncompetitive inhibition of myocardial cytochrome C oxidase in sepsis. Shock 21:110–114

Lew WY, Yasuda S, Yuan T, Hammond HK (1996) Endotoxin-induced cardiac depression is associated with decreased cardiac dihydropyridine receptors in rabbits. J Mol Cell Cardiol 28:1367–1371

Li XQ, Zhao MG, Mei QB, Zhang YF, Guo W, Wang HF, Chen D, Cui Y (2003) Effects of tumor necrosis factor-alpha on calcium movement in rat ventricular myocytes. Acta Pharmacol Sin 24:1224–1230

Linz KW, Meyer R (2000) Profile and kinetics of L-type calcium current during the cardiac ventricular action potential compared in guinea-pigs, rats and rabbits. Pflügers Archiv 439:588–599

Liu MS, Wu LL (1991) Reduction in the Ca^{2+}-induced Ca^{2+} release from canine cardiac sarcoplasmic reticulum following endotoxin administration. Biochem Biophys Res Commun 174:1248–1254

Martin GS, Mannino DM, Eaton S, Moss M (2003) The epidemiology of sepsis in the United States from 1979 through 2000. N Engl J Med 348:1546–1554

Matsuda N, Hattori Y, Akaishi Y, Suzuki Y, Kemmotsu O, Gando S (2000) Impairment of cardiac beta-adrenoceptor cellular signaling by decreased expression of G(s alpha) in septic rabbits. Anesthesiology 93:1465–1473

McDonald TF, Pelzer S, Trautwein W, Pelzer DJ (1994) Regulation and modulation of calcium channels in cardiac, skeletal, and smooth muscle cells. Physiol Rev 74:365–507

Meldrum DR (1998) Tumor necrosis factor in the heart. Am J Physiol 274:R577–595

Merx MW, Weber C (2007) Sepsis and the heart. Circulation 116:793–802

Munt B, Jue J, Gin K, Fenwick J, Tweeddale M (1998) Diastolic filling in human severe sepsis: an echocardiographic study. Crit Care Med 26:1829–1833

Murray DR, Freeman GL (1996) Tumor necrosis factor-alpha induces a biphasic effect on myocardial contractility in conscious dogs. Circ Res 78:154–160

Murray AJ, Anderson RE, Watson GC, Radda GK, Clarke K (2004) Uncoupling proteins in human heart. Lancet 364:1786–1788
Ochi R, Kawashima Y (1990) Modulation of slow gating process of calcium channels by isoprenaline in guinea pig ventricular cells. J Physiol 424:187–204
O'Garra A, Murphy KM (2009) From IL-10 to IL-12: how pathogens and their products stimulate APCs to induce T(H)1 development. Nature Immunol 10:929–932
Opal SM, Fisher CJ Jr, Dhainaut JF, Vincent JL, Brase R, Lowry SF, Sadoff JC, Slotman GJ, Levy H, Balk RA, Shelly MP, Pribble JP, LaBrecque JF, Lookabaugh J, Donovan H, Dubin H, Baughman R, Norman J, DeMaria E, Matzel K, Abraham E, Seneff M (1997) Confirmatory interleukin-1 receptor antagonist trial in severe sepsis: a phase III, randomized, double-blind, placebo-controlled, multicenter trial. The Interleukin-1 Receptor Antagonist Sepsis Investigator Group. Crit Care Med 25:1115–1124
Opal SM, Garber GE, LaRosa SP, Maki DG, Freebairn RC, Kinasewitz GT, Dhainaut JF, Yan SB, Williams MD, Graham DE, Nelson DR, Levy H, Bernard GR (2003) Systemic host responses in severe sepsis analyzed by causative microorganism and treatment effects of drotrecogin alfa (activated). Clin Infect Dis 37:50–58
Oral H, Dorn GW 2nd, Mann DL (1997) Sphingosine mediates the immediate negative inotropic effects of tumor necrosis factor-alpha in the adult mammalian cardiac myocyte. J Biol Chem 272:4836–4842
Osuchowski MF, Welch K, Yang H, Siddiqui J, Remick DG (2007) Chronic sepsis mortality characterized by an individualized inflammatory response. J Immunol 179:623–630
Parker SJ, Watkins PE (2001) Experimental models of gram-negative sepsis. Br J Surg 88:22–30
Parker MM, Shelhamer JH, Bacharach SL, Green MV, Natanson C, Frederick TM, Damske BA, Parrillo JE (1984) Profound but reversible myocardial depression in patients with septic shock. Ann Intern Med 100:483–490
Parker MM, McCarthy KE, Ognibene FP, Parrillo JE (1990) Right ventricular dysfunction and dilatation, similar to left ventricular changes, characterize the cardiac depression of septic shock in humans. Chest 97:126–131
Parrillo JE, Burch C, Shelhamer JH, Parker MM, Natanson C, Schuette W (1985) A circulating myocardial depressant substance in humans with septic shock. Septic shock patients with a reduced ejection fraction have a circulating factor that depresses in vitro myocardial cell performance. J Clin Invest 76:1539–1553
Parrillo JE, Parker MM, Natanson C, Suffredini AF, Danner RL, Cunnion RE, Ognibene FP (1990) Septic shock in humans: advances in the understanding of pathogenesis, cardiovascular dysfunction, and therapy. Ann Intern Med 113:227–242
Patten M, Krämer E, Bünemann J, Wenck C, Thoenes M, Wieland T, Long C (2001) Endotoxin and cytokines alter contractile protein expression in cardiac myocytes in vivo. Pflugers Arch 442:920–927
Piel DA, Gruber PJ, Weinheimer CJ, Courtois MR, Robertson CM, Coopersmith CM, Deutschman CS, Levy RJ (2007) Mitochondrial resuscitation with exogenous cytochrome c in the septic heart. Crit Care Med 35:2120–2127
Piel DA, Deutschman CS, Levy RJ (2008) Exogenous cytochrome C restores myocardial cytochrome oxidase activity into the late phase of sepsis. Shock 29:612–616
Poelaert J, Declerck C, Vogelaers D, Colardyn F, Visser CA (1997) Left ventricular systolic and diastolic function in septic shock. Intensive Care Med 23:553–560
Powers FM, Farias S, Minami H, Martin AF, Solaro RJ, Law WR (1998) Cardiac myofilament protein function is altered during sepsis. J Mol Cell Cardiol 30:967–978
Reilly JM, Cunnion RE, Burch-Whitman C, Parker MM, Shelhamer JH, Parrillo JE (1989) A circulating myocardial depressant substance is associated with cardiac dysfunction and peripheral hypoperfusion (lactic acidemia) in patients with septic shock. Chest 95:1072–1080
Reim D, Westenfelder K, Kaiser-Moore S, Schlautkotter S, Holzmann B, Weighardt H (2009) Role of T cells for cytokine production and outcome in a model of acute septic peritonitis. Shock 31:245–250

Remick DG, Bolgos G, Copeland S, Siddiqui J (2005) Role of interleukin-6 in mortality from and physiologic response to sepsis. Infect Immun 73:2751–2757
Roshon MJ, Kline JA, Thornton LR, Watts JA (2003) Cardiac UCP2 expression and myocardial oxidative metabolism during acute septic shock in the rat. Shock 19:570–576
Rossi MA, Celes MR, Prado CM, Saggioro FP (2007) Myocardial structural changes in long-term human severe sepsis/septic shock may be responsible for cardiac dysfunction. Shock 27:10–18
Rudiger A, Singer M (2007) Mechanisms of sepsis-induced cardiac dysfunction. Crit Care Med 35:1599–1608
Rueckschloss U, Isenberg G (2001) Cytochalasin D reduces Ca^{2+} currents via cofilin-activated depolymerization of F-actin in guinea-pig cardiomyocytes. J Physiol 537:363–370
Schouten M, Wiersinga WJ, Levi M, van der Poll T (2008) Inflammation, endothelium, and coagulation in sepsis. J Leukoc Biol 83:536–545
Schrauwen P, Hesselink M (2002) UCP2 and UCP3 in muscle controlling body metabolism. J Exp Biol 205:2275–2285
Scott MJ, Godshall CJ, Cheadle WG (2002) Jaks, STATs, Cytokines, and Sepsis. Clin Diagn Lab Immunol 9:1153–1159
Sharshar T, Gray F, Lorin de la Grandmaison G, Hopkinson NS, Ross E, Dorandeu A, Orlikowski D, Raphael JC, Gajdos P, Annane D (2003) Apoptosis of neurons in cardiovascular autonomic centres triggered by inducible nitric oxide synthase after death from septic shock. Lancet 362:1799–1805
Shepherd RE, Lang CH, McDonough KH (1987) Myocardial adrenergic responsiveness after lethal and nonlethal doses of endotoxin. Am J Physiol 252:H410–6
Sipido KR, Callewaert G, Carmeliet E (1995) Inhibition and rapid recovery of Ca^{2+} current during Ca^{2+} release from sarcoplasmic reticulum in guinea pig ventricular myocytes. Circ Res 76: 102–109
Solomon MA, Correa R, Alexander HR, Koev LA, Cobb JP, Kim DK, Roberts WC, Quezado ZM, Scholz TD, Cunnion RE et al (1994) Myocardial energy metabolism and morphology in a canine model of sepsis. Am J Physiol 266:H757–H768
Stein B, Frank P, Schmitz W, Scholz H, Thoenes M (1996) Endotoxin and cytokines induce direct cardiodepressive effects in mammalian cardiomyocytes via induction of nitric oxide synthase. J Mol Cell Cardiol 28:1631–1639
Stengl M, Bartak F, Sykora R, Chvojka J, Benes J, Krouzecky A, Novak I, Sviglerova J, Kuncova J, Matejovic M (2010) Reduced L-type calcium current in ventricular myocytes from pigs with hyperdynamic septic shock. Crit Care Med 38:579–587
Stief TW, Ijagha O, Weiste B, Herzum I, Renz H, Max M (2007) Analysis of hemostasis alterations in sepsis. Blood Coagul Fibrinolysis 18:179–186
Suliman HB, Welty-Wolf KE, Carraway M, Tatro L, Piantadosi CA (2004) Lipopolysaccharide induces oxidative cardiac mitochondrial damage and biogenesis. Cardiovasc Res 64:279–288
Sykora R, Chvojka J, Krouzecky A, Radej J, Karvunidis T, Varnerova V, Novak I, Matejovic M (2009) High versus standard-volume haemofiltration in hyperdynamic porcine peritonitis: effects beyond haemodynamics? Intensive Care Med 35:371–380
Takeuchi K, del Nido PJ, Ibrahim AE, Poutias DN, Glynn P, Cao-Danh H, Cowan DB, McGowan FX Jr (1999) Increased myocardial calcium cycling and reduced myofilament calcium sensitivity in early endotoxemia. Surgery 126:231–238
Tang C, Liu MS (1996) Initial externalization followed by internalization of beta-adrenergic receptors in rat heart during sepsis. Am J Physiol 270:R254–R263
Tavernier B, Garrigue D, Boulle C, Vallet B, Adnet P (1998) Myofilament calcium sensitivity is decreased in skinned cardiac fibres of endotoxin-treated rabbits. Cardiovasc Res 38: 472–479
Tavernier B, Mebazaa A, Mateo P, Sys S, Ventura-Clapier R, Veksler V (2001) Phosphorylation-dependent alteration in myofilament Ca^{2+} sensitivity but normal mitochondrial function in septic heart. Am J Respir Crit Care Med 163:362–367

Thongboonkerd V, Chiangjong W, Mares J, Moravec J, Tuma Z, Karvunidis T, Sinchaikul S, Chen ST, Opatrný K, Matejovic M (2009) Altered plasma proteome during an early phase of peritonitis-induced sepsis. Clin Sci 116:721–730

Tracey KJ, Beutler B, Lowry SF, Merryweather J, Wolpe S, Milsark IW, Hariri RJ, Fahey TJ 3rd, Zentella A, Albert JD et al (1986) Shock and tissue injury induced by recombinant human cachectin. Science 234:470–474

Trautwein W, Cavalié A, Flockerzi V, Hofmann F, Pelzer D (1987) Modulation of calcium channel function by phosphorylation in guinea pig ventricular cells and phospholipid bilayer membranes. Circ Res 61:I17–I23

Tschaikowsky K, Hedwig-Geissing M, Schiele A, Bremer F, Schywalsky M, Schüttler J (2002) Coincidence of pro- and anti-inflammatory responses in the early phase of severe sepsis: Longitudinal study of mononuclear histocompatibility leukocyte antigen – DR expression, procalcitonin, C-reactive protein, and changes in T-cell subsets in septic and postoperative patients. Crit Care Med 30:1015–1023

Tsien RW, Bean BP, Hess P, Lansman JB, Nilius B, Nowycky MC (1986) Mechanisms of calcium channel modulation by beta-adrenergic agents and dihydropyridine calcium agonists. J Mol Cell Cardiol 18:691–710

Tsien RW, Hess P, McCleskey EW, Rosenberg RL (1987) Calcium channels: mechanisms of selectivity, permeation, and block. Annu Rev Biophys Biophys Chem 16:265–290

van der Heyden MA, Wijnhoven TJ, Opthof T (2005) Molecular aspects of adrenergic modulation of cardiac L-type Ca^{2+} channels. Cardiovasc Res 65:28–39

Vincent JL, Bakker J, Marécaux G, Schandene L, Kahn RJ, Dupont E (1992) Administration of anti-TNF antibody improves left ventricular function in septic shock patients. Results of a pilot study. Chest 101:810–815

Volders PG, Vos MA, Szabo B, Sipido KR, de Groot SH, Gorgels AP, Wellens HJ, Lazzara R (2000) Progress in the understanding of cardiac early afterdepolarizations and torsades de pointes: time to revise current concepts. Cardiovasc Res 46:376–392

Watts JA, Kline JA, Thornton LR, Grattan RM, Brar SS (2004) Metabolic dysfunction and depletion of mitochondria in hearts of septic rats. J Mol Cell Cardiol 36:141–150

White M, Lawless MW, O'Dwyer MJ, Grealy R, Connell BO, Storceur P, Kelleher D, McManus R, Ryan T (2010) Transforming growth factor beta-1 and interleukin-17 gene transcription in peripheral blood mononuclear cells and the human response to infection. Cytokine 50:322–327

Wu LL, Liu MS (1992) Altered ryanodine receptor of canine cardiac sarcoplasmic reticulum and its underlying mechanism in endotoxin shock. J Surg Res 53:82–90

Wu SN, Lue SI, Yang SL, Hsu HK, Liu MS (1993) Electrophysiologic properties of isolated adult cardiomyocytes from septic rats. Circ Shock 41:239–247

Wu LL, Ji Y, Dong LW, Liu MS (2001a) Calcium uptake by sarcoplasmic reticulum is impaired during the hypodynamic phase of sepsis in the rat heart. Shock 15:49–55

Wu LL, Tang C, Liu MS (2001b) Altered phosphorylation and calcium sensitivity of cardiac myofibrillar proteins during sepsis. Am J Physiol 281:R408–R416

Wu LL, Tang C, Dong LW, Liu MS (2002) Altered phospholamban-calcium ATPase interaction in cardiac sarcoplasmic reticulum during the progression of sepsis. Shock 17:389–393

Ye P, Garvey PB, Zhang P, Nelson S, Bagby G, Summer WR, Schwarzenberger P, Shellito JE, Kolls JK (2001) Interleukin-17 and lung host defense against Klebsiella pneumoniae infection. Am J Respir Cell Mol Biol 25:335–340

Yokoyama T, Vaca L, Rossen RD, Durante W, Hazarika P, Mann DL (1993) Cellular basis for the negative inotropic effects of tumor necrosis factor-alpha in the adult mammalian heart. J Clin Invest 92:2303–2312

Zhong J, Hwang TC, Adams HR, Rubin LJ (1997) Reduced L-type calcium current in ventricular myocytes from endotoxemic guinea pigs. Am J Physiol 273:H2312–H2324

Chapter 5
The Role of Proinflammatory Cytokines in Regulation of Cardiac Bioelectrical Activity: Link to Mechanoelectrical Feedback

Vlad S. Kuzmin, Denis V. Abramochkin, Vadim M. Mitrochin, Bo Tian, Ekaterina Yu. Makarenko, Leonid V. Kovalchuk, Marina V. Khoreva, Anna Nikonova, Leonid Kalugin, Natalia N. Lysenko, Ilya Lozinsky, Alexander Rozanov, Grigory Arutyunov, Irina Kiseleva, and Andre Kamkin

Abstract In this review we cover key issues, concerned with the effects of cytokines in cardiac tissue and cardiac cells. We discuss the effects of proinflammatory cytokines under non-pathological conditions and their mechanisms dependent and independent of nitric oxide. The role of proinflammatory cytokines is considered in acute myocardial infarction and in heart failure. We also describe proinflammatory cytokines as inductors of arrhythmia. We discuss ionic current alternation as possible mechanisms of cytokines action in heart. We consider TNF-α as a possible player in this signaling cascade. It was shown that TNF-α induced alternation of transmembrane action potentials. Influence of TNF-α on transient outward current (I_{to}), I_{Kur}, I_{Kr}, I_{Ks}, I_{K1} is also reported. We discuss the interplay between TNF-α and Ca^{2+} current, influence of TNF-α on SERCA. Then we consider influence of IL-1 on action potentials, I_{Na}, I_{Ca}, I_K. We also address the role of IL-2, IL-6, and IL-11. Finally using TNF-α and IL-6 as an example we discuss the effects of cytokines on mechanoelectrical feedback. Perfusion of cardiac tissue with TNF-α containing solution leads to abnormalities in cardiac electrical activity, majorly to prolongation of APD90 and appearance of hump-like depolarization at APD90 level. After reaching E_c hump-like depolarization transforms into extra-AP, leading to sustained arrhythmias. TNF-α activates NO cardiomyocyte synthases and the rise of intracellular NO levels opens MGCs, which leads to sodium entry into the cell, which depolarizes cellular membrane, shifting resting potential towards E_C. We proposed and proved that TNF-α triggered arrhythmias can be mediated through activation of MGCs. Stretching of preparations removed TNF-α. Perfusion of preparation with IL-6 containing solution leads to fibrillation in response to low levels of stretch. IL-6 mechanisms of action are mediated by NO synthases

A. Kamkin (✉)
Department of Fundamental and Applied Physiology, Russian State Medical University, 117997 Moscow, Russia
e-mail: Kamkin.A@g23.relcom.ru; andre.kamkin@gmail.com

A. Kamkin, I. Kiseleva (eds.), *Mechanical Stretch and Cytokines*,
Mechanosensitivity in Cells and Tissues 5, DOI 10.1007/978-94-007-2004-6_5,

in cardiomyocytes. The circulating levels of TNF-α and IL-6 were found to be significantly higher in patients suffering from atrial fibrillation. It suggests a positive feedback between inflammation and atrial fibrillation. Proinflammatory cytokines are believed to be markers of atrial fibrillation, and more over, the key element in this positive feedback system.

Keywords Heart · Arrhythmia · Proinflammatory cytokines · Cardiomyocytes · Mechanoelectrical feedback · Mechanically gated channels · TNF-α · IL-2 · IL-6 · IL-11 · Nitric oxide · Ionic currents

5.1 Introduction

First cytokine related investigations date back to 1950s, but till present a number of questions in this field remain unanswered. For example till late 2000s it was generally accepted that biological effects of cytokines are known. However recently it was found that their effects expand far beyond the immune system and affect many organs and tissues. Latest studies of cardiovascular pathology genesis lead to emergence of a new related field, which can be called "Cardio-immunology". The role of cytokines in regulation of cardiac function is well described in literature. Unfortunately much fewer studies investigated the cross talk between cytokines effects and effects of myocardium stretch of healthy and pathological tissue. First studies in this field addressed this topic on whole heart level. For example one lab studied the induction of myocardial stretch in an isolated perfused Langendorff preparation by inflation of an intraventricular balloon to an end-diastolic load, which according to their publication resulted in a nearly six fold increase in VEGF message level. They reported a possibility of stretch-induced induction of VEGF expression (Li et al., 1997). Later works focused primarily on cytokine production during stretching of cardiac cells in culture (For example Yokoyama et al., 1999). This field of research remains relatively new and to our knowledge there is only a few number of reviews addressing this topic (for example see Chapter 2 of this Volume). The effects of cytokines on stretched heart and their role in regulation of mechanoelectrical feedback remains unknown. Moreover the role of cytokines in regulation of ion channel activity, their single channel currents and action potentials remain unaddressed. For this reasons we focused our chapter on discussion of the role of main proinflammatory cytokines in regulation of heart bioelectrical activity. In conclusion of our review we analyze the role of a number of cytokines in regulation of mechanoelectrical feedback.

5.2 Physiological Role of Proinflammatory Cytokines

5.2.1 *Effects of Proinflammatory Cytokines under Non-pathological Conditions and Their Mechanisms Dependent and Independent of Nitric Oxide*

The main and the most obvious physiological function of proinflammatory cytokines is induction of inflammation in the case of tissue damage. However, some of them demonstrate pleiotropic nature, for example IFN-γ is considered a proinflammatory cytokine, but possesses antiviral activity and induces activation of cytotoxic T cells (Dinarello, 2000). The basic mechanism of proinflammatory cytokine-mediated promotion of inflammation is up-regulation of several genes encoding the mediators of inflammation, which are usually not produced in healthy persons. For example, some of these genes code for enzymes that increase the synthesis of leukotrienes, prostanoids and NO. Others are genes of chemokines, the small peptides that facilitate the passage of leukocytes from bloodstream into the tissues, and endothelial adhesion molecules, which are essential for leukocytes adhesion to endothelium (Dinarello, 2000). Of course, proinflammatory action of cytokines is expressed mostly in the damaged tissue, however, as far as proinflammatory cytokines can get to the bloodstream and induce fever and other types of systemic response, they may affect all other organs located far from the nidus of inflammation. Heart is one of their possible targets.

That is why cardiotropic effects of proinflammatory cytokines undoubtedly merit extensive investigation. In the last two decades various effects of proinflammatory cytokines, mostly TNF and IL-1, on contractile and electrical activity of myocardium were demonstrated in diverse in vivo and in vitro studies. These effects may be divided in two groups: the early ones begin to develop immediately after cytokine application and completely develop within the first 3 h, while the delayed response is lasting from hours to days (Prabhu, 2004a).

The early response may be either cardiostimulatory or inhibitory depending on the concentration of cytokine and prevailing cellular and biological milieu. For example, low concentrations of TNF-α (200–500 U/ml) cause transient decrease of AP duration and amplitude of peak Ca^{2+} transient without affecting $I_{Ca,L}$ in adult mammalian cardiomyocytes (Yokoyama et al., 1993; Sugishita et al., 1999; Stamm et al., 2001a, b). However, high concentrations ($\geq$10,000 U/ml) were observed to produce either inhibition of $I_{Ca,L}$ accompanied with depression of peak Ca^{2+} transient (Krown et al., 1995) or no changes in Ca^{2+} turnover despite reduced contractility (Goldhaber et al., 1996). In other studies early positive inotropic effects of TNF-α mediated by increase of peak Ca^{2+} transient were observed in adult rat cardiomyocytes (Cailleret et al., 2004; Amadou et al., 2002). Other proinflammatory cytokines: IL-1β (Liu and Schreur, 1995; Schreur and Liu, 1997), IL-6 (Sugishita et al., 1999) and IL-2 (Cao et al., 2003a, b) were shown to suppress contractile activity in moderate concentrations (about 1–5 ng/ml). This effect is mediated by $I_{Ca,L}$ depression in the case of IL-1β.

Alteration of Ca^{2+} cycling is not the only mechanism of immediate cytokine effects. Stimulation of cytokine receptors also leads to rapid induction of NO production by cNOS as well as activation of sphingomyelinase and PLA2-dependent signaling cascades (Prabhu, 2004a; Hedayat et al., 2010). The role of cNOS activation in distribution of early negative inotropic effects was demonstrated for TNF-α (Finkel et al., 1992; Goldhaber et al., 1996), IL-1β (Cain et al., 1999; Kumar et al., 1999), IL-2 (Finkel et al., 1992; McGowan et al., 1994) and IL-6 (Finkel et al., 1992; Kinugawa et al., 1994; Sugishita et al., 1999) in various cardiac preparations and isolated cardiomyocytes. Some of these authors argue that NO-dependent negative inotropic effects of proinflammatory cytokines are produced via activation of soluble guanylate cyclase and subsequent elevation of intracellular cGMP level (Kinugawa et al., 1994; Kumar et al., 1999). On the contrary, several other groups presented serious argumentation against participation of NO-dependent pathway in mediation of acute cytokine effects (Yokoyama et al., 1993; Edmunds et al., 1999; Grandel et al., 2000). Thus, interrelations between cytokines and NO signaling cascade are rather controversial.

It is known that proinflammatory cytokines activate sphingomyelinases in several cell types (Liu et al., 1998; Kolesnick 2002). Sphingomyelinases hydrolyze membrane phospholipide sphingomyeline to form ceramide, which, in turn, can be deacylated by ceramidase to yield sphingosine. This amino alcohol substance and its metabolite sphingosine-1-phosphate act as a second messenger, producing suppression of Ca^{2+} turnover in cardiomyocytes (Dettbam et al., 1994; Friedrichs et al., 2002). Several studies have indicated the possible contribution of sphyngolipid signaling pathway to early negative inotropic effects of TNF-α in adult feline (Oral et al., 1997) and guinea pig cardiomyocytes (Sugishita et al., 1999), rat ventricular myocardium (Hofmann et al., 2003), isolated rat (Edmunds and Woodward, 1998) and rabbit hearts (Stamm et al., 2001a, b). The role of sphingomyelinase signaling cascade in the contractile effects of other proinflammatory cytokines is sparsely studied, although it was proposed for negative inotropic effect of IL-1β in human atrial trabeculae (Cain et al., 1999). Interestingly, activity of neutral sphingomyelinases is inhibited by glutathione, therefore negative inotropic effect of cytokines exerted via activation of sphingomyelinase signaling cascade depends on the intracellular redox state. In vivo administration of glutathione precursor *N*-acetylcysteine blocks early cardiodepressant effects of TNF-α, unmasking the positive inotropic effect of this cytokine (Cailleret et al., 2004).

Effects of proinflammatory cytokines attributed to activation of sphingolipid pathway are at least partly dependent on another second messenger – arachidonic acid. TNF-α and IL-1β stimulate phospholipase A2 via activation of their receptors and therefore increase the intracellular level of arachidonic acid (Jayadev et al., 1994; Liu and McHowat, 1998). This messenger activates neutral sphingomyelinases (Robinson et al., 1997), but also has own positive inotropic effect in cardiomyocytes, which is produced via increase of Ca^{2+} transients (Kang and Leaf, 1994; Damron and Summers, 1997). Therefore, small concentrations of arachidonic acid improve myocardial contractility, while higher concentrations suppress it due to the activation of sphingolipid pathway (Amadou et al., 2002; Damron and Summers, 1997). That is why some researchers explain the dual inotropic effects of TNF-α by such dual action of arachidonic acid (Amadou et al., 2002). Arachidonic acid may also mediate the positive inotropic effect of IL-2 (Bracco et al., 1991).

Delayed effects of proinflammatory cytokines are substantially less controversial than the acute ones. It is generally accepted that sustained application of proinflammatory cytokines, in the first place TNF-α and IL-1β, induces marked decrease of basal myocardial contractility which is aggravated by suppressed responsiveness to stimulation of β-adrenoreceptors (Prabhu, 2004a, b). The latter effect for the first time was demonstrated in rat neonatal myocytes (Gulick et al., 1989; Chung et al., 1990). Authors explained this effect by loss of β-adrenoreceptors ability to activate adenylate cyclase, because cytokines didn't change the density of receptors and on the other hand direct stimulation of adenylate cyclase by forskolin induced normal cAMP accumulation and augmentation of contractile activity. Subsequent studies confirmed that impairment of signal transduction from receptor to adenylate cyclase, which is mediated by G_s protein, is the main reason of decreased β-adrenoreceptors responsiveness (Rozanski and Witt, 1994; Bick et al., 1997).

Several other groups of investigators believe that suppression of β-adrenoreceptors responsiveness is mediated by sustained upregulation of iNOS leading to increased cGMP intracellular content resulting in decrease of cAMP level due to the activation of PDE2 by cGMP (Balligand et al., 1993, 1994; Balligand, 1999; Ungureanu-Longrois et al., 1995, 1997; Joe et al., 1998). It seems likely that both NO-dependent and NO-independent mechanisms of β-adrenoreceptors responsiveness are decreased by proinflammatory cytokines which are present in the myocardium. Moreover, upregulation of iNOS, which is already shown for most of proinflammatory cytokines in the case of their prolonged application (Hedayat et al., 2010), exerts several other crucial consequences such as S-nitrosylation of thiol residues of important regulatory proteins and formation of peroxynitrite, leading together to prolonged depression of contractile function (Campbell et al., 1996; Massion et al., 2003; Prabhu, 2004a). Obviously, increase in generation of reactive oxygen species (ROS), which is also induced in myocardium by different proinflammatory cytokines (Oyama et al., 1998; Cheng et al., 1999; Panas et al., 1998; Ferdinandy et al., 2000), amplifies the latter effect greatly. For example, in isolated rat heart combination of IL-1β, TNF-α and IFN-γ provokes augmentation of xanthine oxidoreductase and NAD(P)H oxidase activity, which is accompanied with enhanced activity of iNOS and leads to sustained decline in contractility (Ferdinandy et al., 2000). It is important to note that oxidative modifications induced by ROS and peroxynitrite tend to be irreversible (Stamler et al., 2001) and are therefore attributed to serious pathological changes in the structure and function of myocardium.

Thus, the modulation of normal heart function by proinflammatory cytokines is extremely complex and is mediated by diverse signaling mechanisms. However, it becomes even more versatile in the case of cardiac disease. Here we briefly discuss the role of proinflammatory cytokines in pathogenesis of myocardial infarction, heart failure and arrhythmias.

5.2.2 Proinflammatory Cytokines in Acute Myocardial Infarction

The onset of myocardial infarction starts from ischemic and subsequent reperfusion injury of myocardium leading to cardiomyocytes necrosis. Proinflammatory cytokines have significant impact on heart function during all these stages. Although they promote myocardial malfunction during and after ischemic injury, cytokines are able to increase resistance of heart to such injury if added prior to ischemia.

It is accepted that application of TNF-α (Nakano et al., 1998; Eddy et al., 1992), IL-1β (Brown et al., 1990; Maulik et al., 1993) and IL-6 (Smart et al., 2006) prior to induction of ischemia-reperfusion injury exerts clear beneficial cardioprotective effects. The mechanisms of such cytoprotective preconditioning are diverse, but in the case of TNF-α and IL-1β they are undoubtedly associated with induction of manganous superoxide dismutase (MnSOD), which neutralizes the cytotoxic oxygen free radicals, which are generated in the myocardium during ischemia-reperfusion (Chen et al., 1998; Wong and Goeddel, 1988; Maulik et al., 1993). The second substantial protective mechanism is concerned with upregulation of various heat-shock proteins, which attenuate the excessive myocardial production of inflammatory cytokines, following the ischemic and reperfusion injury and having strong deleterious impact on cardiac function (Nakano et al., 1996). IL-1β preconditioning promotes accumulation of leukocytes and generation of H_2O_2 in myocardium, leading to moderate oxidative stress and induction of cytoprotective mechanisms, which increase resistance of myocardium to ischemic injury (Brown et al., 1990). IL-6 preconditioning is mediated mainly by inhibition of apoptotic signaling pathways in cardiomyocytes via PI3K/Akt-mediated activation of iNOS (Smart et al., 2006).

It is not surprising that during ischemia and subsequent reperfusion proinflammatory cytokines aggravate irreversible injury of myocardium (Schulz, 2008; Schulz and Heusch, 2009). In ischemic myocardium negative effects of TNF-α, such as induction of apoptosis (Haudek et al., 2007) are produced via activation of first type TNF receptors (Flaherty et al., 2008) and can be prevented by treatment with TNF-α antibodies (Belosjorow et al., 2003) or soluble TNF receptors of first type (Sugano et al., 2004), which are capable of binding and inactivation of TNF-α. IL-1 also enhances apoptosis during ischemia and reperfusion (Haudek et al., 2007). This effect is mediated by activation of NO production (Ing et al., 1999) via stimulation of IL-1 receptors and can be blocked by antagonists of these receptors (Abbate et al., 2008). Similarly to IL-1, IL-6 provokes contractile dysfunction in ischemic myocardium via activation of iNOS mediated by JAK2/STAT3 signaling pathway (Yu et al., 2003a, b). In the last years IL-18 was accepted as a new powerful proapoptotic factor, which plays very negative role during ischemia. This cytokine promotes apoptosis in cardiomyocytes via upregulation of death receptors leading to enhancement of extrinsic proapoptotic signaling (Marino and Cardier, 2003; Chandrasekar et al., 2004), activates caspases and suppresses antiapoptotic PI3K/Akt pathway via induction of phosphatase and tensin homolog expression (Chandrasekar et al., 2006).

After the ischemic lesion the myocardium itself becomes the source of proinflammatory cytokines. Obviously, the increased production of cytokines endangers nearby cells altering cardiomyocytes structure and function. However, infarction leads to increased level of proinflammatory cytokines in blood and, as a result, severe systemic inflammatory response syndrome known as the leading cause of death in patients with acute myocardial infarction. TNF-α, IL-6 and, especially, IL-1β play the key role in its pathogenesis (Debrunner et al., 2008). Nevertheless, we will focus on alteration of heart function caused by proinflammatory cytokines, produced in the heart itself after myocardial infarction.

Firstly, all acute cardiotropic effects of proinflammatory cytokines, described in Section 5.2.1, take place after infarction but in much bigger extent than in normal conditions, because concentration of cytokines in the heart are many times higher. Second, described proapoptotic action of cytokines aggravates after infarction. Moreover, other consequences of inflammation, such as infiltration of leukocytes and increase of ROS production take place in the heart and together with proper cytokines' effects lead to progressive loss of cardiac function. Finally, expression of proinflammatory cytokines provokes hypertrophy of cardiomyocytes and remodeling of extracellular cardiac matrix. These effects potentiate delayed cardiodepressant action of cytokines (see Section 5.2.1) and accelerate development of cardiac failure following acute myocardial infarction (Hedayat et al., 2010).

5.2.3 Role of Proinflammatory Cytokines in Heart Failure

Hypertrophic growth response is shown for all proinflammatory cytokines (Hedayat et al., 2010). Hypertrophic effects of TNF-α, measured in terms of cardiomyocytes size (both length and width), cardiac weight and left ventricular wall thickness (Janczewski et al., 2003) are at least partly mediated by upregulation of angiotensin-converting enzyme and resulting increase in angiotensin II protein level (Flesch et al., 2003). TNF-α stimulates expression of other proinflammatory cytokines in myocardium, which in turn amplify its own hypertrophic action. Hypertrophic effect of IL-1, which was demonstrated in several in vitro (Palmer et al., 1995; Thaik et al., 1995) and in vivo studies (Nishikawa et al., 2006), is NO-independent and appears to be mediated through a tyrosine kinase signaling pathway (Palmer et al., 1995; Thaik et al., 1995). IL-6 also contributes to hypertrophic growth response (Yamauchi-Takihara and Kishimoto, 2000), which is believed to be mediated mainly through JAK/STAT3 signaling pathway (Kunisada et al., 2000). On the contrary to IL-1 and IL-6, hypertrophy of cardiomyocytes induced by IL-18 is mediated through PI3K cascade (Chandrasekar et al., 2005; Colston et al., 2007). Thus, cytokines produced in the heart due to acute infarction induce hypertrophic growth of cardiomyocytes, which primarily helps to regain pumping function, but frequently transforms to maladaptive response leading to progressive heart failure.

Alterations of collagenous extracellular matrix (ECM) providing the physical scaffolding for the spatial organization of cardiomyocytes into cardiac tissue have been demonstrated to play a central role in cardiac remodeling and resulting progressive heart failure (Fedak et al., 2005; Ju and Dixon, 1996). It is shown that TNF-α (Siwik and Colucci, 2004), IL-1β (Siwik et al., 2000) and other proinflammatory cytokines (Hedayat et al., 2010) cause imbalance between synthesis and degradation of ECM through disregulation of degradative enzymes and of matrix metalloproteinases (MMPs) activity. These effects are largely influenced by the duration of exposure, which ranges from increased fibrillar collagen degradation to excessive fibrillar collagen deposition. For example, in short term, activation of MMPs induced by TNF-α leads to enhanced degradation of ECM components which promotes progressive dilation of left ventricle (Sivasubramanian et al., 2001; Li et al., 2000). However, in long term, sustained action of TNF-α results in excessive collagen deposition and increased LV stiffness (Sivasubramanian et al., 2001).

Thus, increased expression of proinflammatory cytokines in the myocardium leads to sustained depression of cardiomyocytes contractile activity (see Section 5.2.1), which is accompanied with even more oppressive irreversible effects: cardiomyocytes hypertrophy and ECM remodeling and results in chronic heart failure.

5.2.4 Proinflammatory Cytokines as Inductors of Arrhythmia

The possibility of proinflammatory cytokines contribution to pathogenesis of cardiac arrhythmias is not so obvious as in the case of myocardial infarction and cardiac failure. Among cardiac arrhythmias, the most common and well-studied is atrial fibrillation (AF). However the specific mechanism of AF initiation remains unknown. It is generally accepted that AF is associated somehow with inflammation. The circulating levels of IL-6 (Marcus et al., 2008), TNF-α (Sata et al., 2004) and IL-18 (Luan et al., 2010) were found to be significantly higher in patients suffering from AF. However, it is unclear whether inflammation is a cause or an effect of AF and, in either case, what mechanistic pathways may be important. One of the hypothesis proposes that AF leads to myocyte calcium overload, promoting atrial myocyte apoptosis, which, in turn, induces local inflammation and complement activation. Tissue damage then ensues and fibrosis sets in (Avilles et al., 2003; Demelis and Panaretou, 2001). Locally released factors of inflammation, including cytokines, can further contribute to membrane dysfunction by inhibiting the calcium cycling machinery. This can eventually lead to the maintenance of AF (Avilles et al., 2003; Demelis and Panaretou, 2001, 2006).

Some experimental data sheds light on mechanism of arrhythmogenic action of proinflammatory cytokines in atrial myocardium. Administrating TNF-α in rats can induce arrhythmias (Krown et al., 1995) and mice with overexpression of TNF-α have a greater incidence of AF (London et al., 2003). The latter is associated with abnormalities in systolic and diastolic Ca^{2+} handling and may be due to them. In particular, enhanced expression of TNF-α is accompanied by increase in frequency and magnitude of spontaneous Ca^{2+} releases from the sarcoplasmic reticulum in working cardiomyocytes (Saba et al., 2004). Such releases may lead to induction of delayed afterdepolarizations, which are well known triggers for AF, via activation of reverse Na^+/Ca^{2+} exchanger. In cardiomyocytes from pulmonary veins, which are believed to be important foci of ectopic beats initiating AF or tachycardia (Chen et al., 1999), TNF-α also markedly alters the calcium homeostasis in different ways, increases amplitudes and frequency of delayed afterdepolarizations and thereby increases pulmonary veins ability to initiate AF (Lee et al., 2007).

Thus, there seems to be a positive feedback between inflammation and AF and proinflammatory cytokines are believed to be not only markers of AF, but the key element in this positive feedback system. Moreover, various anti-inflammatory agents, such as statins or angiotensin-converting enzyme inhibitors, are able to suppress atrial remodeling associated with AF and, therefore decrease the possibility of new AF paroxysms (Issac et al., 2007).

The role of proinflammatory cytokines in ventricular arrhythmias induction was also studied just recently. In dogs, etanercept, which binds and inactivates TNF-α, decreased probability of ventricular tachyarrhythmias after coronary artery ligation (Yu et al., 2005a, b), indicating important role of this cytokine in sudden cardiac death. Mechanisms of proarrhythmic effects of cytokines in ventricles seem to be very similar to shown in atrial myocardium. TNF-α (Duncan et al., 2010; Saba et al., 2004) and IL-1β (Duncan et al., 2010) suppress total release of Ca^{2+} from the sarcoplasmic reticulum induced by AP, but promote local spontaneous Ca^{2+} releases during diastole. These releases lead to initiation of delayed afterdepolarizations via activation of depolarizing Na^+/Ca^{2+} exchanger current.

5.3 Ionic Current Alternation as a Possible Mechanisms of Cytokines Action in Heart

5.3.1 Tumor Necrosis Factor-α (TNF-α)

Tumor necrosis factor-α (TNF-α) is a proinflammatory cytokine that is implicated in a number of pathophysiological events. It was established that TNF-α affects myocardium through several pathways. TNF-α influences both mechanical and bioelectrical activity of myocardium by alternating membrane potassium and calcium currents and cytoplasmic calcium handling. It was also shown that tumor necrosis factor-α is able to induce rapid as well as slow effects in cardiac muscle.

A lot of studies were carried out to clarify mechanisms of TNF-α action in heart, but often, represented results are contradictory, especially in cases of investigation of cardiac bioelectrical activity. It is necessary to mention that in animal models or patients with heart failure serum TNF-α concentrations belong to pictogram range (up to 10 pg/ml) (Cugno et al., 2000; Testa et al., 1996). While, TNF-α concentration range, utilized to treat heart preparation and cells in electrophysiological studies extend from picogram to hundreds of nanogram per milliliter.

Here some aspects of TNF-α influence on action potentials and ionic currents in heart and TNF-α induced contraction reduction are discussed.

5.3.1.1 TNF-α Induced Alternation of Transmembrane Action Potentials

In isolated guinea pigs papillary muscle (Alloatti et al., 1999) TNF-α in concentration of 1–10 ng/ml induce APD decrease by 10–15% after short term (20–25 min) cytokine administration (Fig. 5.1a). Decreasing of TNF-α induced AP duration was concentration- and time-dependent. APD decreasing was accompanied by reduction of the contractile force. However, as Alloatti reports, the resting membrane potential, the overshoot and the maximum rate of depolarization of the APs were not affected by TNF-α. Recovery of mechanical activity and APs parameters was observes after 20 min of washout by solution free from TNF-α.

Effects of TNF-α on APs were completely abrogated by pretreatment of papillary muscle by the NOS inhibitor – L-NAME or PAF-receptor antagonists (PAF – platelet aggregation factor, CV 3988). Also TNF-α-induced APD decreasing was abolished by chemical (Triton X-100) endocardial endothelium removing. Alloatti et al. suggested that both the generation of NO and PAF production contributes to the TNF-induced alteration of APD. It was also speculated that the production of NO is consequent to the production of PAF.

TNF-α influence on APs of isolated canine ventricular cardiomyocytes was opposite to that, observed in guinea pigs papillary muscle. APD was significantly longer in canine cells pretreated with TNF-α (Fig. 5.1b) then in control cells. Alternation of APD occurred only after 10 h of incubation with 10 ng/ml TNF-α (Wang et al., 2004).

Effects of TNF-α were investigated in experiments with isolated rat ventricular cardiomyocytes. In this study it was shown that after 48 h of exposition of

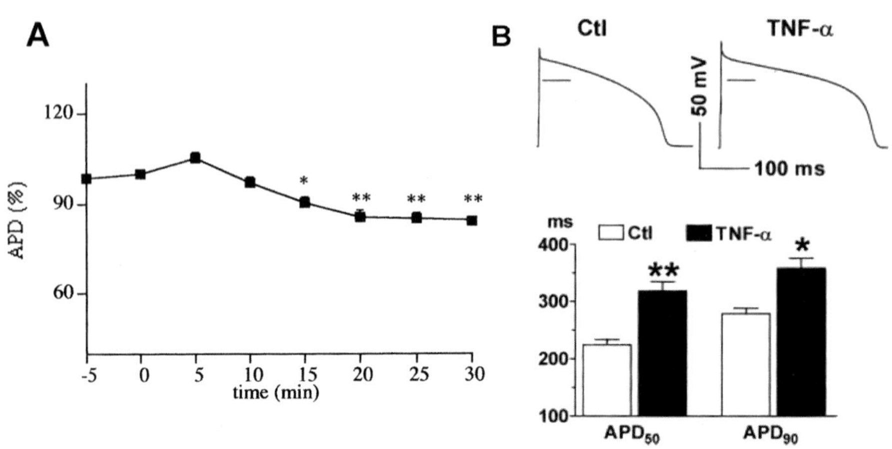

Fig. 5.1 TNF-α alternate action potential duration in heart. (**a**) Effects of TNF-α (10 ng/ml) on action potential duration (APD) in the isolated guinea pig papillary muscle (from Alloatti et al. (1999) with permission of Elsevier and British Cardiac Society from Copyright Clearance Center). (**b**) Effects of TNF-α (10 ng/ml, 6-h incubation at 4°C) on action potential duration in canine ventricular cells (from Wang et al. (2004) with permission of American Society for *Biochemistry and Molecular Biology* via Copyright Clearance Center)

cardiomyocytes to 5 ng/ml TNF-α APD increased more than two times (Fernandez-Velasco et al., 2007).

In ventricular cardiomyocytes, isolated from normal mice, which received chronically NTF-α significant alternation of ionic currents was observed, while AP duration remained unaltered (Grandy and Fiset, 2009). Only slight reduction of AP amplitude was observed. Authors speculate, that the lack of TNF-α-induced APs alterations could be due to counterbalanced alternation of repolarizating and depolarizating currents (resulting in APD preservation).

Transgenic mice that overexpress TNF-α selectively in the heart tissue have been studied as a model of congestive heart failure (Janczewski et al., 2003; Kubota et al., 1997; London et al., 2003; Sivasubramanian et al., 2001). TNF-α- overexpressing mice develop atrial and ventricular arrhythmias. In such mice heart level of TNF-α exceeds cytokine level usually reached in disease state more than ten times (Kubota et al., 1997). Several electrophysiological investigations were realized with such animals. AP duration in ventricles of perfused isolated hearts and in isolated ventricular transgenic mice cardiomyocytes was significantly lengthened (London et al., 2003; Petkova-Kirova et al., 2006) compared to APD in wild type mice.

Thereby, literature data, describing influence of TNF-α to action potential is contradictory. Some authors report decreasing of APs duration, on the contrary to other reports, TNF-α induce prolongation of APs. These discrepancies may occur due to different animals (mice, rat, dog), experimental protocol and wide range of TNF-α concentration used in experiments.

5.3.1.2 TNF-α and Ca^{2+} Current

Krown and coworkers has shown that in isolated rat ventricular cardiomyocytes TNF-α caused significant inhibition of I_{Ca} which mediated by TNF-α receptors TNFRI (Krown et al., 1995). TNF-α inhibition of the I_{Ca} was dose dependent (Fig. 5.2), but significant inhibition of I_{Ca} was observed in presence of very high concentration of TNF-α (>300 ng/ml), which 1000 times exceeds cytokine level observed in the pathophysiological condition (≈0.01–0.05 ng/ml) (Testa et al., 1996). As follows from *I–V* curves TNF-α exerts an inhibitory action at all tested potentials. Only partial recovery of I_{Ca} from inhibition was observed after TNF-α washout. Therefore, authors conclude involvement of intracellular second messengers in inhibitory effects of TNF-α.

Therefore, Krown and colleagues assume I_{Ca} suppression as a significant component of mechanism of TNF-α cardiodepressant action in different pathophysiological conditions.

On the contrary several studies reported lack of TNF-α effects on I_{Ca}. In isolated rat and cat ventricle myocytes treated by TNF-α peak I_{Ca} suppression and alternation of *I–V* relation were not detected (Yokoyama et al., 1993; Fernandez-Velasco et al., 2007).

Same observation was made in cardiomyocytes isolated from NTF-a overexpressing mice (Petkova-Kirova et al., 2006). Measurements of $I_{Ca,L}$ showed no difference in the peak current density between control and TNF-overexpressing

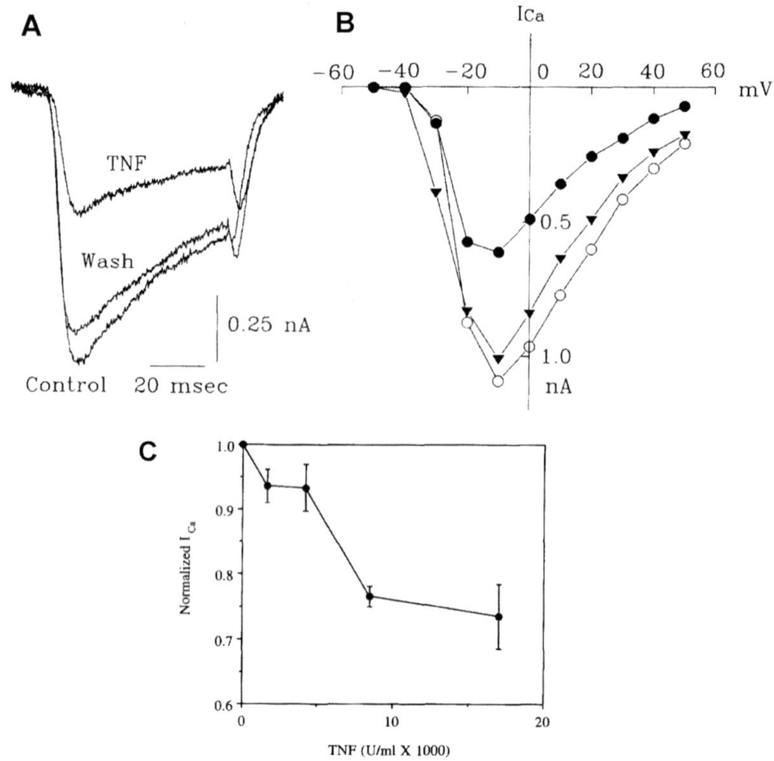

Fig. 5.2 TNF-α alternate cardiac $I_{Ca,L}$. (**a**) L-type channel recordings from a whole cell patch-clamped adult rat cardiac myocyte (5 min, 18,000 U/ml). There was a partial recovery after washout. Control experiments using fatty acid free BSA (data not shown) failed to show a non-specific protein effect on the channel in the range of protein used in the TNF-α experiments. (**b**) Current–voltage relationship for $I_{Ca,L}$ in the same cell prior to TNF-α application (*open circles*), 5 min after TNF-α application (*closed circles*) and after washout (*closed triangles*). (**c**) Average dose-dependent effect of TNF-α on L-type channel current in adult cardiac myocytes demonstrated half-maximal inhibition with 6400 U/ml. Holding potential –50 mV, pulses to 0 mV (from Krown et al. (1995) with permission of Elsevier Limited and *FEBS Letters* via Copyright Clearance Center)

myocytes over the range of potentials from –10 to +50 mV. At more negative potentials $I_{Ca,L}$ was slightly larger in TNF-α myocytes. No differences of inactivation and deactivation kinetics of $I_{Ca,L}$ were detected.

Hereby, ability of TNF-α to alternate I_{Ca} in cardiomyocytes remains questionable.

5.3.1.3 Influence of TNF-α on Transient Outward Current (I_{to})

Under various pathological conditions, including heart failure, prolongation of action potentials, which occurs due to downregulation or suppression of the potassium currents (I_K) (Nuss et al., 1999; Beuckelmann et al., 1993) was observed. Depression of I_K and AP prolongation increases probability of arrhythmogenic events in heart (Nattel et al., 2007). Transient outward current (I_{to}) plays an essential role in early repolarization and regulation of duration of APs in myocardium (Hund and Rudy, 2004). In several studies it was shown that failing hearts (in human and in animals model) demonstrate significantly decreased I_{to} (Benitah et al., 1993; Kaab et al., 1998; Li et al., 2004). Isolated cardiomyocytes from transgenic mice overexpressing TNF-α demonstrate prolonged APs (London et al., 2003). Therefore, hypothesis accounting TNF-α as a mediator of I_{to} reduction was elaborated. In turn, TNF-α could modulate I_{to} through mechanisms involving upregulation of iNOS, NO and oxidant species (Fernandez-Velasco et al., 2007).

In study developed by Fernandez-Velasco isolated rat ventricular cardiomyocytes were exposed to TNF-α in concentration of 1–10 ng/ml for 48 h. After exposure to TNF-α amplitude of I_{to} was significantly reduced. Suppression of I_{to} was accompanied by AP prolongation. Inhibitory effect of TNF-α was concentration-dependent. TNF-α affect amplitude and density of current but not voltage dependence. Inactivating curve for I_{to} in cells treated with TNF-α was shifted to negative potentials. Half-maximal inactivation was –49.9 mV in control and –55.6 mV in cells treated with TNF-α. Authors also showed that exposure to TNF-α leads to lowering of I_{to} channel proteins level.

Thereby, depression of I_{to} is caused by several mechanisms – decreasing of current density, lowering level of channels proteins and shifting of inactivating curve to negative potentials.

The decrease of I_{to} induced by TNF-α was fully reversed in cells pretreated by iNOS inhibitors. Similarly, preliminary exposure of cardiomyocytes with antioxidants also reverses TNF-α induced I_{to} decreasing. Discussing these data, authors support the idea that highly reactive oxidant species, such a peroxynitrite, formed by NO and superoxide anions, play a central role in TNF-α-induced ionic currents suppression. This report was hypothesized that posttranslational modification of I_{to} channels by peroxynitrite (nitrosylation, nitration or oxidation) is the main mechanism of TNF-α action (Fernandez-Velasco et al., 2007).

Investigation provided by Kawada and colleagues was focused on TNF-α effect in cultured cardiomyocytes from neonatal (1-day-old) rats (Kawada et al., 2006). Cultured cells were incubated during 48 h with 50 ng/ml TNF-α. Peak I_{to} was significantly reduced in TNF-α – treated myocytes compared with the control cells (17.7 and 3.2 pA/pF respectively). Expression of I_{to}-related mRNA was also reduced.

Study of mice ventricular cardiomyocytes, similar to results obtained in rat and dog cardiomyocytes, demonstrated TNF-α – induced inhibition of I_{to} (by chronic TNF-α treatment) (Grandy and Fiset, 2009). However, depression of I_{to} was not accompanied by lowering of mRNA or protein expression level.

In transgenic mice overexpressing NTF-a, ventricle cardiomyocytes demonstrate significantly decreased I_{to} (Petkova-Kirova et al., 2006). Level of channels proteins responsible for I_{to} was measured in this study. As in rat ventricular cardiomyocytes, transgenic mice show lowered level of I_{to}-related proteins.

Thereby, in studies utilized different animals and experimental protocols it was shown that TNF-α causes reduction of I_{to}. Such convergent results make possible speculation that preferentially I_{to} depression is responsible for AP alternation and plays an important role in TNF-α-induced heart dysfunctions.

5.3.1.4 TNF-α Induced Alternation of I_{Kur}

In most species, including human, ultrarapid-activating K^+ current (I_{Kur}) was present only in atrial myocardium (Nerbonne and Kass, 2005). Studies focused on electrophysiological effects of NTF-α in heart predominantly deal with ventricular cardiomyocytes. This is, probably, the reason why there is insufficient information about NTF-α influence on the I_{Kur}. TNF-α-induced alternation of I_{Kur} was investigated only in mice. As shown previously, ultrarapid K^+ current (also called $I_{K,slow1}$) is present in adult mice ventricular myocardium (Liu et al., 2011; Brouillette et al., 2004).

In study provided by Grandy and Fiset I_{Kur} parameters were measured in isolated mice ventricular cardiomyocytes after NTF-a chronic (6 week in vivo) administration (Grandy and Fiset, 2009). In cardiomyocytes from TNF-α pretreated mice density of I_{Kur} was significantly smaller than in cell from control animals. I–V curves for I_{Kur} reveal significant difference, especially at positive potentials (Fig 5.3a). Inactivation of I_{Kur}.was similar in myocytes from all animals. Recovery from inactivation was significantly longer in the cells isolated from TNF-α-treated then from control animals. Authors suggested that the prolongation in the recovery from inactivation of I_{Kur} could result in a reduction in current. Also this study revealed that level of mRNA and expression of proteins related to I_{Kur} were not altered by chronic administration of TNFα. Posttranslation modification of channels as result of reactive oxygen species generation in cardiomyocytes is postulated as a main mechanism of TNF-α action.

$I_{K,slow1}$ was measured in transgenic mice, overexpressing TNF-α. In TNF-α cardiomyocytes at potentials positive than –10 mV the density of $I_{K,slow1}$ was significantly smaller than in cells from wild type mice (Fig. 5.3b). The mean $I_{K,slow1}$ density measured at +40 mV in TNF-α cardiomyocytes was suppressed approximately two times in comparison to controls (22.6 and 12.5 pA/pF respectively). In contrast with previous, level of channels proteins responsible for $I_{K,slow1}$ was reduced by more than 40% (Petkova-Kirova et al., 2006).

5.3.1.5 Role of TNF-α in Regulation of K^+ Currents – I_{Kr}, I_{Ks}

Down-regulation of K^+ current, including rapid component of delayed-rectifier current – I_{Kr} observed in congestive heart failure is accompanied by prolongation of APs (Nattel et al., 2007). Pore-forming α-subunit of I_{Kr} is encoded by human "ether-a-go-go"-related gene ("HERG channels"). In human ventricular myocardium I_{Kr} is the main contributor to AP repolarization among delayed currents (Virag et al., 2001). Excessive decreasing of I_{Kr} leads to early afterdepolarization and tachyarrhythmias (Rashmi and Hondeghem, 2005; Priebe and Beuckelmann, 1998).

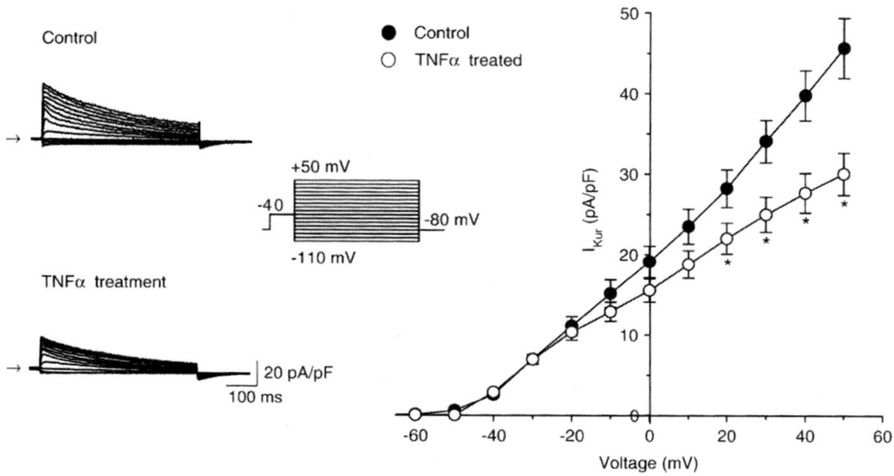

Fig. 5.3 Reduction of I_{Kur} by TNF-α. (**a**) I_{Kur} was markedly reduced in myocytes from TNF-α treated mice compared to myocytes from control mice. *Left*: Representative I_{Kur} traces for a control myocyte (*top*) and a myocyte from a TNF-α treated mouse (*bottom*). I_{Kur} was calculated by subtracting currents recorded in the presence of 100 μmol/l 4-AP from currents recorded in the absence of 4-AP (voltage protocol shown in *inset*). *Right*: The mean *I*/*V* curves show that I_{Kur} was significantly reduced in myocytes from TNF-α treated mice compared to myocytes from control animals at voltages of +20 mV to +50 mV. (**b**) Slowly inactivating delayed rectifier potassium current $I_{K,slow1}$ suppressed in TNF-α overexpressing mice. *Left*: Representative families of $I_{K,slow1}$ currents in control wild-type cells (*top*) and TNF cells (*bottom*). *Right*: Mean *I*/*V* relations for $I_{K,slow1}$ in cells from TNF mice (*squares*) and control wild-type mice (*triangles*) (from Grandy and Fiset (2009) with permission of Academic Press, International Society for Heart Research and *Journal of Molecular and Cellular Cardiology* via Copyright Clearance Center)

As was mentioned above serum level and tissue production of TNF-α in such pathological condition as heart failure is highly increased.

In study provided by Wang et al. (2004) it was shown that short time (15 min) exposure of isolated dog atrial and ventricular cardiomyocytes to TNF-α leads to concentration-dependent decreasing of I_{Kr} (Fig. 5.4a). Amplitude of HERG channels current was decreased by 9 and 35% after exposure to 1 and 100 ng/ml of TNF-α respectively. Effect of NTF-α was larger at more negative potentials. However, I_{Kr} kinetics was unaltered by TNF-α. Depression of I_{Kr} by TNF-α was accompanied by AP duration prolongation (Fig. 5.4b).

In this study it was shown that TNF-α -induced decreasing of I_{Kr} was mediated via TNFR1 receptors. Preincubation of cardiomyocytes with inhibitory anti-TNFR1 antibody for 1 h prevents suppression of I_{Kr} induced by short-time or prolonged TNF-α administration. As in case of I_{to}, author suggests intracellular reactive oxygen species generation as a main mechanism of TNF-α-mediated current suppression. Pretreatment of cardiomyocytes with antioxidants (vitamin E, MnTBAP) for 2 h prevented consequent I_{Kr} reduction induced by TNF-α.

Hatada and colleagues examined effects of NTF-α on delayed-rectifier potassium currents in cardiomyocytes isolated from guinea pig ventricle (Hatada et al., 2006).

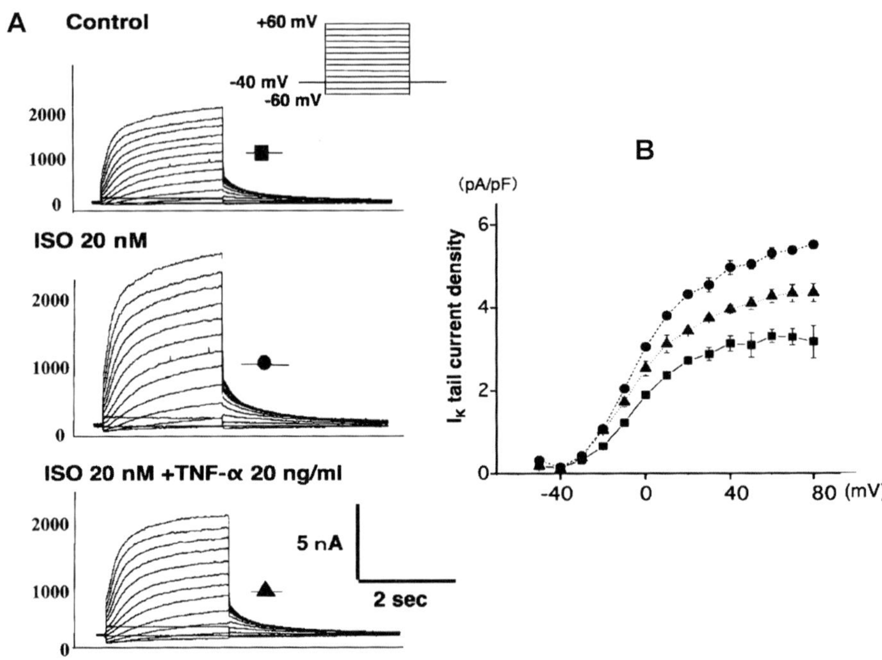

Fig. 5.4 Impairment of I_{Kr} function by TNF-α. (**a**) Three sets of current traces at test potentials between –60 and +60 mV in 10-mV increments. (**b**) Tail current amplitudes plotted against test potentials. *Solid square* indicates the control condition; *solid circle*, exposure to 20 nmol/l ISO; and *solid triangle*, after addition of 20 ng/ml TNF-α (from Hatada (2006) with permission of *Biochemical and Biophysical Research Communications* via Copyright Clearance Center)

In these experiments TNF-α suppressed I_K only after preliminary treatment of cells with isoproterenol, histamine or forskolin. Suppression of I_K in normal condition (at basal level) was not observed. Relating to pharmacological analysis, authors conclude that TNF-α preferentially affects slow component of delayed-rectifier (i.e. I_{Ks}) and that NTF-α inhibits potassium currents by reducing intracellular cAMP level.

5.3.1.6 TNF-α in Regulation of I_{K1}

Scarce data regarding influence of TNF-α on inwardly rectifying K^+ current (I_{K1}) is present to our knowledge. Only in Grandy and Fiset study (see above) analysis of I_{K1} after TNF-α administration was performed. In mice ventricular cardiomyocytes TNF-α has no effects on I_{K1}. The current–voltage (*I–V*) relationships show that the I_{K1} was comparable in cells from control and TNF-α pretreated mice (Fig. 5.5).

5.3.1.7 TNF-α Induced Stimulation of I_{KATP}

One study was reported on influence of TNF-α on ATP-sensitive potassium channels in cardiomyocytes (El-Ani and Zimlichman, 2003). In cardiac cell line H9c2

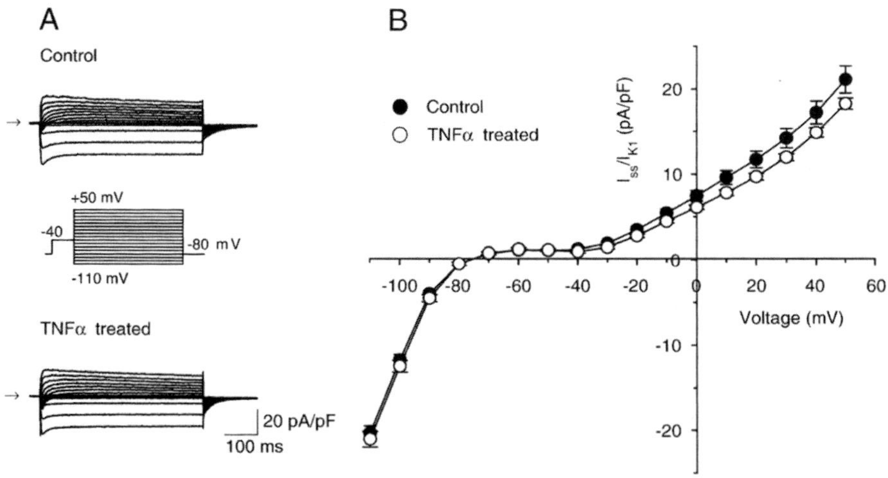

Fig. 5.5 I_{K1} after chronic TNF-α treatment in mice. The steady-state outward K^+ current (I_{ss}) and the inward rectifier K^+ current (I_{K1}) were comparable in ventricular myocytes from control and TNF-α treated mice. (**a**) Representative traces for I_{ss} and I_{K1} from a control myocyte (*top*) and a myocyte from a TNF-α treated mouse (*bottom*). (**b**) Mean *I*/*V* curves show that I_{ss} and I_{K1} density were similar in control myocytes and myocytes from TNF-α treated mice. The voltage protocol is shown in the inset. The current density for I_{K1} was measured between −110 and −40 mV. To record I_{ss}, the currents were initiated in the presence 100 μmol/l 4-AP and test steps were preceded with a 100 ms step to −40 mV (from Grandy and Fiset (2009) with permission of Academic Press, International Society for Heart Research and *Journal of Molecular and Cellular Cardiology* via Copyright Clearance Center)

pretreated with 50 ng/ml TNF-α for 30 min was caused activation of I_{KATP}. Preincubation of cardiomyocytes with glibenclamide, I_{KATP} blocker, inhibited the effects of TNF-α. Activation of I_{KATP} leads to depression of action potential and contractile force of cardiomyocytes (Zingman et al., 2007). Unfortunately, in this study ionic current was measured with use of radioactive isotope – ^{86}Rb and no electrophysiological data were obtained. Thereby, the role of TNF-α-induced I_{KATP} activation in APD regulating remains questionable.

5.3.1.8 TNF-α as a Regulator of Cytoplasmic Calcium Handling; Influence on SERCA

TNF-α affects contractility in the heart. Two main visions of TNF-α-induced negative inotropic action in cardiomyocytes exists. First one proposes that there is a reduced myofilament responsiveness to $[Ca^{2+}]_i$ which is mediated by nitric oxide or sphingosine production (Finkel et al., 1992; Goldhaber et al., 1996; Oral et al., 1997). Second one proposes that intracellular Ca^{2+} transients are mediated via down-regulation of structures (SERCA, NCX, RyR) relating to Ca^{2+} handling (Kao et al., 2010; Janczewski et al., 2003).

In study performed on isolated rabbit and guinea pig ventricular myocytes (Goldhaber et al., 1996) TNF-α affected only cell shortening, but not cytoplasmic calcium transients (amplitude or kinetics). Moreover, TNF-α caused a significant increase in diastolic cell length without any change in diastolic and systolic $[Ca^{2+}]_i$. In this investigation inhibition of NO production by L-NAME prevented TNF-α-induced suppression of myocytes contractile activity. The same observation was made in experiments with isolated hamster papillary muscles (Finkel et al., 1992). Contraction alternation without changing of cytoplasmic calcium level may occur due to a reduction in the responsiveness of the myofilaments to $[Ca^{2+}]_i$. Authors suggest that an NO-mediated increase in cGMP could account for decreased myofilament Ca^{2+} responsiveness.

TNF-α reduces $[Ca^{2+}]_i$ transients in isolated rat and feline cardiomyocytes (Krown et al., 1995; Yokoyama et al., 1993) (Fig. 5.6). TNF-α blocks spontaneous contractions of cultured murine cardiomyocytes (Weisensee et al., 1993). In heart of TNF-α overexpressing transgenic mice ("TNF1.6 mice") authors observed altered intracellular Ca^{2+} transients with decreased amplitude and peak systolic Ca^{2+}, elevated diastolic Ca^{2+}, and slower kinetics (Janczewski et al., 2003; London et al., 2003). Also in transgenic mice twitch contractions and response to the β-adrenergic stimulation of myocytes were decreased (Janczewski et al., 2003). However, mechanisms of cytoplasmic Ca^{2+} transients alternation remain unrevealed.

SERCA2a (sarcoplasmic reticulum Ca^{2+}-ATPase) plays an essential role in the cytoplasmic Ca^{2+} regulation in cardiomyocytes. Heart failure is accompanied with reduction of SERCA2a level (Meyer et al., 1995). Several studies suggested that key mechanisms of TNF-α-induced contraction dysfunction is the down-regulation of SERCA.

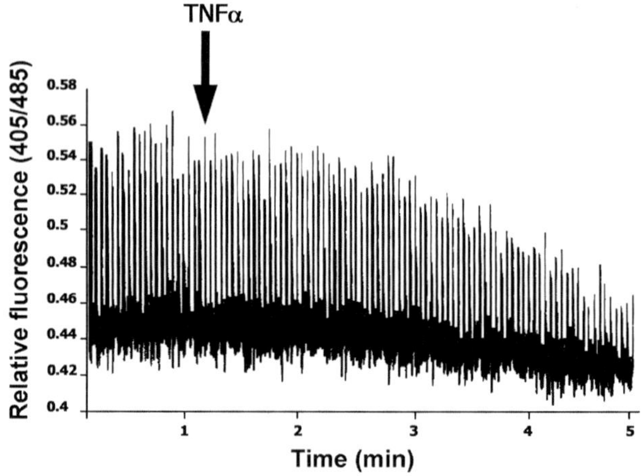

Fig. 5.6 Impairment of Ca^{2+} transients in cardiomyocites by TNF-α. Indo 1 Ca^{2+} transients of an adult rat ventricular myocyte are shown. Ca^{2+} transient measurements were made at room temperature from the cytoplasmic regions of individual cells electrically paced at 0.3 Hz. At the indicated times, 18,000 U/ml recombinant murine TNF-α was added (from Krown et al. (1995) with permission of Elsevier Limited and *FEBS Letters* via Copyright Clearance Center)

TNF-α has been shown to down-regulate the SERCA2a in the expressions of RNA and proteins level in isolated rabbit cardiomyocytes (Lee et al., 2007). Expression of SERCA mRNA and protein were markedly decreased in mouse cardiac muscle cell line – HL-1 treated by TNF-α (50 ng/ml for 24 h) (Kao et al., 2010). Down-regulation of SERCA was accompanied by reduction of $[Ca^{2+}]_i$ transients amplitude.

On the one hand in isolated ventricular myocytes from "TNF1.6 mice" transcripts encoding SERCA 2 and phospholamban (PLB) were significantly reduced in comparison with wild type mice (Kubota et al., 1997; Janczewski et al., 2003). Despite reduced mRNA expression SERCA and PLB protein levels were not different in transgenic and wild type mice. Similarly to previous, the molecules related to Ca^{2+}-handling, such as ryanodine receptor, SERCA and NCX (Na/Ca-exchanger) did not show any specific changes in rat cultured cardiomyocytes after treatment with 50 ng/ml TNF-α (Kawada et al., 2006). Thereby, mechanism of TNF-α-induced cytoplasmic calcium transients altern .tion remain unclear.

5.3.2 *Interleukin IL-1*

5.3.2.1 I_{Na}

Alterations of Na^+ channel function can contribute to rhythm disturbances (Grandy et al., 2010). Disease-mediated alterations in Na^+ current could be partially responsible for arrhythmias that have been observed HIV patients (Sani and Okeahialam, 2005). Similarly HIV patients and CD4C/HIV mice have elevated levels of the proinflammatory cytokine IL-1β (Monsuez et al., 2007; Grandy et al., 2010). In ventricular myocytes isolated from HIV mice Na^+ peak current was significantly reduced. It is possible that alterations in serum proinflammatory cytokines could mediate the reduction of Na^+ current and the corresponding changes in the AP waveform as observed in ventricular myocytes from CD4C/HIV mice. Overall, the results suggest that elevated levels of IL-1β may result in a reduction in Na^+ current, altering the action potential upstroke and increasing the risk of arrhythmias (Grandy et al., 2010).

5.3.2.2 $I_{Ca,L}$

IL-1β has been suggested to play a role in impaired cardiac performance under these conditions. This suggestion has been supported by in vitro studies showing that IL-1β decreases cardiac contractility (Evans et al., 1993; Hosenpud et al., 1989; Weisensee et al., 1993).

To determine whether cytokines alter the electrical properties of heart cells, the effects of human recombinant IL-1β were examined in excised tissues and dissociated myocytes from guinea pig ventricles (Li and Rozanski, 1993). In excised papillary muscles, IL-1β significantly prolonged action potential duration (measured at 90% repolarization) by 24.2 ± 2.2 (SEM) ms and effective refractory period by 22.9 ± 2.3 ms. Other measured variables were not affected. Treatment of muscles with cyclo-oxygenase inhibitors, indomethacin (10^{-5} M) or acetyl salicylic acid (2×10^{-4} M), abolished the prolongation of action potential duration elicited by IL-1β. However, the effects of IL-1β were also blocked by the lipoxygenase inhibitor nordihydroguaiaretic acid (2×10^{-5} mol/l) or by treating tissues with the leukotriene receptor blocker, ICI 1986 I5 (10^{-8} mol/l). In isolated myocytes, 1 ng/ml IL-1β increased I_{Ca} density in 44 of 78 cells by 33.6 ± 7.5% during voltage steps from –40 to 0 mV. Thus IL-1β modifies electrical properties of cardiac cells via lipid second messengers generated by cyclo-oxygenase and lipoxygenase pathways. Voltage clamp analyses suggest that these effects are mediated, at least in part, by changes in the conductance of Ca^{2+} channels (Li and Rozanski, 1993). In guinea pig ventricular myocytes, IL-1 has been reported to increase $I_{Ca,L}$ (Li and Rozanski 1993). However, these authors in that study GTP was not included in the pipette solution, and only 56% of ventricular myocytes responded to IL-l exposure (Liu and Schreur, 1995). When GTP was included in the pipette solution, authors found that IL-1β inhibited I_{Ca} in all experiments. Therefore, in the presence of GTP, IL-1β suppressed the cardiac $I_{Ca,L}$ by a receptor-coupled G protein-mediated mechanism (Liu and Schreur, 1995). IL-1β attenuates cardiac L-type Ca^{2+} currents ($I_{Ca,L}$) via pertussis toxin-insensitive G protein rather than by stimulation of cGMP production. IL-1β caused a concentration-dependent decrease in the peak $I_{Ca,L}$ (Ba^{2+} as the charge carrier). IL-1β did not significantly alter the voltage dependence of the peak $I_{Ca,L}$ nor the steady-state inactivation and activation, but did slightly slow the rate of inactivation (Liu and Schreur, 1995).

Many studies have suggested that nitric oxide generation via IL-1β-mediated induction of nitric oxide synthase plays a role in the observed effects of IL-1β on cardiac contractility (for example, Evans et al., 1993; Harding et al., 1995). In ferret papillary muscles, IL-1β suppressed the contraction by increasing nitric oxide (NO) production via a dexamethasone-sensitive pathway that consequently increased cellular cGMP (9). However, the NO-mediated negative inotropic effect was not observed when hamster papillary muscles were exposed to IL-1α (Finkel et al., 1992). In cells internally dialyzed with pipette solutions without GTP, the hydrolysis of which is essential for cGMP production, the absence of GTP blocks the suppression of $I_{Ca,L}$ by IL-1β. These results could be explained by an increase in cell cGMP level stimulated by NO synthesis. The reduction of $I_{Ca,L}$ by IL-1β may account for the decreased myocardial contractility associated with pathophysiological conditions, such as immunologically mediated cardiac disorders and myocardial infarction.

Modulation of the β-adrenergic control of the cardiac L-type Ca^{2+} current (I_{Ca}) by human recombinant IL-1β was examined in guinea pig ventricular myocytes using the whole cell voltage clamp technique (Rozanski and Witt, 1994b). I_{Ca} was evoked in Cs^{+}-loaded myocytes by depolarizing pulses from a holding potential of –40 mV. In the presence of an acidic external solution (pH 5.8), the response of I_{Ca} to isoproterenol was markedly decreased compared with control myocytes studied at pH 7.4. However, when cells were pretreated with 1 ng/ml IL-lβ and then exposed to acid media, β-responsiveness was significantly increased compared with untreated cells. Despite this effect of IL-lβ, maximum I_{Ca} density with 0.01 and 1 μM isoproterenol was still 51 and 58%, respectively, less than that measured at pH 7.4. The enhanced β-responsiveness produced by IL-1β was eliminated by adding amiloride to block Na^+/H^+ exchange or protein kinase C inhibitors staurosporine (10 nM) and calphostin C (50 nM). However, a direct activator of protein kinase C, phorbol 12-myristate 13-acetate, did not mimic the effects of the cytokine. These data demonstrate that IL-lβ partially restores the β-adrenergic control of cardiac Ca^{2+} channels suppressed under acidic conditions. Moreover, they suggest that IL-lβ acts by enhancing Na^+/H^+ exchange through a second messenger pathway that may involve protein kinase C. These cellular mechanisms may play a role in altering ventricular function during cytokine-mediated inflammatory processes that are initiated by myocardial ischemia (Rozanski and Witt, 1994b). Isoproterenol exposed to myocytes pretreated with 1 ng/ml IL-1β evoked a significantly smaller increase in I_{Ca} density compared with control cells. This IL-l-mediated decrease in β-responsiveness was usually observed with pretreatment periods of >l h and varied as a function of the L-arginine concentration of the pretreatment medium. It was prevented by (i) IL-1 receptor antagonist, (ii) substituting D-arginine for L-arginine, or (iii) incubating cells with the nitric oxide synthase inhibitor NG-monomethyl-L-arginine. Thus the present data illustrate that IL-1 significantly alters the β-adrenergic control of cardiac Ca^{2+} channels by cellular mechanisms that involve the activation of nitric oxide synthase. These mechanisms may play a role in altering ventricular function during cytokine-mediated inflammatory processes affecting the heart (Rozanski and Witt, 1994a).

The possible mechanism by which IL-1β affects β-adrenergic responsiveness of L-type Ca^{2+} current ($I_{Ca,L}$) was examined in adult rat ventricular myocytes by use of whole cell patch-clamp techniques (Liu et al., 1999). In the presence of isoproterenol, exposure for 3 min to IL-1β suppressed the isoproterenol-activated $I_{Ca,L}$. In the presence of IL-1β, the response of $I_{Ca,L}$ to isoproterenol was decreased, and the EC_{50} for isoproterenol stimulation was increased. However, IL-1β had no effect on basal and isoproterenol-enhanced cAMP content. When $I_{Ca,L}$ was activated by extracellular application of forskolin or 8-(4-chlorophenylthio)-cAMP, a membrane-permeable cAMP analog, or by intracellular dialysis with cAMP, IL-1β had little effect on $I_{Ca,L}$. In contrast, in the presence of cAMP, IL-1β still suppressed the Iso-enhanced $I_{Ca,L}$. These results show that the IL-1β -induced decrease in β-adrenergic responsiveness of $I_{Ca,L}$ does not result from inhibition of β-adrenoceptor binding, adenylyl cyclase activity, or cAMP-mediated pathways, suggesting a cAMP-independent mechanism (Liu et al., 1999).

5.3.2.3 Na^+–K^+ ATPase

Heart failure coupled with ischemia and myocardial hypertrophy were associated in turn with a decrease in the concentration of myocardial Na^+–K^+ ATPase (Bundgaard and Kjeldsen, 1996) an enzyme located in the T tubules and peripheral sarcolemma (McDonough et al., 1996) and responsible for the generation and maintenance of the electrochemical gradient for Na^+ and K^+. Many studies have shown that heart diseases are always accompanied with high levels of IL-1β and a decrease in Na^+–K^+ ATPase concentrations (Harada et al., 1999; Kreydiyyeh et al., 2004). Because in heart failure, changes in the Na^+–K^+ ATPase activity and expression were noted to accompany increases in circulating levels of IL-1β, the presence of a cause effect relationship between the cytokine and the pump has been investigated (Kreydiyyeh et al., 2004). IL-1 β reduced the protein expression of the α1 subunit of the Na^+–K^+ ATPase in right and left ventricular homogenates and exerted a significant inhibition of the ATPase activity which correlated with its down-regulatory effect and could thus be ascribed to the decrease in the number of the catalytic a1 subunits. Because the ventricular homogenate is a mixture of different types of cells (endothelial cells, macrophages, fibroblasts), the observed change in the ATPase activity and expression may not reflect necessarily changes in cardiomyocytes. To clarify this point cardiomyocytes were isolated from rats treated with IL-1β, and the activity and protein expression of the ATPase was studied in a homogenate of these isolated cells. IL-1β caused down-regulation of the pump in isolated cardiomyocytes and inhibited its activity. The results suggest that the cytokine acts directly on the cardiomyocytes. Effect of IL-1β is mediated through mitogen-activated protein kinases (MAPK).

Such an inhibition of the Na^+–K^+ ATPase by IL-1β is expected to decrease the Na^+ gradient and Na^+-dependent electrolyte movements that regulate excitation–contraction coupling in cardiac myocytes especially calcium movements. Cytosolic Ca^{2+} increases during contractions and is restored back to normal levels upon relaxation via the Na^+–Ca^{2+} exchanger whose activity is dependent on the Na^+ gradient established by the Na^+–K^+ pump. An inhibition of the pump is thus expected to dissipate the Na^+ gradient, inhibit the Na^+–Ca^{2+} antiporter, increase cytosolic Ca^{2+} and by so doing alter the systolic–diastolic cycle and delay relaxation. In fact, inhibition of the Na^+–K^+ ATPase has been reported in some cases of heart failure (Paganelli et al., 2001; Long, 2001) and was associated with increases in the levels of pro-inflammatory cytokines. Similarly increases in cytosolic Ca^{2+} were reported to occur in association with increases in the levels of IL-1β (Kato et al., 1987). The demonstrated inhibitory effect of IL-1β on the pump may thus provide a possible explanation of the previously reported increase in cytosolic Ca^{2+} levels (Bick et al., 1997, 1999) that occur in myocytes following IL-1β treatment and whose cause remained controversial.

The inhibition of the pump is expected also to alter the function of the Na^+/H^+ exchanger which regulates intracellular pH and the activity of which is dependent on the Na^+ gradient generated by the Na^+–K^+ ATPase. IL-1β is thus expected to increase intracellular acidity, a process that has already been reported in the ischemic heart where pH_i and pH_o were noted to reach values as low as 6.0 and 6.5, respectively (Karmazym et al., 1999; Park et al., 1999).

5.3.3 Interleukin-2

IL-2 has been known as a mediator which plays one of the key roles in the immune system via stimulating and coordinating immune responses (Arai et al., 1990). IL-2 and its receptor have been intensively studied on the molecular level. Apart from the cells belonging to the immune system, IL-2 binding sites have been found in various tissues, including rat brain (Araujo et al., 1989), isolated human and rats muscular cells and some other (Smith et al., 1989). Moreover, physiological effects of IL-2, suppressing long-term potentiation of hippocampal neurons and prolonging the open time of extrajunctional acetylcholine receptors in muscle have been described (Trotter et al., 1988).

IL-2 occurs in increased amounts during inflammatory neuronal diseases, such as multiple sclerosis and polyradiculoneuritis. Since the latter is characterized by a severe paralysis, the question arose whether IL-2 may be responsible for a reduction of cellular excitability by blocking excitatory Na^+ channels. A preliminary account of some of the results has appeared (Brinkmeier et al., 1992).

Despite direct action as an immune system mediator some other indirect effects of IL-2 such as sodium current inhibition, sarcoplasmic reticulum activity mediation have been described. IL-2 may act either through own IL-2 receptor or through direct interaction with other receptors, including k-opioid receptor (Cao et al., 2003a, b) and sodium channels (Brinkmeier et al., 1992). It has been shown that treatment with IL-2 increasing in a dose-dependent manner the activity of sarcoplasmic reticulum Ca^{2+}-ATPase, but the sarcolemmal Ca^{2+}-ATPase activity is not changing (Cao et al., 2003a, b).

Molecular mechanisms of sarcoplasmic reticulum mediation include cAMP. Treatment with IL-2 at 200 U/ml decreasing the intracellular cAMP concentration in the isolated rat heart, which suggest that cAMP is a target component of IL-2 in the heart. It is also known that cardiac k-opioid receptor stimulation inhibits adenylyl cyclase via Gi/o proteins, leading to a decreased intracellular cAMP level (Zhang and Wong, 1998). Results of decreased sensitivity of SR Ca^{2+}-ATPase to free calcium by IL-2 action, indicates that the activation of k-opioid receptor by IL-2 results in a decreased intracellular cAMP level, which may be responsible for reduced sensitivity of SR Ca^{2+}-ATPase to calcium by affecting the degree of cAMP-dependent phosphorylation of phospholamban.

IL-2 may also act without involvement of second messenger system, directly to the sodium channels in human muscle cells. Clear results demonstrate an inhibitory effect of human IL-2 on the excitatory Na^+-channels in skeletal muscle. With IL-2 concentration as those found in body fluids of patients suffering from inflammatory neuronal diseases (Lorenzon et al., 1991) of those used to induce T-lymphocyte activation in vitro. Doubts in direct action of the IL-2 was solved in experiments with anti-IL-2 receptor antibody. This preincubation did not prevent the inhibitory effect of IL-2 on the Na^+-channels on the Na^+-currents. Furthermore, it is shown that the cytokine blocks the voltage-dependent muscular Na^+-channels by keeping the channels in the state of fast inactivation. An IL-2 receptor and a second messenger system are not likely to be involved in this reaction. Thus, the effect of IL-2 is comparable to the action of local anesthetics on Na^+-channels (Kaspar et al., 1994).

The same results of the IL-2 was shown on sodium channels in human cardiomyocytes. It is shown that IL-2 has class I antiarrhythmic-like effect with extracellular cytokine concentrations that are also usually reached when patients are under high-dose rIL-2 therapy. According to (Vaughan Williams, 1984) class I antiarrhythmics block sodium channels, shift the steady state inactivation curve in the negative direction, delay recovery of the sodium channels from block and are use dependent. All these properties are possessed by rIL-2.

Human IL-2 consists of 133 amino acids with a molecular weight of 15,000 Da in its active form. Its three-dimensional structure is known. Carboxy-peptidase removes the amino acids one after the other, starting at the C-terminal. Only arginine, proline and hydroxyproline are resistant to its action. The sequence of IL-2 contains an arginine at location 120, so only the last 13 amino acids can be removed. Unless removal of these amino acids changes the structure of the molecule, the sodium channel-binding site of rIL-2 is presumably located within the last 13 amino acids of the molecule. The receptor binding site of IL-2 is located within the first 54 amino acids, near the N-terminus, because blocking of amino acids 59–133 with monoclonal antibodies does not inhibit receptor binding (Proebstle et al., 1995).

Affinity quantitative comparison with established antiarrhythmic drugs shows that the blocking potency of rIL-2 on sodium channels is quite high.

5.3.4 Interleukin-6

IL-6 along with TNF-α is known as one of the main pro-inflammatory cytokines, especially during acute myocardial infarction (AMI). IL-6 exerts acute negative inotropic action without detectable reduction in I_{ca} under basal conditions. Expression of IL-6 and IL-8 genes is increasing in patients assigned to take calcium blockers.

IL-6 exerts an acute negative inotropic action without detectable reduction in I_{Ca} under basal conditions. This sounds as if IL-6 directly inhibited EC-coupling instead of blocking I_{Ca}. However, inhibition of NOS (by L-NMMA) abolished the negative inotropic action of IL-6 (Sugishita et al., 1999). An acute inhibitory effect of NO on I_{Ca} has been shown to be mediated through cGMP-dependent protein kinase in mammalian hearts (Thakkar et al., 1988; Levi et al., 1989; Mery et al., 1991; Ono et al., 1991). The effects of cGMP have been shown to be mediated by phosphodiesterase-dependent mechanisms in lower species such as frogs while cGMP-dependent protein kinase (cGMP-PK) may play a key role in the action of cGMP in rat hearts. Thus, IL-6 may act on L-type calcium channels mediated by activation of cGMP-PK in the ventricular myocytes of the guinea-pig, as used in this study. The NO/cGMP pathway appears to be involved in the reversal of cAMP-stimulated I_{Ca} in guinea-pig ventricular myocytes (Ono et al., 1991; Levi et al., 1994; Wahler et al., 1995). IL-1 also decreased α-adrenergic control of I_{Ca} through NO production in adult guinea-pig ventricular myocytes (Rozanski et al., 1994).

Four different Ca^{2+}-channel blocker, amlodipine, felodipine, isradipine and manidipine, are capable of inducing the activation of genes coding for IL-6 and EL-8 in cultured human VSMC and fibroblasts. The effect occurs at nanomolar concentrations which can be achieved in vivo during therapy. The novel activity of the dihydropyridines cannot be substituted for by two other blood pressure lowering agents such as β-receptor antagonists or diuretics. This suggests an apparently specific activity of the class of drugs affecting inflammatory mechanisms. However, the exact mechanism by which the drugs stimulate remains unclear. Two possible modes of action can be hypothesized; (i) the binding site of the potential operated Ca^{2+}-channels contain an unknown signal transducing element which can be activated by interaction with Ca^{2+}-channel blockers, or (ii) the induction of transcription of IL genes by Ca^{2+}-channel blockers is mediated via an unknown membranous binding site for Ca^{2+}-channel blockers which is linked to a signaling pathway. Further studies should unravel the mechanisms of Ca^{2+}-channel blockers on the interaction of the two ILs with their various receptors (Roedler et al., 1994).

Concentration of pro-inflammatory cytokines (IL-6, TNF-α and hs-CRP) and anti-inflammatory cytokine IL-10 were elevated in AMI patients in comparison with the group with stable angina pectoris, but statistically significant difference was found in IL-6 and hs-CRP level. IL-6:IL-10 ratio is higher in AMI which emphasis the hypothesis of pro-inflammatory to anti-inflammatory cytokine misbalance in AMI. IL-10 may be useful as a marker of myocardial reperfusion in AMI. There is positive linear correlation between IL-6 and IL-10 in acute myocardial infarction and negative linear correlation between HDL and IL-6 (Dizdarević-Hudić et al., 2009).

It is also well known that the cardiac effects of NO are complex and somewhat controversial, probably due to the distinct concentration-dependent sensitivities of its targets and distinguishable signaling mechanisms. NO is known to target the L-type Ca^{2+} channel and SR in cardiac myocytes through signaling pathways such as cGMP, and/or by directly modifying the proteins (such as by S-nitrosylation/oxidation) (Campbell et al., 1996; Hare, 2003; Massion et al., 2003). Studies in adult ferret ventricular myocytes showed that SIN-1, a NO donor, induces biphasic and bimodal changes in basal $I_{Ca,L}$ (Campbell et al., 1996), which were not observed in frog (Mery et al., 1993), guinea-pig (Wahler and Dollinger, 1995) or rat (Abi-Gerges et al., 2001) ventricular myocytes. We also observed no significant change in basal $I_{Ca,L}$ in ARVM upon acute exposure to 0.1 mM SIN-1 (S.J. Liu, unpublished data: In Yu et al., 2005a, b). Similarly, SIN-1 (up to 0.1 mM) has no significant effect on contractile function of papillary muscle isolated from adult rat (Wyeth et al., 1996). Although we found that cumulative NO production in cell lysates and in culture medium is 2.5- to 3-fold of the basal level after 2 h of IL-6 treatment, it is not surprising that $I_{Ca,L}$ is not significantly altered in our experimental conditions. However, the possibility for alteration in $I_{Ca,L}$ during longer exposure (e.g. >24 h) to IL-6 cannot be excluded.

In summary, exposure to IL-6 for 2 or more hours results in a negative inotropy probably by inhibiting SR Ca^{2+} reuptake in concomitance with increases in NO in ARVM. The IL-6-induced decrease in PLB phosphorylation is also in accord with its reduction of SR function. The increase in NO production results primarily from expression/activation of iNOS that is mediated via IL-6-induced JAK2/STAT3 activation as shown previously (Yu et al., 2003a, b). Inhibition of iNOS/NO abolishes the IL-6-induced decrease in SR function. Therefore, this leads to our conclusion that IL-6 suppresses SR function via iNOS after chronic exposure. Moreover, such cardiac effects of IL-6 are sustained even after removal of IL-6. This suggests that the transient elevation of IL-6 observed in plasma or ventricular muscle of patients with many cardiac diseases can cause prolonged effects on contractility. An in vivo study in transgenic mice showed that cardiac-specific overexpression of iNOS causes an increase in NO_x in hearts (2.5-fold above the wild-type controls) but no severe cardiac dysfunction (Heger et al., 2002). One interpretation for this seeming inconsistency is that the increase in plasma and heart NO_x in the transgenic models is moderate so the associated systemic adaptation during development in these transgenic mice results in less detrimental functional consequences. A recent in vivo study reported that IL-6 administration causes heart failure in a dose-dependent manner (Janssen et al., 2005). Thus, IL-6 might play an important role in the pathogenesis of associated chronic heart diseases. Treatment to antagonize the induction and activation of iNOS in the early stage of IL-6-associated heart dysfunction might be beneficial (Yu et al., 2005a, b).

5.3.5 Interleukin-11

There are clinical reports of sudden cardiac fibrillation in cancer patients who were treated with recombinant human IL-11. The question was either that effect is a direct arrhythmogenic action of IL-11 or that was indirect system side effect. The results from (Sartiani et al., 2002) which was performed on single human atrial myocytes using standard patch-clamp technique clearly show the lack of effect on $I_{Ca,L}$, I_K, I_{to}, in agreement with the observation that rhIL-11 does not influence the atrial action potential configuration and particularly it's duration.

Shortening of atrial action potential is generally associated with shortening of atrial refractory period and hence facilitates induction of atrial flutter/fibrillation (Narayan et al., 1997). Conversely, lengthening of action potential duration is a perquisite for the appearance of early after depolarization, which may trigger arrhythmias (Janse, 1992). Calcium handling by sarcoplasmic reticulum and Na/Ca exchanger is crucial for the changes in atrial action potential duration and the appearance of delayed afterdepolarizations (Tavi et al., 1996, 1998; Benardeau et al., 1996). To avoid any perturbation by the patch pipette of the intracellular milieu and calcium homeostasis, which could have hindered the development of Ca-dependent electrical alterations, action potentials were measured with the perforated-patch technique. Thus, our results allow inferring that superfusion with rhIL-11 does not have direct, acute electrophysiological effects on action potential parameters measured in atrial myocytes. Furthermore, acute exposure of atrial myocytes to rhIL-11 does not affect the pacemaker current I_f, nor did it modify the resting potential in these cells. Taken together, these data suggest that it is unlikely that rhIL-11, at least by its direct effects, may cause afterdepolarizations, thought to be the trigger for atrial fibrillation, which is maintained by re-entry (Janse, 1997). We did not find a single direct electrophysiological alteration indicative that rhIL-11 may acutely influence atrial electrical activity. The only electrophysiologic effect of rhIL-11, however not statistically significant, was a small reduction in the I_f amplitude.

I_f is an inward current that is responsible for the diastolic depolarization phase and automatic activity in pacemaker cells (Difrancesco, 1993). This current has been recently described and characterized in human atrial myocytes (Hoppe et al., 1998; Pino et al., 1998; Porciatti et al., 1997). An increase of I_f amplitude in association with a reduction in the inward rectifier I_{K1} or in association with the presence of delayed afterdepolarizations due to a cellular calcium overload has been suggested to potentially trigger premature depolarizations able to generate atrial fibrillation if an appropriate substrate for re-entry is present (Opthof, 1988; Pino et al., 1998). A reduction in I_f amplitude should reduce the likelihood for generation of such premature depolarizations.

Thus, it is possible that acute atrial dilatation and consequent atrial stretch rather than direct action on the myocytes by rhIL-11 could be the mechanism by which rhIL-11 might affect cellular atrial electrophysiology and induce atrial arrhythmias. The experimental study in rats presented in the accompanying paper supports this hypothesis. In summary, it seems reasonable to conclude that the lack of effect of rhIL-11 on AP and ionic currents during acute exposure of human atrial myocytes seems to exclude a direct effect of rhIL-11 in the genesis of atrial arrhythmias in patients.

In conclusion, it is very likely that effect of IL-11 is mediated via blockade of sodium channels, water accumulation, edema, atrium dilatation and further mechanically induced fibrillation.

5.4 Cytokines and Mechanosensitivity of the Heart

In recent years many groups investigated the role of cytokines in generation of cardiac stretch effects. This studies employed one of two approaches. Firstly – investigation of cytokine production during stretch of issue or single cells. Those studies are discussed in review of Kovalchuk et al. (Chapter 2, this volume). The second approach – investigation of cytokine effects on cells and tissues during their stretching. We will discuss their findings in this section.

However there is a number of studies addressing cytokine production during tissue stretching (for review see Chapters 2 and 10 of this Volume), investigation of the effects of exogenous cytokines on cells, for example in heart, are limited to those, which were described in previous sections of this chapter. On the other hand there are practically no studies of the regulation of mechanoelectrical feedback in heart by cytokines. Therefore their effects on stretched cardiac tissue remain unknown. For the purpose of investigation of the role of cytokines in regulation of mechanoelectrical feedback we used microelectrode registration of bioelectrical activity of the fragment of the right atria during discrete tissue stretching (Kamkin et al., 2000a).

Several intracellular electrophysiological alterations in the healthy and diseased heart, which were ascribed to mechanoelectrical feedback were reported for the first time about 10 years ago (Kiseleva et al., 2000; Kamkin et al., 2000a). Later it was showed that stretch of isolated cardiomyocytes induced mechanosensitive whole-cell currents which lead to membrane depolarization (Kamkin et al., 2000b, 2003a, b; Zeng et al., 2000; Zhang et al., 2000; Isenberg et al., 2003; Kondratev and Gallitelli, 2003) and elicited stretch-induced depolarizations (SID), observed as prolongation of APD90 (stretch-induced depolarizations: SID) or hump-like SID, that appeared in repolarization phase at APD50 or APD90 level (Kamkin et al., 2003b). Membrane depolarization and SID in action potentials provoke extra-action potentials when these hump-like SID reach a threshold potential. This finding correlated with previous studies of step-wise application of stretch to atrial and ventricular tissue from healthy animals and animals, suffering from infarct, which reported SIDs, emerging as prolongation of APD90 and as typical hump-like SIDs, leading to extra-action potentials (extra-AP), paroxysmal tachycardia and even fibrillation (Kiseleva et al., 2000; Kamkin et al., 2000a). Later they were studied in detail (Lozinsky and Kamkin, 2010; Ward et al., 2010). Thus, activation of mechanically gated channels (MGCs) and the increase of mechanosensitive whole-cell currents, which during cardiomyocyte stretching leads to ion entry carrying the positive charge together compose the cellular mechanisms underlying different mechanically induced alterations and abnormalities.

It is known that certain cytokines induce signaling, which includes for example, transducing signals within intracellular compartments, such as in the case of modification of transcription factors in mitochondria and the nucleus (see for review Calabrese et al., 2009). One of those signaling cascades includes NO-dependent pathway, which is well described in many publications. In the heart NO-synthases play the key role in NO level regulation.

Besides, in response to stretch in cardiac cells NO-synthase activity significantly increases (reviewed in Shah and MacCarthy, 2000; Casadei and Sears, 2003; Seddon et al., 2007). Recently it was shown that NO and NO-synthases as NO source directly regulates mechanically gated channels (MGCs) conductivity. Experimental data in this study was obtained from the isolated ventricular myocytes of mouse, rat and guinea pig by means of patch-clamp method in whole-cell configuration. The data demonstrated that NO donors lead to MGCs activation and appearance of MG-like currents in undeformed ventricular myocytes, whereas, it can also lead to the inactivation of the stretched cells with activation MGCs and inhibit the conductivity of these channels (Kazanski et al., 2010a, 2011). NO scavenger PTIO causes inactivation of all MGCs. In undeformed cells conductance through MGCs is blocked and their activation in the controlled, and on the background of stretch cell the PTIO causes complete MG-current inhibition. Non-selective inhibitors of NO-synthases L-NAME or LNMMA resulted in complete blocking of MGCs. In ventricular myocytes of wild-type mice, NOS1–/– and NOS2–/– the stretching of the cells triggered typical MG-current. On the contrary, in cells from NOS3–/– mice stretch does not cause MG-current. The results discussed on the channel level testify to the NO role in activation and inactivation of MGCs in cardiomyocytes and demonstrate that NOS3 dominates as NO source (Kazanski et al., 2010b, 2011). Therefore the appearance of hump-like SIDs and extra-AP during stretching of atrial tissue stripes is determined by MGCs functioning, which are modulated by NO and NO-synthases.

5.4.1 Tumor Necrosis Factor-α

TNF-α has been implicated in the pathogenesis of cardiovascular diseases, including heart failure, myocarditis, acute myocardial infarction and sepsis-related cardiac dysfunction (Levine et al., 1990; Low-Friedrich et al., 1992; Yokoyama et al., 1993; Latini et al., 1994; Packer, 1995; Kumar et al., 1996). TNF-α also has been shown to be related to the occurrence of cardiac arrhythmias. Mice with a higher TNF-α expression have a greater incidence of atrial and ventricular arrhythmias (London et al., 2003). Administrating TNF-α in rats can induce arrhythmias with a loss of the myocyte inotropy (Krown et al., 1995). TNF-α changes the L-type calcium currents ($I_{Ca,L}$), and calcium transient in ventricular myocytes (Krown et al., 1995; Goldhaber et al., 1996; Cailleret et al., 2004).

The direct effects of proinflammatory cytokine TNF-α on the contractility of mammalian heart were studied a lot of years ago (Finkel et al., 1992). It was shown, that TNF-α inhibited contractility of isolated hamster papillary muscles in a concentration-dependent, reversible manner. The authors demonstrated, that the nitric oxide synthase inhibitor NG-monomethyl-L-arginine (L-NMMA) blocked these negative inotropic effects. L-Arginine reversed the inhibition by L-NMMA. Removal of the endocardial endothelium did not alter these responses. These findings demonstrate that TNF-α regulated the contractility and the direct negative inotropic effect of cytokines is mediated through a myocardial nitric oxide synthase. The negative inotropic effect of TNF-α was observed in isolated rabbit cells (Goldhaber et al., 1996). This effect was apparent within 20 min of exposure to the cytokine, was similar to that described by Finkel et al. (1992) in a multicellular preparation.

Previous studies have shown that nitric oxide (NO) has important regulatory effects on the cardiovascular system (Moncada et al., 1991; Kelly et al., 1996). NO has been shown to have a role in the development of triggered arrhythmias generated by Ca^{2+}overload (Kubota et al., 2000). Previous study in vivo also showed that NO could suppress trigged activity induced ventricular tachycardia (Chen et al., 2001). It is known that pulmonary veins contain endothelium and smooth muscle which may produce NO through the enzyme of eNOS or iNOS. In addition, cardiac myocytes also express eNOS activity (Massion and Balligand, 2003; Balligand et al., 1994). NO has been shown to regulate PV arrhythmogenesis through mechanoelectrical feedback (Hu et al., 2009). Because PVs was known to induce atrial arrhythmia through the enhancement of triggered activity, it is possible that NO may play a critical role in the pulmonary veins arrhythmogenic activity.

It was shown that cardiac cells themselves produce different quantities of TNF-α during prolonged stretch or during cyclic application of mechanical stretch. Acute stretch caused by volume overload of aorto-caval fistula induces a variety of myocardial responses. These responses can be part of myocardial inflammation dictated by TNF-α, which is elevated after acute aorto-caval fistula. Western blot demonstrated increased local production of TNF-α in the left ventricle and cardiomyocytes. TNF-α produced by cardiomyocytes mediates a predominant inflammatory response to stretch in the early volume overload in the aorto-caval fistula rat, suggesting an important role of TNF-α in initiating pathophysiological response of myocardium to volume overload (Chen et al., 2010). In addition to production of TNF-α by cardiomyocytes during their stretching, cardiac fibroblasts also produce TNF-α during stretch. To determine whether mechanical stress affect the production of TNF-α in the heart, neonatal rat cardiac myocytes and fibroblasts were cultured separately and treated for 6 h with cyclic mechanical stretch. Mechanical stretch stimulated the production of TNF-α in cardiac fibroblasts. Thus mechanical stress, can stimulate production of TNF in cardiac fibroblasts (Yokoyama et al., 1999).

Thus, TNF-α has been shown to be related to the occurrence of cardiac arrhythmias. Moreover, TNF-α regulated the contractility and the direct negative inotropic effect of TNF-α is mediated through a myocardial nitric oxide synthase. This data and studies reporting that NO directly regulates mechanically gated channels (MGCs) (Kazanski et al., 2010a, b, 2011) allowed us to propose that in heart TNF-α would be an effective regulator of MGCs operating and therefore will regulate mechanoelectrical feedback.

In order to test this hypothesis we used the standard microelectrode technique in combination with stretching of the tissue stripe of the right rat atria. Tissue was perfused with standard physiological solution, containing 50 ng of TNF-α per milliliter. Perfusion of tissue with TNF-α lead to appearance of abnormalities in development of bioelectrical activity observed as prolongation of APD90 and appearance of hump-like depolarization at APD90 level. After reaching E_c hump-like depolarization transformed into single extra-APs, with following development of sustained arrhythmia. Abnormalities at APD50 level were also observed.

Since TNF-α activates NO synthases of cardiomyocytes (Finkel et al., 1992; Goldhaber et al., 1996; Nakayama et al., 2009; Bougaki et al., 2010) and rise of the concentration of the intracellular NO activates MGCs (Kazanski et al., 2010a, b, 2011), which allows Na ions entry into the cell leading to cellular depolarization and shift of resting potential to E_C we proposed, that TNF-α induced arrhythmias can be linked with MGCs activation. This is the most simple mechanism, explaining the appearance of arrhythmias after TNF-α application.

Since the blocker of MGCs Gd^{3+} at 40 μmol/l eliminates not only hump-like depolarization, but arrhythmias, triggered by TNF-α, as well, it was proposed that TNF-α triggered abnormalities are linked with MGCs activity, which was triggered by increasing of intracellular NO concentration, caused by TNF-α application. In this case they would be similar to stretch induced depolarizations (SID). For testing the hypothesis about the role of NO in regulation of MGCs we used patch-clamp method in whole-cell configuration in combination with stretching of isolated cells (Kamkin et al., 2000b, 2003a; Lozinsky and Kamkin, 2010). It was shown that NO donor SNAP and DEO-NO open MGCs in unstretched cells, and closes them in stretched cells. Therefore MGCs activity is determined by NO concentration. It is known that stretching of cells and tissues stimulates NO synthases activity, which rises intracellular NO concentration (Shah and MacCarthy, 2000; Seddon et al., 2007). Besides that under control conditions tissue stretch induces hump-like depolarizations, which are originating from SIDs (Kamkin et al., 2000a). That's why we conducted a series of experiments in which we stretched atrial tissue preparations during registration of TNF-α induced hump-like depolarizations or arrhythmias. Tissue stretch completely eliminated mechanoinduced electrical abnormalities, originating in cardiomyocytes in response to TNF-α application. Therefore, TNF-α induces electrical abnormalities in cardiomyocytes, while tissue stretching eliminates them. This effects is likely to be mediated via NO signaling cascades.

If TNF-α acts via an increase in NO concentration after activation of NO-synthase then application of NO donor, for example SNAP sold lead to similar changes in APD90 with following generation of arrhythmias. In order to prove that in absence of any changes in preload we perfused tissue with standard saline solution, containing 3×10^{-4} mol/l of SNAP. Already after 10 min of perfusion we observed appearance of hump-like depolarizations, which later transformed into sustained paroxysmal arrhythmic events, mediated by altered or even normal action potentials. Therefore SNAP caused abnormalities such as hump-like depolarizations, similar to those induced by TNF-α. In order to answer the question whether abnormalities and arrhythmias, induced by SNAP, are similar to those stretch-induced depolarizations (SID), we conducted two series of experiments with 40 μmol/l of Gd^{3+}. In the first series we added Gd^{3+} to perfusion solution after appearance of SNAP induced hump-like depolarizations or SNAP induced arrhythmias. It was shown that hump-like depolarization in abnormal patterns is completely blocked by 40 μmol/l of Gd^{3+} within 10 min after its application. SNAP induced arrhythmias were also blocked by 40 μmol/l of Gd^{3+}. In another series Gd^{3+} was added to the perfusion solution together with SNAP. In this case we did not observe any abnormal action potentials. Since discrete tissue stretching during TNF-α perfusion led to complete elimination of arrhythmias and then blocked electrical abnormalities and taking into consideration that TNF-α effects can be mediated via NO concentration increase we conducted a series of experiments in which we triggered SNAP induced SIDs and arrhythmias, and then stretched cells for further activation of NO-synthases. After appearance of SNAP induced abnormalities or arrhythmias the tissue was stretched. This stretch completely arrhythmias and SI-like depolarization triggered by SNAP application.

In conclusion it was shown that application of SNAP during observation of TNF-α induced abnormalities completely abolishes those abnormalities. For investigation of the functional role of NO-synthases in mediation of TNF-α effects we used their non-selective inhibitor L-NAME, with which we preincubated tissue during 30 min. After that we applied TNF-α for 30 more minutes in combination with L-NAME. Pre incubation of tissue with L-NAME containing solution did not lead to any abnormalities in action potentials even after 30 min of perfusion. More importantly after such pre-incubation TNF-α, applied in combination with L-NAME did not trigger appearance or development of hump-like depolarizations. Neither stretch lead to any hump-like depolarizations.

Therefore arrhythmogenic effect of TNF-α is mediated via MGCs activity and is NO dependent. TNF-α activates NO synthases, which leads to NO concentration increase triggering MGCs activation and Na^+ influx to the cell which depolarizes cellular membrane triggering arrhythmias. Tissue stretch completely abolishes mechanoinduced electrical abnormalities, triggered by application of TNF-α to cardiomyocytes probably via additional activation of NO synthases and rise of intracellular NO concentration. Such NO concentration rise would lead to MGCs closing. This data revealing TNF-α NO mediated effects completely correlate with data from patch clamp experiments conducted in combination with single cardiomyocyte stretch. In unstretched cardiomyocytes NO donors trigger MGCs activation, which is abolished by cardiomyocyte stretch.

Thus, these experimental data sheds light on mechanism of arrhythmogenic action of TNF-α in atrial myocardium. We are trying to reveal the link between arrhythmogenic effect of TNF-α and its underlying mechanisms, which we described earlier, data about TNF-α application triggered arrhythmias in animals and clinical studies reporting levels of TNF-α from patients with cardiac arrhythmias and fibrillation. Administrating TNF-α in rats can induce arrhythmias (Krown et al., 1995) and mice with overexpression of TNF-α have a greater incidence of AF (London et al., 2003). The circulating levels of TNF-α (Sata et al., 2004) was found to be significantly higher in patients suffering from AF. In our study we were working with patients' congestive heart failure (New York Heart Association: Class II, Class III and Class IV). According to NYHA, at congestive heart failure arrhythmias and fibrillation are not considered to be symptoms of the decease. Data from Table 5.1 demonstrates the link between concentration levels of TNF-α and observation of ventricular arrhythmias. Table shows that elevation of TNF-α concentration correlates with increasing percentage of patients with clinical arrhythmic events.

5.4.2 Interleukin-6

IL-6 has been shown to be related to the occurrence of cardiac arrhythmias. IL-6 has been shown to significantly correlate with increased left atrial size (Psychari et al., 2005), an important risk factor for developing atrial fibrillation. In addition, polymorphisms in the promoter region of the *IL-6* gene have been associated with post-operative atrial fibrillation (Bittar et al., 2005; Gaudino et al., 2003). IL-6 has not previously been shown to be independently associated with spontaneous (non-post-operative) atrial fibrillation. Polymorphisms of the IL-6 gene were not studied

Table 5.1 Correlation of concentration of TNF-α with frequency of occurrence of ventricular arrhythmias

	Control $n = 55$	NYHA Class II, $n = 56$	NYHA Class III, $n = 62$	NYHA Class IV, $n = 42$	P
TNF-α, pg/ml	4.1±2.6	12.1±7.6	24.1±10.1	38.5±12.4	<0.01
Ventricular arrhythmia, n; %	4 7.3%	8 14.3%	23 37%	32 76%	<0.01

in these patients. Recent studies showed that atrial fibrillation is associated with elevated IL-6 levels and the IL-6 -174CC genotype (Marcus et al., 2008).

In vivo study also showed that elevated IL-6 levels were observed 1–2 h after aortic declamping in cardiopulmonary bypass (Wan et al., 1996). The direct cardiac effect of IL-6 during this period of time remains undefined. De novo synthesis and activation of iNOS induced by IL-6 can be detected in adult rat ventricular myocytes as early as 2 h after exposure (Yu et al., 2003a, b). This study also demonstrated that the IL-6-elicited iNOS activation and decrease in postrest potentiation of contraction in adult rat ventricular myocytes are mediated by activation of Janus kinase (JAK)2/signal transducer and activator of transcription (STAT)3, the upstream mediators of IL-6 signaling (Yu et al., 2003a, b). However, whether iNOS is the downstream mediator of IL-6-induced negative inotropy remains undefined.

In vitro studies on cardiac contractile function have shown that exposure to IL-6 for 2–3 min decreased contractility in papillary muscle isolated from hamster heart (Finkel et al., 1992). Studies with single myocytes also showed that IL-6 suppressed peak systolic $[Ca^{2+}]_i$ and cell shortening within 5 min in chick embryonic cardiac myocytes (Kinugawa et al., 1994) and in adult guinea-pig ventricular myocytes (Sugishita et al., 1999). This acute negative inotropic effect of IL-6, accompanied by an increase in cell cGMP production (Kinugawa et al., 1994), was blocked by NG-monomethyl-L-arginine (L-NMMA), an inhibitor of nitric oxide synthase (NOS) (Finkel et al., 1992; Kinugawa et al., 1994; Sugishita et al., 1999). Thus, acute IL-6-induced suppression of cardiac contractility and $[Ca^{2+}]_i$ have been suggested to be mediated by a NO-dependent pathway via activation of NOS (Finkel et al., 1992; Kinugawa et al., 1994; Sugishita et al., 1999), probably a constitutive endothelial isoform (eNOS) (Kinugawa et al., 1994). After 2 h incubation IL-6 decreases contractility and $Ca^{2+}{}_o$ responsiveness with no significant effect on $I_{Ca,L}$ in ARVM. (IL-6 reduces PRP, caffeine responsiveness and the phosphorylation of PLB. The negative inotropic effect is sustained for at least 1 h after removal of IL-6 and IL-6 induces production of NO via activation of iNOS, with inhibition of iNOS abolishing IL-6-induced changes in PRP, caffeine- and $Ca^{2+}{}_o$-responsiveness, and PLB phosphorylation (Yu et al., 2005a, b).

In general, IL-6 has been shown to be related to the occurrence of cardiac arrhythmias. Moreover, IL-6 regulated the contractility and the direct negative inotropic effect of IL-6 is mediated through a myocardial nitric oxide synthase. This data and studies reporting that NO directly regulates activity of mechanically gated channels (MGCs) (Kazanski et al., 2010a, b, 2011) allowed us to suppose that in heart IL-6 would be an effective regulator of MGCs activity and therefore will regulate mechanoelectrical feedback.

For testing this hypothesis we used standard microelectrode technique in combination with stretching of right rat atria cardiac tissue preparation. Under control conditions (Kamkin et al., 2000a), during perfusion with standard saline solution stretching of tissue by 1.7 mN caused prolongation of APD90 and appearance of hump-like depolarizations at APD90 level (Fig. 5.7a). Perfusion of tissue with IL-6

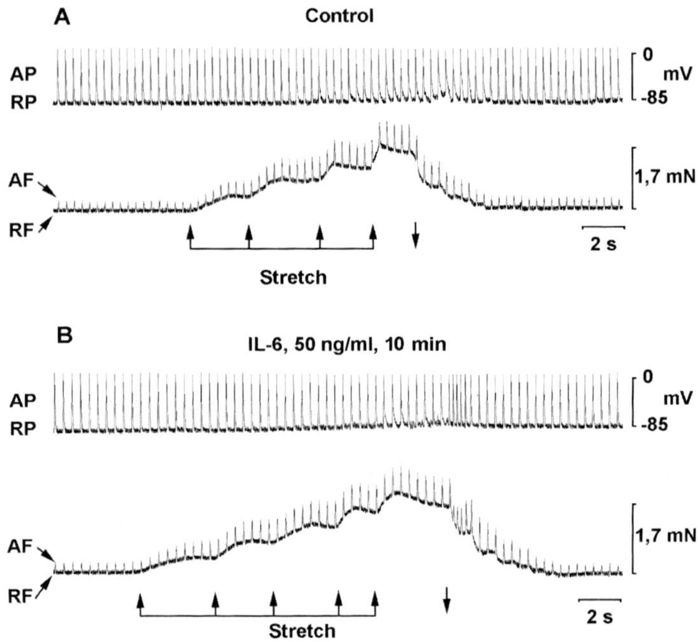

Fig. 5.7 Response of atrial cardiomyocyte membrane potential to stretch applied by the digital micromanipulator. (**a**) Control conditions. Stepwise increases in resting force due to stretch (indicated by ↑) up to 1.7 mN significantly increased APD90. Removal of stretch (indicated by ↓) reversed these effects. (**b**) IL-6 in concentration of 50 ng/ml, 10 min of perfusion. Stepwise increases in resting force due to stretch (indicated by ↑) up to 1.7 mN produced stretch-induced depolarization and tachyarrhythmia. Note: resting force (RF), active force (AF), resting potential (RF), active potential (AP) (the data obtained from the Department of Professor Andre Kamkin)

containing solution (50 ng/ml) lead to appearance of abnormalities in bioelectrical activity, majorly observed as APD90 prolongation. After 7–10 min of perfusion tissue stretch by 1.5–1.7 mN trigged hump-like depolarizations at APD90 level. After reaching E_c hump-like depolarization transformed into single extra-action potential later leading to typical tachycardia, which could be abolished by stretch (Fig. 5.7b).

Continuous perfusion of the preparation with IL-6 containing solution for periods of time longer than 20 min leads to fibrillation triggered by extremely low levels of stretch. Figure 5.8a shows a trace of registration of the resting force (RF), active force (AF), resting potential (RF) and active potential (AP) in control conditions during perfusion of the fragment of right rat atria with standard saline solution. Stretching of tissue by 1.7 mN triggered APD90 prolongation. Figure 5.8b shows effects of tissue stretch after 23 min of perfusion with IL-6 containing solution (50 ng/ml). Even small stretch by 0.3 mN triggered fibrillation, observed as typical bioelectrical activity and absence of AF. Stretch elimination stopped fibrillation. IL-6 effects and their underlying mechanisms are similar to those of TNF-α.

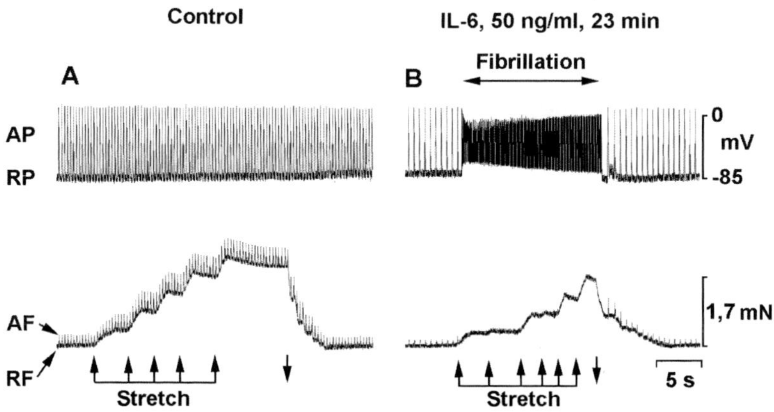

Fig. 5.8 Example of response of cardiomyocyte membrane potential to stretch applied by the digital micromanipulator of right atrial tissue. (**a**) Control conditions. Stepwise increases in resting force due to stretch (indicated by ↑) up to 1.7 mN produced stretch-induced depolarization (SID) near APD90. (**b**) IL-6 in concentration of 50 ng/ml, 10 min of perfusion. Stepwise increases in resting force due to stretch (indicated by ↑) only up to 0.3 mN produced fibrillation. Release of the stretch (↓) resulted in complete reversibility of stretch-induced effects. Note: resting force (RF), active force (AF), resting potential (RF), active potential (AP) (the data obtained from the Department of Professor Andre Kamkin)

5.5 Conclusion and Perspectives

All together the data from literature testifies towards hypothesis that studied cytokines regulate action potentials of cardiac cells influencing a number of ion fluxes including I_{Na}, I_{Ca}, I_{to}, I_{Kur}, I_{Kr}, I_{Ks}, I_{K1}. Besides that and possibly even more importantly those compounds regulate mechanoelectrical feedback in heart via activation of NO synthases in cardiomyocytes, primarily NOS3 as is shown on Fig. 5.9 (Prabhu, 2004a). NO imparts cardiac effects via two general signaling modalities: (i) the activation of soluble guanylate cyclase and the formation of cGMP, which in turn activates protein kinase G (PKG) and PKG-dependent phosphorylation events; and (2) the direct oxidation of thiol residues on critical regulatory proteins (S-nitrosylation) (Kojda and Kottenberg, 1999; Massion et al., 2003; Prabhu, 2004a). Rise of intracellular NO concentration opens MGCs, which allows sodium entry into the cell which in turn depolarizes them, shifting resting membrane potential towards E_{C} (Kazanski et al., 2011). This can directly or indirectly lead to arrhythmias triggered by elevated levels of cytokines. A large number of studies, performed in a variety of experimental models, have examined whether NO, derived from NOS3, contributes to early cytokine-induced dysfunction. We already highlighted that cytokine levels go up in different pathological states, including infarct and hypertrophy of cardiac tissue. In combination with data regarding correlation between such electrophysiological manifestations of cardiac pathology as arrhythmias and fibrillation and cytokine blood levels it is possible to propose cytokines as triggers of cardiac arrhythmias and fibrillation. Besides pure theoretical

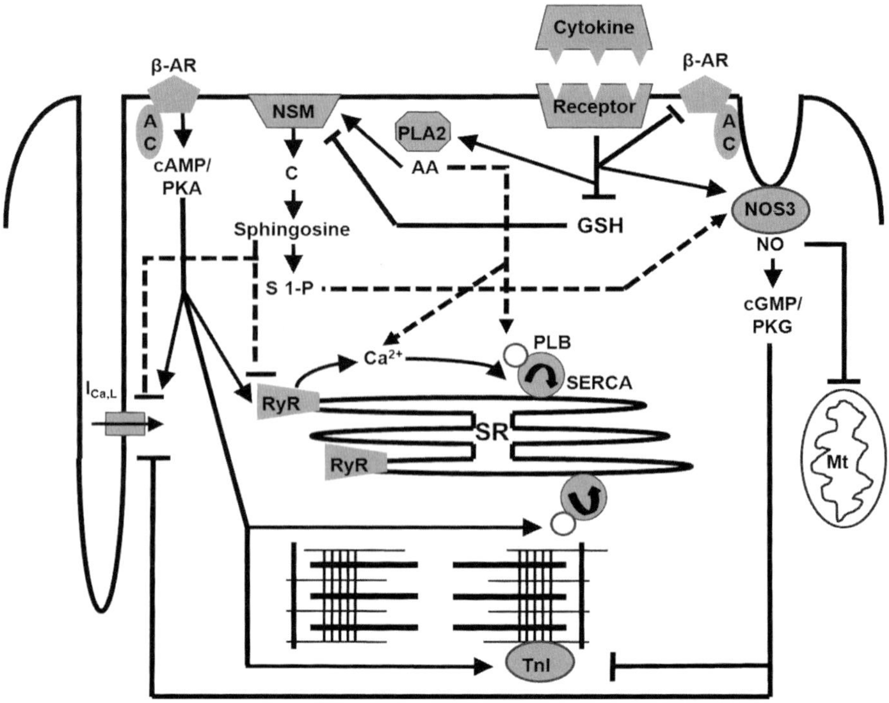

Fig. 5.9 Cellular mechanisms underlying immediate cytokine-induced contractile effects. Proinflammatory cytokines modulate several signaling pathways that can ultimately impact contractile function. Rapid activation of NOS3 generates NO, with depression of Ca^{2+} transients and myofilament Ca^{2+} sensitivity via the production of cGMP and cGMP-dependent activation of PKG. cGMP/PKG also blunts cAMP/PKA-mediated effects, and NO can also directly depress mitochondrial (Mt) respiration, thus altering mechanoenergetics. In low concentrations, NO can also exert a positive inotropic effect. Cytokines such as TNF-α also deplete GSH and activate membrane-associated neutral sphingomyelinase (NSM), which in turn increases ceramide (C) and sphingosine levels. Sphingosine blunts the inward Ca^{2+} current ($I_{Ca,L}$) and Ca^{2+} release from the ryanodine receptor (RyR) of the SR. Ceramide also blunts $I_{Ca,L}$. Cytokines also activate cytosolic PLA2 to generate AA, which can, under appropriate conditions, augment the Ca^{2+} transient and contractility. Cytokines may also acutely impair β-AR coupling to AC via unclear mechanisms. Cross-talk likely occurs between these various pathways. Free fatty acids such as AA have been shown to activate NSM, and sphingosine 1-phosphate (S 1-P) can phosphorylate and activate cNOS. GSH, in turn, inhibits NSM activation. Thus, the ambient oxidant stress and metabolic state, along with the cytokine dose and applicable reflex effects, influences the relative activation of these pathways, resulting in responses that are either cardiostimulatory (with low concentrations and short exposure duration) or cardiodepressant (with higher concentrations and longer exposure time). PLB indicates phospholamban; SERCA, SR Ca^{2+} ATPase; TnI, troponin I (from Prabhu (2004a) with permission of Lippincott Williams & Wilkins, *Circulation Research* and American Heart Association via Copyright Clearance Center)

importance such point of view can have clinical application. It would be a valuable approach for creation of test systems for early prediction of cardiac arrhythmias and fibrillation.

Acknowledgements This work was supported by the Russian Foundation for Basic Research (grant no. 09-04-01277-a). Department of Fundamental and Applied Physiology (Professor and Chairman – Andre Kamkin) was supported by Ministry of Education and Science of the Russian Federation. The Order of Ministry of Education and Science of the Russian Federation No. 743 from 01 July 2010, Supplement, Event 4.4, the Period of Financing 2010–2019.

References

Abbate A, Salloum FN, Vecile E et al (2008) Anakinra, a recombinant human interleukin-1 receptor antagonist, inhibits apoptosis in experimental acute myocardial infarction. Circulation 117:2670–2683

Abi-Gerges N, Fischmeister R, Mery PF (2001) G protein mediated inhibitory effect of a nitric oxide donor on the L-type Ca^{2+} current in rat ventricular myocytes. J Physiol (London) 531:117–130

Alloattia G, Pennaa C, De Martinoa A et al (1999) Role of nitric oxide and platelet-activating factor in cardiac alterations induced by tumor necrosis factor-a in the guinea-pig papillary muscle. Cardiovasc Res 41:611–619

Amadou A, Nawrocki A, Best-Belpomme M et al (2002) Arachidonic acid mediates dual effect of TNF-α on Ca^{2+} transients and contraction of adult rat cardiomyocytes. Am J Physiol 282:C1339–C1347

Arai K, Nishida J, Hayashida K et al (1990) Coordinate regulation of immune and inflammatory responses by cytokines. Rinsho Byori 38(4):347–353

Araujo DM, Lapchak PA, Collier B, Quirion R (1989) Localization of interleukin-2 immunoreactivity and interleukin-2 receptors in the rat brain: interaction with the cholinergic system. Brain Res 498(2):257–266

Aviles R, Martin D, Apperson-Hansen C et al (2003) Inflammation as a risk factor for atrial fibrillation. Circulation 108:3006–3010

Balligand JL (1999) Regulation of cardiac β-adrenergic response by nitric oxide. Cardiovasc Res 43:607–620

Balligand JL, Ungureanu D, Kelly RA, Kozaik L, Pimental D, Michel T, Smith TW (1993) Abnormal contractile function due to induction of nitric oxide synthesis in rat cardiac myocytes follows exposure to activated macrophage-conditioned medium. J Clin Invest 91(5):2314–2319

Balligand JL, Ungureanu-Longrois D, Simmons WW et al (1994) Cytokine-inducible nitric oxide synthase (iNOS) expression in cardiac myocytes. Characterization and regulation of iNOS expression and detection of iNOS activity in single cardiac myocytes *in vitro*. J Biol Chem 269:27580–27588

Belosjorow S, Bolle I, Duschin A et al (2003) TNF-alpha antibodies are as effective as ischemic preconditioning in reducing infarct size in rabbits. Am J Physiol 284:H927–H930

Benardeau A, Hatem SN, Rucker-Martin C et al (1996) Contribution of Na^{+}/Ca^{2+} exchange to action potential of human atrial myocytes. Am J Physiol 271:H1151–H1161

Benitah JP, Gomez AM, Bailly P et al (1993) Heterogeneity of the early outward current in ventricular cells isolated from normal and hypertrophied rat hearts. J Physiol (London) 469:111–138

Beuckelmann DJ, Näbauer M, Erdmann E (1993) Alterations of K^{+} currents in isolated human ventricular myocytes from patients with terminal heart failure. Circ Res 73(2):379–385

Bick RJ, Liao JP, King TW, LeMaistre A, McMillin JB, Buja LM (1997) Temporal effects of cytokines on neonatal cardiac myocyte calcium transients and adenylate cyclase activity. Am J Physiol 272:H1937–H1944

Bick RJ, Wood DE, Poindexter B, McMillin JB, Karoly A, Wang D et al (1999) Cytokines increase neonatal cardiac myocyte calcium concentrations: the involvement of nitric oxide and cyclic nucleotides. J Interferon Cytokine Res 19:645–653

Bittar MN, Carey JA, Barnard J et al (2005) Interleukin 6 G-174C polymorphism influences outcome following coronary revascularization surgery. Heart Surg Forum 8(3):E140–145

Bougaki M, Searles RJ, Kida K, Yu J, Buys ES, Ichinose F (2010) NOS3 protects against systemic inflammation and myocardial dysfunction in murine polymicrobial sepsis. Shock 34(3): 281–290

Bracco MM, Fink SB, Finiasz MR et al (1991) Positive inotropic effect of interleukin-2. Role of phospholipases and protein kinase C. Int J Immunopharm 13:509–515

Brinkmeier H, Kaspar A, Wiethoelter H, Ruedel R (1992) Interleukin-2 inhibits sodium currents in human muscle cells. Eur J Physiol 420:621–623

Brouillette J, Clark RB, Giles WR, Fiset C (2004) Functional properties of K^+ currents in adult mouse ventricular myocytes. J Physiol (London) 559(Pt 3):777–798

Brown JM, White CW, Terada LS et al (1990) Interleukin 1 pretreatment decreases ischemia/reperfusion injury. PNAS 87:5026–5030

Bundgaard H, Kjeldsen K (1996) Human myocardial Na,K-ATPase concentration in heart failure. Mol Cell Biochem 163–164:277–283

Cailleret M, Amadou A, Andrieu-Abadie N, Nawrocki A, Adamy C, Ait-Mamar B, Rocaries F, Best-Belpomme M, Levade T, Pavoine C, Pecker F (2004) N-Acetylcysteine prevents the deleterious effect of tumor necrosis factor-(alpha) on calcium transients and contraction in adult rat cardiomyocytes. Circulation 109(3):406–411

Cain BS, Meldrum DR, Dinarello CA et al (1999) Tumor necrosis factor-α and interleukin-1β synergistically depress human myocardial function. Crit Care Med 27:1309–1318

Calabrese V, Cornelius C, Rizzarelli E, Owen JB, Dinkova-Kostova AT, Butterfield DA (2009) Nitric oxide in cell survival: a janus molecule. Antioxid Redox Signal 11:2717–2739

Campbell DL, Stamler JS, Strauss HC (1996) Redox modulation of L-type calcium channels in ferret ventricular myocytes – dual mechanism regulation by nitric oxide and S-nitrosothiols. J Gen Physiol 108:277–293

Cao C, Xia Q, Bruce Iain C, Zhang X, Fu C, Chen JZ (2003a) Interleukin-2 increases activity of sarcoplasmic reticulum Ca^{2+}-ATPase, but decreases its sensitivity to calcium in rat cardiomyocytes. J Pharmacol Exp Ther 306:572–580

Cao CM, Xia Q, Bruce IC, Shen YL, Ye ZG, Lin GH, Chen JZ, Li GR (2003b) Influence of interleukin-2 on Ca^{2+} handling in rat ventricular myocytes. J Mol Cell Cardiol 35:1491–1503

Casadei B, Sears CE (2003) Nitric-oxide-mediated regulation of cardiac contractility and stretch responses. Prog Biophys Mol Biol 82(1–3):67–80

Chandrasekar B, Vemula K, Surabhi RM et al (2004) Activation of intrinsic and extrinsic proapoptotic signaling pathways in interleukin-18-mediated human cardiac endothelial cell death. J Biol Chem 279:20221–20233

Chandrasekar B, Mummidi S, Claycomb WC, Mestril R, Nemer M (2005) Interleukin-18 is a pro-hypertrophic cytokine that acts through a phosphatidylinositol 3-kinase-phosphoinositide dependent kinase-1-Akt-GATA4 signaling pathway in cardiomyocytes. J Biol Chem 280: 4553–4567

Chandrasekar B, Valente AJ, Freeman GL et al (2006) Interleukin-18 induces human cardiac endothelial cell death via a novel signaling pathway involving NF-kappaB-dependent PTEN activation. Biochem Biophys Res Commun 339:956–963

Chen Z, Siu B, Ho YS et al (1998) Overexpression of MnSOD protects against myocardial ischemia/reperfusion injury in transgenic mice. J Mol Cell Cardiol 30:2281–2289

Chen SA, Hsieh MH, Tai CT et al (1999) Initiation of atrial fibrillation by ectopic beats originating from the pulmonary veins: electrophysiological characteristics, pharmacological responses, and effects of radiofrequency ablation. Circulation 100:1879–1886

Chen YJ, Tsai CF, Chiou CW, Chan P, Chen SA (2001) Effect of nitric oxide on strophanthidin-induced ventricular tachycardia. Pharmacology 62:213–217

Chen Y, Pat B, Zheng J, Cain L, Powell P, Shi K, Sabri A, Husain A, Dell'italia LJ (2010) Tumor necrosis factor-alpha produced in cardiomyocytes mediates a predominant myocardial inflammatory response to stretch in early volume overload. J Mol Cell Cardiol 49(1):70–78

Cheng XS, Shimokawa H, Momii H et al (1999) Role of superoxide anion in the pathogenesis of cytokine-induced myocardial dysfunction in dogs in vivo. Cardiovasc Res 42:651– 659

Chung MK, Gulick TS, Rotondo RE et al (1990) Mechanism of cytokine inhibition of beta-adrenergic agonist stimulation of cyclic AMP in rat cardiac myocytes. Impairment of signal transduction. Circ Res 67:753–763

Colston JT, Boylston WH, Feldman MD, Jenkinson CP, de la Rosa SD, Barton A, Trevino RJ, Freeman GL, Chandrasekar B (2007) Interleukin-18 knockout mice display maladaptive cardiac hypertrophy in response to pressure overload. Biochem Biophys Res Commun 354:552–558

Cugno M, Agostoni P, Brunner HR et al (2000) Plasma bradykinin levels in human chronic congestive heart failure. Clin Sci (London) 99:461–466

Damron DS, Summers BA (1997) Arachidonic acid enhances contraction and intracellular Ca^{2+} transients in individual rat ventricular myocytes. Am J Physiol 272:H350–H359

Debrunner M, Schuiki E, Minder E et al (2008) Proinflammatory cytokines in acute myocardial infarction with and without cardiogenic shock. Clin Res Cardiol 97:298–305

Dernellis J, Panaretou M (2001) C-reactive protein and paroxysmal atrial fibrillation: evidence of the implication of an inflammatory process in paroxysmal atrial fibrillation. Acta Cardiol 56:375–380

Dernellis J, Panaretou M (2006) Effects of C-reactive protein and the third and fourth components of complement (C3 and C4) on incidence of atrial fibrillation. Am J Cardiol 97:245–248

Dettbarn CA, Betto R, Salviati G et al (1994) Modulation of cardiac sarcoplasmic reticulum ryanodine receptor by sphingosine. J Mol Cell Cardiol 26:229–242

DiFrancesco D (1993) Pacemaker mechanisms in cardiac tissue. Annu Rev Physiol 55:455–472

Dinarello CA (2000) Proinflammatory cytokines. Chest 118:503–508

Dizdarević-Hudić L, Kušljugić Z, Baraković F, Brkić S, Sabitović D, Jahić E, Isabegović M, Smajić E, Hudić I, Divković K (2009) Correlation between interleukin 6 and interleukin 10 in acute myocardial infarction. Bosn J Basic Med Sci 9(4):301–306

Duncan DJ, Yang Z, Hopkins PM et al (2010) TNF-α and IL-1β increase Ca^{2+} leak from the sarcoplasmic reticulum and susceptibility to arrhythmia in rat ventricular myocytes. Cell Calcium 47:378–386

Eddy LJ, Goeddel DV, Wong GH (1992) Tumor necrosis factor alpha pretreatment is protective in a rat model of myocardial ischemia-reperfusion injury. Biochem Biophys Res Commun 184:1056–1059

Edmunds NJ, Woodward B (1998) Effects of tumour necrosis factor-α on the coronary circulation of the rat isolated perfused heart: a potential role for thromboxane A2 and sphingosine. Br J Pharmacol 124:493–498

Edmunds NJ, Lal H, Woodward B (1999) Effects of tumour necrosis factor-α on left ventricular function in the rat isolated perfused heart: possible mechanisms for a decline in cardiac function. Br J Pharmacol 126:189–196

El-Ani D, Zimlichman R (2003) TNFalpha stimulated ATP-sensitive potassium channels and attenuated deoxyglucose and Ca uptake of H9c2 cardiomyocytes. Ann N Y Acad Sci 1010:716–720

Evans HG, Lewis MJ, Shah AH (1993) Interleukin-1β modulates myocardial contraction via a dexamethasone sensitive production of nitric oxide. Cardiovasc Res 27:1486–1490

Fedak PW, Verma S, Weisel RD, Li RK (2005) Cardiac remodeling and failure from molecules to man (part II). Cardiovasc Pathol 14:49–60

Ferdinandy P, Danial H, Ambrus I et al (2000) Peroxynitrite is a major contributor to cytokine-induced myocardial contractile failure. Circ Res 87:241–247

Fernandez-Velasco M, Ruiz-Hurtado G, Hurtado O et al (2007) TNF-α downregulates transient outward potassium current in rat ventricular myocytes through iNOS overexpression and oxidant species generation. Am J Physiol 293:H238–H245

Finkel MS, Oddis CV, Jacob TD, Watkins SC, Hattler BG, Simmons RL (1992) Negative inotropic effects of cytokines on the heart mediated by nitric oxide. Science 257:387–389
Flaherty MP, Guo Y, Tiwari S et al (2008) The role of TNF-alpha receptors p55 and p75 in acute myocardial ischemia/reperfusion injury and late preconditioning. J Mol Cell Cardiol 45: 735–741
Flesch M, Hoper A, Dell'Italia L et al (2003) Activation and functional significance of the renin-angiotensin system in mice with cardiac restricted overexpression of tumor necrosis factor. Circulation 108:598–604
Friedrichs GS, Swillo RE, Jow B et al (2002) Sphingosine modulates myocyte electrophysiology, induces negative inotropy, and decreases survival after myocardial ischemia. J Cardiovasc Pharmacol 39:18–28
Gaudino M, Andreotti F, Zamparelli R et al (2003) The -174G/C interleukin-6 polymorphism influences postoperative interleukin-6 levels and postoperative atrial fibrillation. Is atrial fibrillation an inflammatory complication? Circulation 108(10 Suppl 1):II195–II199
Goldhaber JI, Kim KH, Natterson PD, Lawrence T, Ph Y, Weiss JN (1996) Effects of TNF-α on $[Ca^{2+}]_i$ and contractility in isolated adult rabbit ventricular myocytes. Am J Physiol 271:H1449–H1455
Grandel U, Fink L, Blum A et al (2000) Endotoxin-induced myocardial tumor necrosis factor-α synthesis depresses contractility of isolated rat hearts: evidence for a role of sphingosine and cyclooxygenase-2-derived thromboxane production. Circulation 102:2758–2764
Grandy SA, Fiset C (2009) Ventricular K^+ currents are reduced in mice with elevated levels of serum TNFα. J Mol Cell Cardiol 47:238–246
Grandy SA, Brouillette J, Fiset C (2010) Reduction of ventricular sodium current in a mouse model of HIV. J Cardiovasc Electrophysiol 21:916–922
Gulick T, Chung MK, Pieper SJ et al (1989) Interleukin 1 and tumor necrosis factor inhibit cardiac myocyte β-adrenergic responsiveness. PNAS 86:6753–6757
Harada E, Nakagawa O, Yoshimura M et al (1999) Effect of interleukin-1 beta on cardiac hypertrophy and production of natriuretic peptides in rat cardiocyte culture. J Mol Cell Cardiol 31:1997–2006
Harding P, Carretero OA, LaPointe MC (1995) Effects of interleukin-1β and nitric oxide on cardiac myocytes. Hypertension 25:421–430
Hare JM (2003) Nitric oxide and excitation–contraction coupling. J Mol Cell Cardiol 35:719–729
Hatada K, Washizuka T, Horie M et al (2006) Tumor necrosis factor-alpha inhibits the cardiac delayed rectifier K current via the asphingomyelin pathway. Biochem Biophys Res Commun 344(1):189–193
Haudek SB, Taffet GE, Schneider MD, Mann DL (2007) TNF provokes cardiomyocyte apoptosis and cardiac remodeling through activation of multiple cell death pathways. J Clin Invest 117:2692–2701
Hedayat M, Mahmoudi MJ, Rose NR, Rezaei N (2010) Proinflammatory cytokines in heart failure: double-edged swords. Heart Fail Rev 15:543–562
Heger J, Goedecke A, Flögel U, Merx MW, Molojavyi A, Kühn-VeltenW N, Schrader J (2002) Cardiac-specific overexpression of inducible nitric oxide synthase does not result in severe cardiac dysfunction. Circ Res 90:93–99
Hofmann U, Domeier E, Frantz S et al (2003) Increased myocardial oxygen consumption by TNF-α is mediated by a sphingosine signaling pathway. Am J Physiol 284:H2100–H2105
Hoppe UC, Beuckelmann DJ (1998) Characterization of the hyperpolarization-activated current in isolated human atrial myocytes. Cardiovasc Res 38:788–801
Hosenpud JD, Campbell SM, Mendelson DJ (1989) Interleukin-l-induced myocardial depression in an isolated beating heart preparation. J Heart Transplant 8:460–464
Hu Y, Chen YC, Cheng CC, Higa S, Chen YJ, Chen SA (2009) Fluvastatin reduces pulmonary vein spontaneous activity through nitric oxide pathway. J Cardiovasc Electrophysiol 20:200–206
Hund TJ, Rudy Y (2004) Rate dependence and regulation of action potential and calcium transient in a canine cardiac ventricular cell model. Circulation 110:3168–3174

Ing DJ, Zang J, Dzau VJ et al (1999) Modulation of cytokine-induced cardiac myocyte apoptosis by nitric oxide, Bak, and Bcl-x. Circ Res 84:21–33
Isenberg G, Kazanski V, Kondratev D, Gallitelli MF, Kiseleva I, Kamkin A (2003) Differential effects of stretch and compression on membrane currents and $[Na^{+}]_{C}$ in ventricular myocytes. Prog Biophys Mol Biol 82:43–56
Issac TT, Dokainish H, Lakkis NM (2007) Role of inflammation in initiation and perpetuation of atrial fibrillation. J Am Coll Cardiol 50:2021–2028
Janczewski AM, Kadokami T, Lemster B et al (2003) Morphological and functional changes in cardiac myocytes isolated from mice overexpressing TNF-α. Am J Physiol 284:H960–H969
Janse MJ (1992) The premature beat. Cardiovasc Res 26:89–100
Janse MJ (1997) Why does atrial fibrillation occur? Eur Heart J 18(suppl):C12–C18
Janssen SPM, Gayan-Ramirez G, Van Den Bergh A, Herijgers P, Maes K, Verbeken E, Decramer M (2005) Interleukin-6 causes myocardial failure and skeletal muscle atrophy in rats. Circulation 111:996–1005
Jayadev S, Linardic CM, Hannun YA (1994) Identification of arachidonic acid as a mediator of sphingomyelin hydrolysis in response to tumor necrosis factor-α. J Biol Chem 269:5757–5763
Joe EK, Schussheim AE, Longrois D et al (1998) Regulation of cardiac myocyte contractile function by inducible nitric oxide synthase (iNOS): mechanisms of contractile depression by nitric oxide. J Mol Cell Cardiol 30:303–315
Ju H, Dixon IM (1996) Extracellular matrix and cardiovascular diseases. Can J Cardiol 12: 1259–1267
Kaab S, Dixon J, Duc J et al (1998) Molecular basis of transient outward potassium current downregulation in human heart failure: a decrease in Kv4.3 mRNA correlates with a reduction in current density. Circulation 98:1383–1393
Kamkin A, Kiseleva I, Wagner KD, Leiterer KP, Theres H, Scholz H, Günther J, Lab MJ (2000a) Mechano-electric feedback in right atrium after left ventricular infarction in rats. J Mol Cell Cardiol 32:465–477
Kamkin A, Kiseleva I, Isenberg G (2000b) Stretch-activated currents in ventricular myocytes: amplitude and arrhythmogenic effects increase with hypertrophy. Cardiovasc Res 48: 409–420
Kamkin A, Kiseleva I, Isenberg G (2003a) Ion selectivity of stretch-activated cation currents in mouse ventricular myocytes. Eur J Physiol 446(2):220–231
Kamkin A, Kiseleva I, Wagner KD, Bohm J, Theres H, Günther J, Scholz H (2003b) Characterization of stretch-activated ion currents in isolated atrial myocytes from human hearts. Eur J Physiol 446(3):339–346
Kang JX, Leaf A (1994) Effects of long-chain polyunsaturated fatty acids on the contraction of neonatal rat cardiac myocytes. PNAS 91:9886–9890
Kao YH, Chen YC, Cheng CC et al (2010) Tumor necrosis factor-alpha decreases sarcoplasmic reticulum Ca^{2+}-ATPase expressions via the promoter methylation in cardiomyocytes. Crit Care Med 8(1):217–222
Karmazym M, Gan XT, Humphreys RA, Yoshida H, Kusumoto K (1999) The myocardial Na^{+}-H^{+} exchange structure, regulation, and its role in heart disease. Circ Res 85:777–786
Kaspar A, Brinkmeier H, Riidel R (1994) Local anaesthetic-like effect of interleukin-2 on muscular Na^{+}-channels: no evidence for involvement of the IL-2 receptor. Eur J Physiol 426:61–67
Kato D, Cragoe EJ, Smith TW (1987) Relations among sodium pump inhibition, Na-Ca and Na-H exchange activities, and Ca-H interaction in cultured chick heart cells. Circ Res 60: 185–193
Kawada H, Niwano S, Niwano H et al (2006) Tumor necrosis factor-alpha downregulates the voltage gated outward K_ current in cultured neonatal rat cardiomyocytes: a possible cause of electrical remodeling in diseased hearts. Circulation 70:605–609
Kazanski VE, Kamkin AG, Makarenko EY, Lysenko NN, Sutiagin PV, Bo T, Kiseleva IS (2010a) Role of nitric oxide in activity control of mechanically gated ionic channels in cardiomyocytes: NO-donor study. Bull Exp Biol Med 150(1):1–5 English

Kazanski VE, Kamkin AG, Makarenko EY, Lysenko NN, Sutiagin PV, Kiseleva IS (2010b) Role of nitric oxide in the regulation of mechanosensitive ionic channels in cardiomyocytes: Contribution of NO-synthases. Bull Exp Biol Med 150(2):263–267 English

Kazanski V, Kamkin A, Makarenko E, Lysenko N, Lapina N, Kiseleva I (2011) The role of nitric oxide in regulation of mechanically gated channels in the heart. In: Kamkin A, Kiseleva I (eds) Mechanosensitivity in cells and tissues 4. Mechanosensitivity and mechanotransduction. Springer, New York, NY, pp 109–140

Kelly RA, Balligand JL, Smith TW (1996) Nitric oxide and cardiac function. Circ Res 79:363–380

Kinugawa K, Takahashi T, Kohmoto O, Yao A, Aoyagi T, Momomura S, Hirata Y, Serizawa T (1994) Nitric oxide-mediated effects of interleukin-6 on $[Ca^{2+}]_i$ and cell contraction in cultured chick ventricular myocytes. Circ Res 75:285–295

Kiseleva I, Kamkin A, Wagner KD, Theres H, Ladhoff A, Scholz H, Günther J, Lab MJ (2000) Mechano-electric feedback after left ventricular infarction in rats. Cardiovasc Res 45:370–378

Kojda G, Kottenberg K (1999) Regulation of basal myocardial function by NO. Cardiovasc Res 41:514–523

Kolesnick R (2002) The therapeutic potential of modulating the ceramide/sphingomyelin pathway. J Clin Invest 110:3–8

Kondratev D, Gallitelli MF (2003) Increments in the concentration of sodium and calcium in cell compartments of stretched mouse ventricular myocytes. Cell Calcium 34:193–203

Kreydiyyeh SI, Abou-Chahine C, Hilal-Dandan R (2004) Interleukin-1beta inhibits Na^+-K^+ ATPase activity and protein expression in cardiac myocytes. Cytokine 26(1):1–8

Krown KA, Yasui K, Brooker MJ, Dubin AE, Nguyen C, Harris GL, McDonough PM, Glembotski CC, Palade PT, Sabbadini RA (1995) TNF alpha receptor expression in rat cardiac myocytes: TNF alpha inhibition of L-type Ca^{2+} current and Ca^{2+} transients. FEBS Lett 376(1–2): 24–30

Kubota T, McTiernan CF, Frye CS et al (1997) Dilated cardiomyopathy in transgenic mice with cardiac-specific overexpression of tumor necrosis factor-α. Circ Res 81:627–635

Kubota I, Han X, Opel DJ, Zhao YY, Baliga R, Huang P, Fishman MC, Shannon RP, Michel T, Kelly RA (2000) Increased susceptibility to development of triggered activity in myocytes from mice with raged disruption of endothelial nitric oxide synthase. J Mol Cell Cardiol 32: 1239–1248

Kumar A, Thota V, Dee L, Olson J, Uretz E, Parrillo JE (1996) Tumor necrosis factor-alpha and interleukin 1-beta are responsible for the in vitro myocardial cell depression induced by human septic shock serum. J Exp Med 183(3):949–958

Kumar A, Brar R, Wang P et al (1999) Role of nitric oxide and cGMP in human septic serum induced depression of cardiac myocyte contractility. Am J Physiol 276:R265–R276

Kunisada K, Negoro S, Tone E et al (2000) Signal transducer and activator of transcription 3 in the heart transduces not only a hypertrophic signal but a protective signal against doxorubicin-induced cardiomyopathy. PNAS 97:315–319

Latini R, Bianchi M, Correale E, Dinarello CA, Fantuzzi G, Fresco C, Maggioni AP, Mengozzi M, Romano S, Shapiro L, Sironi M, Tognoni G, Turato R, Ghezzi P (1994) Cytokines in acute myocardial infarction: selective increase in circulating tumor necrosis factor, its soluble receptor, and interleukin 1 receptor antagonist. J Cardiovasc Pharmacol 23(1):1–6

Lee SH, Chen YC, Chen YJ et al (2007) Tumor necrosis factor-alpha alters calcium handling and increases arrhythmogenesis of pulmonary vein cardiomyocytes. Life Sci 80(19):1806–1815

Levi RC, Alloatti G, Fischmeister R (1989) Cyclic GMP regulates the Ca channel current in guinea pig ventricular myocytes. Eur J Physiol 413:685–687

Levi RC, Alloatti G, Penna C, Gallo MP (1994) Guanylate-cyclase-mediated inhibition of cardiac ICa by carbachol and sodium nitroprusside. Pflugers Arch 426(5):419–426

Levine B, Kalman J, Mayer L, Fillit HM, Packer M (1990) Elevated circulating levels of tumor necrosis factor in severe chronic heart failure. N Engl J Med 323(4):236–241

Li Y-H, Rozanski GJ (1993) Effects of human recombinant interleukin-1 on electrical properties of guinea pig ventricular cells. Cardiovasc Res 27:525–530

Li J, Hampton T, Morgan JP, Simons M (1997) Stretch-induced VEGF expression in the heart. J Clin Invest 100(1):18–24

Li YY, Feng YQ, Kadokami T et al (2000) Myocardial extracellular matrix remodeling in transgenic mice overexpressing tumor necrosis factor alpha can be modulated by anti-tumor necrosis factor alpha therapy. PNAS 97:12746–12751

Li GR, Lau CP, Leung TK, Nattel S (2004) Ionic current abnormalities associated with prolonged action potentials in cardiomyocytes from diseased human right ventricles. Heart Rhythm 1:460–468

Liu SJ, McHowat J (1998) Stimulation of different phospholipase A2 isoforms by TNF-α and IL-1β in adult rat ventricular myocytes. Am J Physiol 275:H1462–H1472

Liu S, Schreur KD (1995) G protein-mediated suppression of L-type Ca^{2+} current by interleukin-1β in cultured rat ventricular myocytes. Am J Physiol 268:C339–C349

Liu B, Andrieu-Abadie N, Levade T et al (1998) Glutathione regulation of neutral sphingomyelinase in tumor necrosis factor-α-induced cell death. J Biol Chem 273:11313–11320

Liu SJ, Zhou W, Kennedy RH (1999) Suppression of β-adrenergic responsiveness of L-type Ca^{2+} current by IL-1β in rat ventricular myocytes. Am J Physiol 276:H141–H148

Liu J, Kim KH, London B et al (2011) Dissection of the voltage-activated potassium outward currents in adult mouse ventricular myocytes: $I_{to,f}$, $I_{to,s}$, $I_{K,slow1}$, $I_{K,slow2}$, and I_{ss}. Basic Res Cardiol 106:189–204

London B, Baker LC, Lee JS, Shusterman V, Choi BR, Kubota T, McTiernan CF, Feldman AM, Salama G (2003) Calcium-dependent arrhythmias in transgenic mice with heart failure. Am J Physiol 284:H431–H441

Long CS (2001) The role of interleukin-1 in the failing heart. Heart Fail Rev 6:81–94

Lorenzon P, Ruzzier F, Caratsch CG et al (1991) Interleukin-2 lengthens extrajunctional acetylcholine receptor channel open time in mammalian muscle cells. Pflugers Arch 419(3-4): 380–385

Low-Friedrich I, Weisensee D, Mitrou P, Schoeppe W (1992) Cytokines induce stress protein formation in cultured cardiac myocytes. Basic Res Cardiol 87(1):12–18

Lozinsky I, Kamkin A (2010) Mechanosensitive alterations of action potentials and membrane currents in healthy and diseased cardiomyocytes: cardiac tissue and isolated cell. In: Kamkin A, Kiseleva I (eds) Mechanosensitivity in cells and tissues 3. Mechanosensitivity of the heart. Springer, New York, NY, pp 185–238

Luan Y, Guo Y, Li S et al (2010) Interleukin-18 among atrial fibrillation patients in the absence of structural heart disease. Europace 12:1713–1718

Marcus GM, Whooley MA, Glidden DV, Pawlikowska L, Zaroff JG, Olgin JE (2008) Interleukin 6 and atrial fibrillation in patients with coronary disease: data from the heart and soul study. Am Heart J 155(2):303–309

Marino E, Cardier JE (2003) Differential effect of IL-18 on endothelial cell apoptosis mediated by TNF-alpha and Fas (CD95). Cytokine 22:142–148

Massion PB, Balligand JL (2003) Modulation of cardiac contraction, relaxation and rate by the endothelial nitric oxide synthase (eNOS): lessons from genetically modified mice. J Physiol (London) 546(1):63–75

Massion PB, Feron O, Dessy C, Balligand JL (2003) Nitric oxide and cardiac function: ten years after, and continuing. Circ Res 93:388 –398

Maulik N, Engelman RM, Wei Z et al (1993) Interleukin-1 alpha preconditioning reduces myocardial ischemia reperfusion injury. Circulation 88:387–394

McDonough AA, Zhang Y, Shin V, Frank JS (1996) Subcellular distribution of sodium pump isoform subunits in mammalian cardiac myocytes. Am J Physiol 270:C1221–C1227

McGowan FX, Takeuchi K, del Nido PJ et al (1994) Myocardial effects of interleukin-2. Transplant Proc 26:209–210

Mery PF, Lohmann SM, Walter U, Fischmeister R (1991) Ca^{2+} current is regulated by cyclic GMP-dependent protein kinase in mammalian cardiac myocytes. Proc Natl Acad Sci USA 88:1197–1201

Mery PF, Pavoine C, Belhassen L, Pecker F, Fischmeister R (1993) Nitric oxide regulates cardiac Ca^{2+} current. Involvement of cGMP-inhibited and cGMP-stimulated phosphodiesterases through guanylyl cyclase activation. J Biol Chem 268:26286–26295

Meyer M, Schillinger W, Pieske B et al (1995) Alterations of sarcoplasmic reticulum proteins in failing human dilated cardiomyopathy. Circulation 92:778–784

Moncada S, Palmer RMJ, Higgs EA (1991) Nitric oxide physiology, pathophysiology. Pharmacol Rev 433:109–142

Monsuez JJ, Escaut L, Teicher E, Charniot JC, Vittecoq D (2007) Cytokines in HIV-associated cardiomyopathy. Int J Cardiol 2:150–157

Nakano M, Knowlton AA, Yokoyama T et al (1996) Tumor necrosis factor-alpha-induced expression of heat shock protein 72 in adult feline cardiac myocytes. Am J Physiol 270:H1231–H1239

Nakano M, Knowlton AA, Dibbs Z, Mann DL (1998) Tumor necrosis factor-alpha confers resistance to hypoxic injury in the adult mammalian cardiac myocyte. Circulation 97: 1392–1400

Nakayama H, Bodi I, Correll RN, Chen X, Lorenz J, Houser SR, Robbins J, Schwartz A, Molkentin JD (2009) alpha1G-dependent T-type Ca^{2+} current antagonizes cardiac hypertrophy through a NOS3-dependent mechanism in mice. J Clin Invest 119(12):3787–3796

Narayan SM, Cain ME, Smith JM (1997) Atrial fibrillation. Lancet 350:943–950

Nattel S, Maguy A, Le Bouter S, Yeh YH (2007) Arrhythmogenic ion-channel remodeling in the heart: heart failure, myocardial infarction, and atrial fibrillation. Physiol Rev 87:425–456

Nerbonne JM, Kass RS (2005) Molecular physiology of cardiac repolarization. Physiol Rev 85:1205–1253

Nishikawa K, Yoshida M, Kusuhara M et al (2006) Left ventricular hypertrophy in mice with a cardiac-specific overexpression of interleukin-1. Am J Physiol 291:H176–H183

Nuss HB, Kaab S, Kass DA et al (1999) Cellular basis of ventricular arrhythmias and abnormal automaticity in heart failure. Am J Physiol 277:H80–H91

Ono K, Trautwein W (1991) Potentiation by cyclic GMP of b-adrenergic effect on Ca^{2+} current in guinea pig ventricular cells. J Physiol (London) 443:387–404

Opthof T (1988) The membrane current (I_f) in human atrial cells: implications for atrial arrhythmias. Cardiovasc Res 38:537–540

Oral H, Dorn GWII, Mann DL (1997) Sphingosine mediates the immediate negative inotropic effects of tumor necrosis factor-α in the adult mammalian cardiac myocyte. J Biol Chem 272:4836–4842

Oyama J, Shimokawa H, Momii H (1998) Role of nitric oxide and peroxynitrite in the cytokine-induced sustained myocardial dysfunction in dogs in vivo. J Clin Invest 101:2207–2214

Packer M (1995) Is tumor necrosis factor an important neurohormoral mechanism in chronic heart failure? Circulation 92(6):1379–1382

Paganelli F, Mougenot R, Maixentt JM (2001) Defective activity and isoform of the Na,K-ATPase in the dilated cardiomyopathic hamster. Cell Mol Biol (Noisy-le-grand) 47:255–260

Palmer JN, Hartogensis WE, Patten M et al (1995) Interleukin-1 beta induces cardiac myocyte growth but inhibits cardiac fibroblast proliferation in culture. J Clin Invest 95:2555–2564

Panas D, Khadour FH, Szabo C, Schulz R (1998) Proinflammatory cytokines depress cardiac efficiency by a nitric oxide-dependent mechanism. Am J Physiol 275:H1016–H1023

Park CO, Xiao XH, Allen DG (1999) Changes in intracellular sodium and pH in the rat heart during ischaemia: role of the Na^+/H^+ exchanger. Am J Physiol 276(5Pt2):H1581–H1590

Petkova-Kirova PS, Gursoy E, Mehdi H et al (2006) Electrical remodeling of cardiac myocytes from mice with heart failure due to the overexpression of tumor necrosis factor-α. Am J Physiol 290(5):H2098–H2107

Pino R, Cerbai E, Calamai G et al (1998) Effect of 5-HT4 receptor stimulation on the pacemaker current I_f in human isolated atrial myocytes. Cardiovasc Res 40:516–522

Porciatti F, Pelzmann B, Cerbai E et al (1997) The pacemaker current I_f in single human atrial myocytes and the effect of beta-adrenoceptor and A1-adenosine receptor stimulation. Br J Pharmacol 122:963–969

Prabhu SD (2004a) Cytokine-induced modulation of cardiac function. Circ Res 95:1140–1153
Prabhu SD (2004b) Nitric oxide protects against pathological ventricular remodeling: reconsideration of the role of NO in the failing heart. Circ Res 94:1155–1157
Priebe L, Beuckelmann DJ (1998) Simulation study of cellular electric properties in heart failure. Circ Res 82:1206–1223
Proebstle T, Mitrovics M, Scneider M, Hombach V, Ruedel R (1995) Recombinant interleukin-2 acts like a class I antiarrhythmic drug on human cardiac sodium channels. Eur J Physiol 429:462–469
Psychari SN, Apostolou TS, Sinos L, Hamodraka E, Liakos G, Kremastinos DT (2005) Relation of elevated C-reactive protein and interleukin-6 levels to left atrial size and duration of episodes in patients with atrial fibrillation. Am J Cardiol 95(6):764–767
Rashmi RS, Hondeghem LM (2005) Refining detection of drug-induced proarrhythmia: QT interval and TRIaD. Heart Rhythm 2:758–772
Robinson BS, Hii CS, Poulos A, Ferrante A (1997) Activation of neutral sphingomyelinase in human neutrophils by polyunsaturated fatty acids. Immunology 91:274–280
Roedler S, Roth M, Nauck M, Tamm M, Block LH (1994) Ca^{2+}-channel blockers modulate the expression of interleukin-6 and interleukin-8 genes in human vascular smooth muscle cells. J Mol Cell Cardiol 27:2295–2302
Rozanski GJ, Witt RC (1994a) IL-1 inhibits β-adrenergic control of cardiac calcium current: role of L-arginine/nitric oxide pathway. Am J Physiol 267:H1753–H1758
Rozanski GJ, Witt RC (1994b) Interleukin-1 enhances β-responsiveness of cardiac L-type calcium current suppressed by acidosis. Am J Physiol 267:H1361–H1367
Saba S, Janczewski AM, Baker LC et al (2004) Atrial contractile dysfunction, fibrosis, and arrhythmias in a mouse model of cardiomyopathy secondary to cardiac-specific overexpression of tumor necrosis factor-α. Am J Physiol 289:H1456–H1467
Sani MU, Okeahialam BN (2005) QTc interval prolongation in patients with HIV and AIDS. J Natl Med Assoc 12:1657–1661
Sartiani L, Paoli PD, Lonardo G, Pino R, Conti AA, Cerbai E, Pelleg A, Belardinelli L, Mugelli A (2002) Does recombinant human interleukin-11 exert direct electrophysiologic effects on single human atrial myocytes? J Cardiovasc Pharmacol 39:425–434
Sata N, Hamada N, Horinouchi T et al (2004) C-reactive protein and atrial fibrillation: Is inflammation a consequence or cause of atrial fibrillation? Japan Heart J 45:441–445
Schreur KD, Liu S (1997) Involvement of ceramide in inhibitory effect of IL-1β on L-type Ca^{2+} current in adult rat ventricular myocytes. Am J Physiol 272:H2591–H2598
Schulz R (2008) TNF-α in myocardial ischemia/reperfusion: damage vs. protection. J Mol Cell Cardiol 45:712–714
Schulz R, Heusch G (2009) Tumor necrosis factor-α and its receptors 1 and 2. Yin and Yang in myocardial infarction? Circulation 119:1355–1357
Seddon M, Shah AM, Casadei B (2007) Cardiomyocytes as effectors of nitric oxide signalling. Cardiovasc Res 75(2):315–326
Shah AM, MacCarthy PA (2000) Paracrine and autocrine effects of nitric oxide on myocardial function. Pharmacol Ther 86(1):49–86
Sivasubramanian N, Coker ML, Kurrelmeyer KM et al (2001) Left ventricular remodeling in transgenic mice with cardiac restricted overexpression of tumor necrosis factor. Circulation 104:826–831
Siwik DA, Colucci WS (2004) Regulation of matrix metalloproteinases by cytokines and reactive oxygen/nitrogen species in the myocardium. Heart Fail Rev 9:43–51
Siwik DA, Chang DL, Colucci WS (2000) Interleukin-1beta and tumor necrosis factor-alpha decrease collagen synthesis and increase matrix metalloproteinase activity in cardiac fibroblasts in vitro. Circ Res 86:1259–1265
Smart N, Mojet MH, Latchman DS et al (2006) IL-6 induces PI 3-kinase and nitric oxide dependent protection and preserves mitochondrial function in cardiomyocytes. Cardiovasc Res 69: 164–177

Smith MD, Haynes DR, Roberts-Thomson PJ (1989) Interleukin 2 and interleukin 2 inhibitors in human serum and synovial fluid. I. Characterization of the inhibitor and its mechanism of action. J Rheumatol 16(2):149–157

Stamler JS, Lamas S, Fang FC (2001) Nitrosylation. The prototypic redox-based signaling mechanism. Cell 106:675–683

Stamm C, Cowan DB, Friehs I et al (2001a) Rapid endotoxin-induced alterations in myocardial calcium handling: obligatory role of cardiac TNF-alpha. Anesthesiology 95: 1396–1405

Stamm C, Friehs I, Cowan DB et al (2001b) Inhibition of tumor necrosis factor-α improves postischemic recovery of hypertrophied hearts. Circulation 104:I350–I355

Sugano M, Hata T, Tsuchida K et al (2004) Local delivery of soluble TNF-alpha receptor 1 gene reduces infarct size following ischemia/reperfusion injury in rats. Mol Cell Biochem 266: 127–132

Sugishita K, Kinugawa K, Shimizu T, Harada K, Matsui H, Takahashi T, Serizawa T, Kohmoto O (1999) Cellular basis for the acute inhibitory effects of IL-6 and TNF-α on excitation-contraction coupling. J Mol Cell Cardiol 31:1457–1467

Tavi P, Laine M, Weckstrom M (1996) Effect of gadolinium on stretch-induced changes in contraction and intracellularly recorded action- and after potentials of rat isolated atrium. Br J Pharmacol 118:407–413

Tavi P, Han C, Weckstrom M (1998) Mechanisms of stretch induced changes in $[Ca^{2+}]_i$ in rat atrial myocytes: role of increased troponin C affinity and stretch-activated ion channels. Circ Res 83:1165–1177

Testa M, Yeh M, Lee P et al (1996) Circulating levels of cytokines and their endogenous modulators in patients with mild to severe congestive heart failure due to coronary artery disease or hypertension. J Am Coll Cardiol 28:964–971

Trotter JL, Clifford DB, Anderson CB et al (1988) Elevated serum interleukin-2 levels in chronic progressive multiple sclerosis. N Engl J Med 318(18):1206

Thaik CM, Calderone A, Takahashi N, Colucci WS (1995) Interleukin-1 beta modulates the growth and phenotype of neonatal rat cardiac myocytes. J Clin Invest 96:1093–1099

Thakkar J, Tang SB, Sperelakis N, Wahler GM (1988) Inhibition of cardiac slow action potentials by 8-bromo-cyclic GMP occurs independent of changes in cyclic AMP levels. Can J Physiol Pharmacol 66:1092–1095

Ungureanu-Longrois D, Balligand JL et al (1995) Contractile responsiveness of ventricular myocytes to isoproterenol is regulated by induction of nitric oxide synthase activity in cardiac microvascular endothelial cells in heterotypic primary culture. Circ Res 77:486–493

Ungureanu-Longrois D, Bezie Y, Perret C, Laurent S (1997) Effects of exogenous and endogenous nitric oxide on the contractile function of cultured chick embryo ventricular myocytes. J Mol Cell Cardiol 29:677–687

Vaughan Williams EM (1984) A classification of antiarrhythmic actions reassessed after a decade of new drugs. J Clin Pharmacol 24:129–147

Virag L, Iost N, Opincariu M et al (2001) The slow component of the delayed rectifier potassium current in undiseased human ventricular myocytes. Cardiovasc Res 49:790–797

Wahler GM, Dollinger SJ (1995) Nitric oxide donor SIN-1 inhibits mammalian cardiac calcium current through cGMP-dependent protein kinase. Am J Physiol 268:C45–C54

Wan S, DeSmet JM, Barvais L, Goldstein M, Vincent JL, LeClerc JL (1996) Myocardium is a major source of proinflammatory cytokines in patients undergoing cardiopulmonary bypass. J Thorac Cardiovasc Surg 112:806–811

Wang J, Wang H, Zhang Y, Gao H, Nattel S, Wang Z (2004) Impairment of HERG K^+ channel function by tumor necrosis factor-alpha: role of reactive oxygen species as a mediator. J Biol Chem 279(14):13289–13292

Ward M-L, David G, Allen DG (2010) Stretch-activated channels in the heart: contribution to cardiac performance. In: Kamkin A, Kiseleva I (eds) Mechanosensitivity in cells and tissues 3. Mechanosensitivity of the heart. Springer, New York, NY, pp 141–167

Weisensee D, Bereiter-Hahn J, Schoeppe W, Low-Friedrich I (1993) Effects of cytokines on the contractility of cultured cardiac myocytes. Int J Immunopharmacol 15:581–587

Wong GH, Goeddel DV (1988) Induction of manganous superoxide dismutase by tumor necrosis factor: possible protective mechanism. Science 242:941–944

Wyeth RP, Temma K, Seifen E, Kennedy RH (1996) Negative inotropic actions of nitric oxide require high doses in rat cardiac muscle. Eur J Physiol 432:678–684

Yamauchi-Takihara K, Kishimoto T (2000) Cytokines and their receptors in cardiovascular diseases – role of gp130 signalling pathway in cardiac myocyte growth and maintenance. Int J Exp Pathol 81:1–16

Yokoyama T, Vaca L, Rossen RD, Durante W, Hazarika P, Mann DL (1993) Cellular basis for the negative inotropic effects of tumor necrosis factor-a in the adult mammalian cardiac myocytes. J Clin Invest 92(5):2303–2312

Yokoyama T, Sekiguchi K, Tanaka T, Tomaru K, Arai M, Suzuki T, Nagai R (1999) Angiotensin II and mechanical stretch induce production of tumor necrosis factor in cardiac fibroblasts. Am J Physiol 276(6 Pt 2):H1968–H1976

Yu X, Kennedy RH, Liu SJ (2003a) JAK2/STAT3, not ERK1/2, mediates interleukin-6-induced activation of inducible nitric oxide synthase and decrease in contractility of adult ventricular myocytes. J Biol Chem 278:16304–16309

Yu X-W, Kennedy RH, Liu SJ (2003b) JAK2/STAT3, not ERK1/2, mediates interleukin-6-induced activation of inducible nitric-oxide synthase and decrease in contractility of adult ventricular myocytes. J Biol Chem 278:16304–16308

Yu X-W, Chen Q, Kennedy RH, Liu SJ (2005a) Inhibition of sarcoplasmic reticular function by chronic interleukin-6 exposure via iNOS in adult ventricular myocytes. J Physiol (London) 566(2):327–340

Yu X, Patterson E, Huang S et al (2005b) Tumor necrosis factor α, rapid ventricular tachyarrhythmias, and infarct size in canine models of myocardial infarction. J Cardiovasc Pharmacol 45:153–159

Zeng T, Bett GCL, Sachs F (2000) Stretch-activated whole cell currents in adult rat cardiac myocytes. Am J Physiol 278:H548–H557

Zhang WM and Wong TM (1998) Suppression of cAMP by phosphoinositol/Ca^{2+} pathway in the cardiac kappa-opioid receptor. Am J Physiol 274(1 Pt 1):C82–C87

Zhang YH, Youm JB, Sung HK, Lee SH, Ryu SY, Lee S-H, Ho W-K, Earm YE (2000) Stretch-activated and background non-selective cation channels in rat atrial myocytes. J Physiol (London) 523:607–619

Zingman LV, Alekseev AE, Hodgson-Zingman DM, Terzic A (2007) ATP-sensitive potassium channels: metabolic sensing and cardioprotection. J Appl Physiol 103(5):1888–1893

Chapter 6
An Anti-inflammatory Cytokine Interleukin-13: Physiological Role in the Heart and Mechanoelectrical Feedback

Denis V. Abramochkin, Ekaterina Yu. Makarenko, Vadim M. Mitrochin, Bo Tian, Vlad S. Kuzmin, Leonid Kalugin, Leonid V. Kovalchuk, Marina V. Khoreva, Anna Nikonova, Ilya Lozinsky, and Andre Kamkin

Abstract The role of cytokines in responses not associated with inflammation as well as their involvement in regulation of non-haematopoietic cell activity is intensively studied during last decades. Control of heart activity can be carried out by pro-inflammatory cytokines. During the last decades considerable attention was drawn to the role of cytokines in physiological reactions not related to inflammation as well as their involvement in regulation of non-haematopoietic cell activity. An involvement of pro-inflammatory cytokines in the control of heart activity is thoroughly discussed in many publications. On the other hand such experimental data for the anti-inflammatory cytokines are currently absent. This review briefly summarizes existing evidences of involvement of anti-inflammatory cytokine interleukin-13 in the control of heart functioning and presents our latest findings on IL-13 influence on cardiomyocytes activity. According to our data, application of the IL-13 led to moderate acute changes in electrical activity of cardiomyocytes. At the same time it did not cause any electrical abnormalities, which is opposite to inflammatory cytokines application effects. Application of IL-13 reduced the effect of the mechanical stretch application on electrical activity of cardiomyocytes. Negative inotropic effect of anti-inflammatory IL-13 contrasts with positive inotropic effect of most pro-inflammatory cytokines. Special attention is given to possible mechanisms of IL-13-signaling and its influence on cardiac function in norm and pathology.

Keywords Heart · Cardiomyocytes · Stretch · Hump-like depolarization · Stretch-induced depolarization · APD90 · Anti-inflammatory cytokines · Interleukin-13

A. Kamkin (✉)
Department of Fundamental and Applied Physiology, Russian State Medical University, 117997 Moscow, Russia
e-mail: Kamkin.A@g23.relcom.ru; andre.kamkin@gmail.com

A. Kamkin, I. Kiseleva (eds.), *Mechanical Stretch and Cytokines*,
Mechanosensitivity in Cells and Tissues 5, DOI 10.1007/978-94-007-2004-6_6,

6.1 Introduction

Cytokines are polypeptide factors acting via specific membrane-bound receptors. They transduce a variety of physiological responses. They are used by cells of different types, including all resident cell types in the myocardium.

There is considerable data about influence of inflammatory cytokines on cardiac functions. On the other hand effects of anti-inflammatory cytokines are far less investigated. Interleukin-13, a most recently characterized anti-inflammatory cytokine, is produced predominately by T-helper type 2 cells (Lee et al., 1986; Minty et al., 1993). Its receptors are widely expressed, and consequently, this cytokine has many diverse effects on different cell types (Akaiwa et al., 2001; Lowenthal et al., 1988). IL-13 has recently been shown to possess a broad spectrum of biologic activities including action on non-hematopoietic cells. It shares many functional and structural properties with IL-4 but there are studies reporting that IL-13 has some unique functions in vivo independent of IL-4 (McKenzie and Zurawski, 1995; Wynn, 2003). Apart from the common receptor with IL-4, IL-13 has an additional receptor IL-13Rα2, which possesses antagonistic decoy functions in addition to unique signaling functions (Fichtner-Feigl et al., 2006).

This chapter briefly summarizes the data about the effects of anti-inflammatory cytokine interleukin-13 in the heart and presents our latest findings regarding IL-13 influence on cardiomyocytes activity.

6.2 Interleukin-13 in Normal and Failing Heart

There is no published data regarding direct effects of IL-13 on cardiac function. Nonetheless, this cytokine has some physiological effects that may further lead to modulation of cardiac activity. Interleukin-13 evidently inhibits the production of pro-inflammatory mediators by monocytes and macrophages, including IL-1, IL-6, IL-8, TNF, and IL-12 (de Vries, 1998). Pro-inflammatory cytokines in turn can impact myocardial function via effects on both myocytes contractility and the extracellular matrix (reviewed in Prabhu, 2004). In addition pro-inflammatory cytokines may exert pro-necrotic as well as pro-apoptotic action, being the primary reason for cardiomyocytes death in heart pathologies (reviewed in Pulkki, 1997). Decrease of pro-inflammatory cytokines level in cardiac pathologies prevents progressive cardiomyocyte loss, in particular by inhibiting cytokine-induced apoptosis.

A number of the cell types represented in heart is susceptible to IL-13. IL-4/IL-13 receptors are immunohistochemically detected in the different cells in the heart. According to data of Akaiwa and colleagues (2001), cardiomyocytes, smooth muscle of vessels and endothelial cells in the heart were positively stained by anti-IL-4-receptors antibody whereas intramuscular connective tissue and fat tissue in the epicardium were negative. In another study, positive immunohistochemical staining for IL-13Rα2 was observed in endomyocardial biopsy specimens from the failing human heart. Positive immunohistochemical staining for IL-13Rα1 was observed in myocardial specimens from both failing human hearts and control subjects (Nishimura et al., 2007). These findings suggest that interleukin-13 acts as a physiological regulator in normal and damaged heart.

The role of the immune system in the development and prognosis of cardiovascular diseases drawn great interest in recent years. In contrast to pro-inflammatory cytokines, there is limited information about the role of anti-inflammatory cytokines in cardiovascular disease. However results of the recent investigations indicate that IL-13 might act in an autoctine and paracrine manner to upregulate cardiomyocytes during at least some cardiological dysfunction (Nishimura et al., 2007, 2008).

Clinical data reveal correlation of alterations in interleukin-13 blood level with different heart pathologies. Plasma IL-13 levels were increased in chronical heart failure patients including ischemic heart disease, valvular heart disease, dilated cardiomyopathy and hypertrophic cardiomyopathy (Nishimura et al., 2007). Moreover, elevated serum IL-13 level in patient with dilated cardiomyopathy is correlated with electrocardiographic left ventricular end diastolic dimension and tissue fibrosis (Ohtsuka et al., 2005). In contrast, serum IL-13 level is reduced in patients with acute myocardial infarction (Jafarzadeh et al., 2009). According to recent data of Llaguno and colleagues (2011), serum IL-13 concentrations in patients with cardiac clinical forms of Chagas disease is increased. It's interesting to note that IL-4 concentration appeared to be decreased that further highlights the difference in physiological role of these cytokines. The same study reported direct correlations between IL-13 concentration and different spectral parameters of electrocardiogram related to the sympatho-vagal balance and negative correlation with parameters related to the nonlinear global cardiac function were demonstrated. IL-13 serum level positively correlated with percentage of area corresponding to the spectral bands of low-frequency components of ECG and negatively with percentage of recurrence, determinism, and Shannon entropy. Therefore, it becomes evident that IL-13 may not only be involved in normal heart work but also in development of pathologies.

6.3 Influence of Interleukin-13 on Cardiomyocytes Activity Under Normal and Stretching Conditions

Obviously mechanical stretch is a significant stimulus in the heart. Among many possible mediators, cytokines have been reported to be produced by cardiomyocytes during stretch exposure. Mechanical stretch induces interleukin-13 production in cardiomyocytes, which upregulates IL-13 receptor 2 expression. Thus, cyclic biomechanical stretch leads to more than fivefold elevation of IL-13Rα2 mRNA levels. In addition IL-13 protein release is significantly increased in the supernatant of cultured rat cardiomyocytes stretched for 24 h (Nishimura et al., 2008).

As mentioned above there is no data concerning influence of IL-13 on activity of intact or stretch-exposed cardiomyocytes. To address this issue we investigated the influence of interleukin-13 on electrical activity of rat cardiomyocytes under normal and stretching conditions. This was performed by means of standard microelectrode technique. Application of stretch stimulus was performed as previously described (Kamkin et al., 2000; Lozinsky and Kamkin, 2010). Interleukin-13 (Sigma) was added to perfusion solution (yielding concentration of 50 ng/ml).

It is important to note that IL-13 application did not cause any bioelectrical abnormalities in contrast to inflammatory cytokines, i.e. TNFα, IL-6 (Chapter 5, this volume). Application of the IL-13 led to moderate changes in bioelectrical activity in non-stretched cardiomyocytes. IL-13 caused the changes of the form, duration and rate of the action potential (Fig. 6.1). During 20–30 min period after application of IL-13 depolarization waves at resting potential level were observed. At the same time IL-13 application decreased firing rate in 65% of the cells (of registrations) by 18.3%. Negative inotropic effect of anti-inflammatory IL-13 contrasts with positive inotropic effect of most pro-inflammatory cytokines (reviewed in Prabhu, 2004).

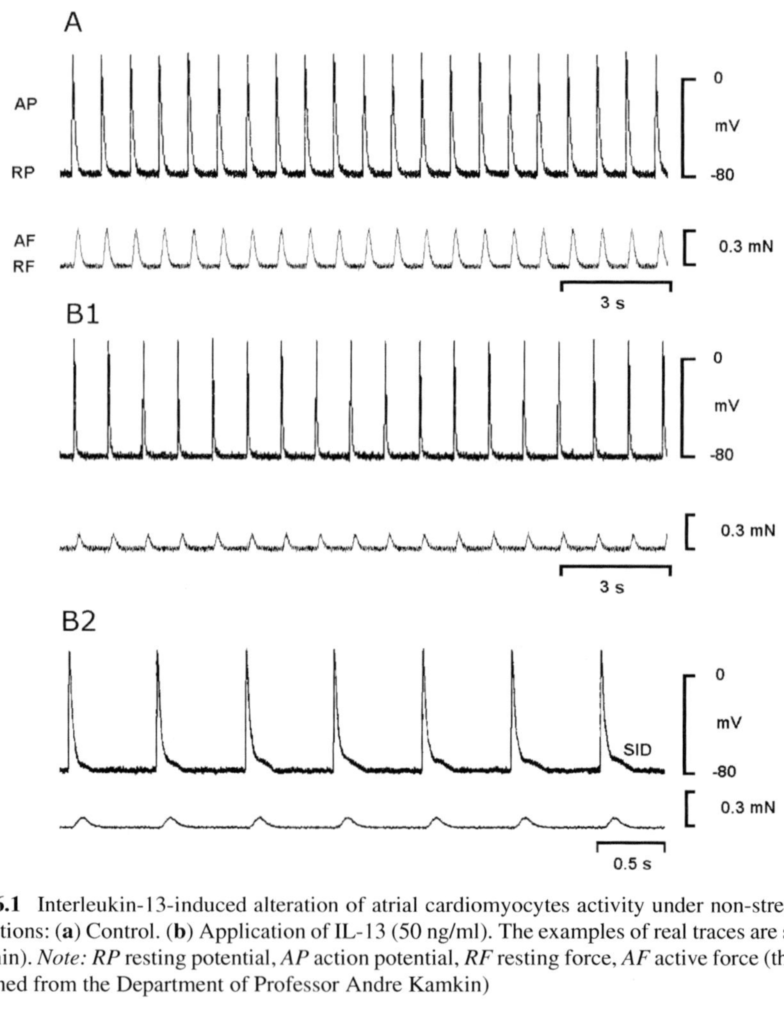

Fig. 6.1 Interleukin-13-induced alteration of atrial cardiomyocytes activity under non-stretching conditions: (**a**) Control. (**b**) Application of IL-13 (50 ng/ml). The examples of real traces are shown (24 min). *Note: RP* resting potential, *AP* action potential, *RF* resting force, *AF* active force (the data obtained from the Department of Professor Andre Kamkin)

Mechanical stretch induced hump-like depolarization (stretch-induced depolarization) and increased the value of APD90, in a manner similar to previously described (for details see Kamkin et al., 2000; Lozinsky and Kamkin, 2010).

IL-13 application had more pronounced effect on stretched cardiomyocytes. IL-13 decreased firing rate of 75% of the stretched cells by 13.2%. In 65% of the cells IL-13 caused marked reduction in duration of action potential: APD90 was decreased by 31.5%, APD50 – by 41%. Therefore, application of IL-13 reduced the effect of the mechanical stretch on bioelectrical activity of cardiomyocytes. Interestingly, this effect is similar to the effect of inflammation cytokines (Chapter 5, this volume). Below we discuss possible mechanisms of the observed effects.

6.4 Cell Signaling of Interleukin-13 and Possible Mechanism of Heart Function Regulation

The biologic responses induced by IL-13 require a complex interaction of signaling pathways and regulators. IL-13 has two cognate receptors, IL-13Rα1 and IL-13α2, which are members of the hematopoietin receptor superfamily (type I cytokine receptor family) and share 37% homology at the aminoacid sequence. Type I cytokine receptors are defined by several features, including four conserved cysteine residues, a W-S-X-W-S motif, fibronectin type II modules in the extracellular domain, and proline-rich box regions in the intracellular domain that are important for binding of Janus tyrosine kinases (JAK) (Leonard and Lin, 2000). The receptor complexes for interleukin-4 and interleukin-13 are intertwined systems that are likely regulated at multiple levels, including differential regulation of expression of the various components or preferential association of certain components, or both (reviewed in Jiang et al., 2000; Hershey, 2003).

Signaling through IL-4Rα/IL-13Rα1 is thought to occur through IL-4Rα because both IL-4 and IL-13 stimulation of the complex result in activation of signaling intermediates characterizing IL-4 response, including phosphorylation of IL-4Rα, insulin receptor substrates (IRS), JAK, and Tyk (Wills-Karp, 2001; Welham et al., 1995). The JAK/STAT is a direct signaling pathway, linking cytokines directly to the nucleus. Activation of JAKs results in phosphorylation of the cytoplasmic tyrosines in IL-4Rα, leading to the recruitment of signal transducer and activator of transcription-6 (STAT6) to the receptor, followed by STAT6 phosphorylation and activation. Activated STAT6 dimers then translocate to the nucleus, bind specific canonic DNA elements, and initiate transcription of downstream genes (Hershey, 2003). Insulin receptor substrate 1 (IRS-1) and a homologous protein IRS-2 (also known as 4-phosphotyrosine substrate) are recruited to phosphorylated IL-4Rα. That leads to phosphorylation and activation of IRS-1 and IRS-2. Two pathways have been implicated in signaling downstream of IRS-1 and IRS-2: the inositol-3-phosphate (IP3) kinase and Ras/mitogen-activated protein (MAP) kinase pathways (Jiang et al., 2000).

Intracellular IL-13 pathways are under complicated influences. A number of negative regulators of the JAK/STAT signaling pathway have been described.

Suppressor of cytokine signaling-1 (SOCS-1) is a molecule that can inhibit the activation of signaling through the inhibition of JAKs. The IL-4Rα–associated tyrosine kinase Fes was recently shown to be important in regulation of the IRS-1/IRS-2 pathway. Latest findings testify towards a role of Fes as a signaling intermediate between JAK1 and phosphorylation of IRS-1 and IRS-2. Fes has also been implicated as a regulator of JAK/STAT signaling (reviewed in Jiang et al., 2000; Hershey, 2003).

Experimental data suggests that JAK/STAT signaling is important for cardiac cells. It has been reported that specific expression of STAT3 in heart cells causes cardiohypertrophy in mice (Kunisada et al., 2000). Mechanical stretch being important physiological stimulus for heart cells causes activation of multiple second messenger systems. JAK/STAT pathway, a major signal transduction cascade of the cytokine superfamily, regulates cardiomyocyte activity under stretch exposure. The JAK/STAT pathway was activated by mechanical stretch in normal (Sadoshima and Izumo, 1997a, b; Pan et al., 1999) and pathological conditions (Mascareno et al., 2001). Moreover, certain level of intracellular calcium was necessary for stretch-induced activation of this pathway (Pan et al., 1999). Nishimura and colleagues (2008) showed that IL-13 treatment as well as mechanical stretch upregulated IL-13Rα2 mRNA expression, which was suppressed by STAT6 siRNA. This data suggest key role of intracellular JAK/STAT signaling in the regulation of the cardiomyocytes activity in different physiological conditions.

The regulation of nitric oxide synthase (NOS) activity is another intracellular mechanism of IL-13 signaling. As a vast majority of T-helper type 2 cytokines IL-13 causes a decrease in the expression of NOS and production of reactive oxygen species and short-lived nitrogenous intermediates (Doherty et al., 1993; Berkman et al., 1996; Bogdan et al., 1997). The inhibition of NOS mRNA expression by IL-13 is blocked by cycloheximide, suggesting that de novo protein synthesis is required for this effect (Berkman et al., 1996). In addition to reduction in NOS mRNA level, the expression of arginase 1 and 2 was is decreased (Chibana et al., 2008). Since these enzymes modulate the cellular levels of arginine, the substrate for NOS, IL-13 signaling down-regulates NOS-pathway at the multiple points.

Numerous experimental and clinical data conclusively established a key role of NOS-pathway, particularly of nitric oxide, in regulation of heart function in normal and pathological conditions. The effects of NO on myocardial functions including its pronounced influence on myocardium contraction and heart rhythm are described in another study (Kazanski et al., 2011). Moreover, a number of experiments conducted on whole hearts allows to suggest that NO and NO-synthases directly regulate the conductivity of mechanically gated channels (Kazanski et al., 2010a, b).

The regulation of cardiac activity by cytokines may be exerted not only by intracellular mechanisms but also via indirect modulation of extracellular regulatory pathways. We suggest that one of the most important mechanisms of IL-13 mediated regulation of cardiac activity is its influence on the concentration of other cytokines. Interleukin-13 inhibits the production of pro-inflammatory mediators, including IL-1, IL-6, IL-8, TNF, and IL-12 (de Vries, 1998). Pro-inflammatory cytokines in turn can impact myocardial function via effects on both myocyte contractility and the extracellular matrix (reviewed in Prabhu, 2004; for details see Chapters 2 and 5, this volume). In addition pro-inflammatory cytokines may exert pro-necrotic as well as pro-apoptotic action on cardiac myocytes in heart pathologies (reviewed in Pulkki, 1997). The signals leading to the death of the cardiomyocytes may carried by second messengers: nitric oxide (Kelly et al., 1996), reactive oxygen species (Gottlieb et al., 1994), calcium overload (Goldhaber et al., 1996).

Decrease of serum concentration of pro-inflammatory cytokines in cardiac pathologies may prevent progressive cardiomyocyte loss, in particular via inhibition of cytokine-induced apoptosis. A number of publications support the hypothesis of defensive role of IL-13 for cardiac tissue. In contrast to IL-4, IL-13 protects cardiac tissues in experimental pathologies. Elnaggar and colleagues (2005) demonstrated marked amelioration of rat experimental autoimmune myocarditis by delivery of an IL-13-Ig fusion gene vector. In the paper of Cihakova and colleagues (2008) also shown that IL-13 protected against myocarditis, induced by immunization of mice with cardiac myosin or by viral infection. IL-13-knockout mice displayed severe cardiac infiltration involving more than 50% of heart tissue in most mice, resulting in increased fibrosis, cardiac dysfunction, and increased mortality (Cihakova et al., 2008).

It is worth mentioning that cardioprotective action of interleukin-13 can be inverted, especially under prolonged exposure to this protein. IL-13 is known to be an important inducer of tissue fibrosis (Wynn, 2003). Myocardial fibrosis is critical in the progression of many cardiac diseases, for instance in heart failure. Cieslik and colleagues (2011) carried out a complex investigation of progressive fibrosis in the aging mouse heart and involvement of cytokines in progress of this pathology. In vitro studies demonstrated that IL-13 markedly enhanced monocyte–fibroblast transformation. In addition, they reported time-dependent increase in IL-13 mRNA in the aging mouse heart. Therefore, a potential modulatory role for IL-13 in age-dependent cardiac interstitial fibrosis may be suggested.

6.5 Conclusion and Perspectives

The regulation of cardiomyocyte activity by interleukin-13 involves intracellular signaling cascades and extracellular regulatory mechanisms (most important are represented on Fig. 6.2). The full spectrum of mechanisms includes short-term moderate acute effects on the electrical activity of cardiomyocytes and long-term indirect mechanisms of modulation of other regulatory factors including pro-inflammatory cytokines.

According to clinical and experimental data interleukin-13 possesses most prominent cardioprotective action (comparing with other anti-inflammatory cytokines, for instance, IL-4). It has cardiotropic effect and protects cardiac cells against pro-inflammatory cytokine–induced death. However, during the progress of heavy pathologies, associated with marked and sustained increase of IL-13 level, this cytokine may be a pro-fibrotic agent and may be related to pathogenesis of

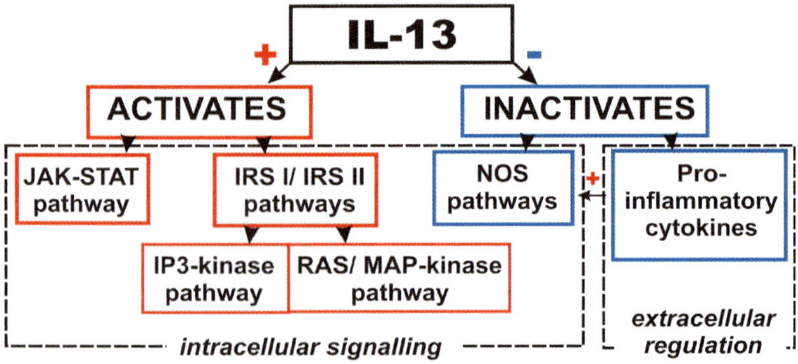

Fig. 6.2 Interleukin-13-mediated pathways that may be involved in the regulation of heart functioning. *Abbreviations*: *JAK* Janus tyrosine kinases; *STAT* signal transducer and activator of transcription; *IRS* insulin receptor substrate; *IP3-kinase* inositol-3-phosphate kinase; *Ras/MAP kinase* Ras/mitogen-activated protein kinase; *NOS* nitric oxide synthase

cardiac fibrosis. This bimodal effect is not unusual and is typical for biologically active regulators like, for example, nitric oxide.

Acknowledgements This work was supported by the Russian Foundation for Basic Research (grant No. 09-04-01277-a). Department of Fundamental and Applied Physiology (Professor and Chairman – Andre Kamkin) was supported by Ministry of Education and Science of the Russian Federation. The Order of Ministry of Education and Science of the Russian Federation №743 from 01 July 2010, Supplement, Event 4.4, the Period of Financing 2010–2019.

References

Akaiwa M, Yu B, Umeshita-Suyama R, Terada N, Suto H, Koga T, Arima K, Matsushita S, Saito H, Ogawa H, Furue M, Hamasaki N, Ohshima K, Izuhara K (2001) Localization of human interleukin 13 receptor in non-haematopoietic cells. Cytokine 13(2):75–84

Berkman N, Robichaud A, Robbins RA, Roesems G, Haddad EB, Barnes PJ, Chung KF (1996) Inhibition of inducible nitric oxide synthase expression by interleukin-4 and interleukin-13 in human lung epithelial cells. Immunology 89(3):363–367

Bogdan C, Thuring H, Dlaska M, Rollinghoff M, Weiss G (1997) Mechanism of suppression of macrophage nitric oxide release by IL-13: influence of the macrophage population. J Immunol 159:4506–4513

Chibana K, Trudeau JB, Mustovitch AT, Hu H, Zhao J, Balzar S, Chuw HW, Wenzel SE (2008) IL-13 induced increases in nitrite levels are primarily driven by increases in inducible nitric oxide synthase as compared with effects on arginases in human primary bronchial epithelial cells. Clin Exp Allergy 38:936–946

Cieslik KA, Taffet GE, Carlson S, Hermosillo J, Trial J, Entman ML (2011) Immune-inflammatory dysregulation modulates the incidence of progressive fibrosis and diastolic stiffness in the aging heart. J Mol Cell Cardiol 50(1):248–256

Cihakova D, Barin JG, Afanasyeva M, Kimura M, Fairweather D, Berg M, Talor MV, Baldeviano GC, Frisancho S, Gabrielson K, Bedja D, Rose NR (2008) Interleukin-13 protects against experimental autoimmune myocarditis by regulating macrophage differentiation. Am J Pathol 172(5):1195–1208

de Vries JE (1998) The role of IL-13 and its receptor in allergy and inflammatory responses. J Allergy Clin Immunol 102:165–169

Doherty TM, Kastelein R, Menon S, Andrade S, Coffman RL (1993) Modulation of murine macrophage function by IL-13. J Immunol 151:7151

Elnaggar R, Hanawa H, Liu H, Yoshida T, Hayashi M, Watanabe R, Abe S, Toba K, Yoshida K, Chang H, Minagawa S, Okura Y, Kato K, Kodama M, Maruyama H, Miyazaki J, Aizawa Y (2005) The effect of hydrodynamics-based delivery of an IL-13-Ig fusion gene for experimental autoimmune myocarditis in rats and its possible mechanism. Eur J Immunol 35(6): 1995–2005

Fichtner-Feigl S, Strober W, Kawakami K, Puri RK, Kitani A (2006) IL-13 signaling through the IL-13alpha2 receptor is involved in induction of TGF-beta1 production and fibrosis. Nat Med 12:99–106

Goldhaber JL, Kim KH, Natterson PD, Lawrence T, Vang P, Weiss JN (1996) Effects of TNF-alpha on $[Ca^{2+}]$, and contractility in isolated adult rabbit ventricular myocytes. Am J Physiol 271:H1449–H1455

Gottlieb RA, Burleson KO, Kloner RA, Babior BM, Engler RL (1994) Reperfusion injury induces apoptosis in rabbit cardiomyocytes. J Clin Invest 94:1621–1628

Jafarzadeh A, Esmaeeli-Nadimi A, Nough H, Nemati M, Rezayati MT (2009) Serum levels of interleukin (IL)-13, IL-17 and IL-18 in patients with ischemic heart disease. Anadolu Kardiyol Derg 9(2):75–83

Jiang H, Harris MB, Rothman P (2000) IL-4/IL-13 signaling beyond JAK/STAT. J Allergy Clin Immunol 105(6 Pt 1):1063–1070

Kamkin A, Kiseleva I, Wagner KD, Leiterer KP, Theres H, Scholz H, Günther J, Lab MJ (2000) Mechano-electric feedback in right atrium after left ventricular infarction in rats. J Mol Cell Cardiol 32:465–477

Kazanski VE, Kamkin AG, Makarenko EY, Lysenko NN, Sutiagin PV, Bo T, Kiseleva IS (2010a) Role of nitric oxide in activity control of mechanically gated ionic channels in cardiomyocytes: NO-donor study. Bull Exp Biol Med 150(1):1–5 (English)

Kazanski VE, Kamkin AG, Makarenko EY, Lysenko NN, Sutiagin PV, Kiseleva IS (2010b) Role of nitric oxide in the regulation of mechanosesnsitive ionic channels in cardiomyocytes: Contribution of NO-synthases. Bull Exp Biol Med 150(2):263–267 (English)

Kazanski V, Kamkin A, Makarenko E, Lysenko N, Lapina N, Kiseleva I (2011) The role of nitric oxide in the regulation of mechanically gated channels in the heart. In: Kamkin A, Kiseleva I (eds) Mechanosensitivity in cells and tissues 4. Mechanosensitivity and mechanotransduction. Springer, New York, NY, pp 109–140

Kelly RA, Balligand J-L, Smith TW (1996) Nitric oxide and cardiac function. Circ Res 79:363–380

Khurana Hershey GK (2003) IL-13 receptors and signaling pathways: an evolving web. J Allergy Clin Immunol 111:677–690

Kunisada K, Negoro S, Tone E, Funamoto M, Osugi T, Yamada S, Okabe M, Kishimoto T, Yamauchi-Takihara K (2000) Signal transducer and activator of transcription 3 in the heart transduces not only a hypertrophic signal but a protective signal against doxorubicin-induced cardiomyopathy. Proc Natl Acad Sci USA 97:315–319

Lee F, Yokota T, Otsuka T, Meyerson P, Villaret D, Coffman R, Mosmann T, Rennick D, Roehm N, Smith C, Zlotnik A, Arai K (1986) Isolation and characterization of a mouse interleukin cDNA clone that expresses B-cell stimulatory factor 1 activities and T-cell- and mast-cell-stimulating activities. Proc Natl Acad Sci USA 83:2061

Leonard WJ, Lin JX (2000) Cytokine receptor signaling pathways. J Allergy Clin Immunol 105:877–888

Llaguno M, Pertili LA, Silva MV, Bunazar P, Reges AM, Faleiros AC, Lages-Silva E, Rodrigues Junior V, Silva VJ, Filho DC (2011) The relationship between heart rate variability and serum cytokines in chronic chagasic patients with persistent parasitemia. Pacing Clin Electrophysiol 1:1–12

Lowenthal JW, Castle BE, Christiansen J, Schreurs J, Rennick D, Arai N, Hoy P, Takebe Y, Howard M (1988) Expression of high affinity receptors for murine interleukin 4 (BSF-1) on hemopoietic and nonhemopoietic cells. J Immunol 140(2):456–464

Lozinsky I, Kamkin A (2010) Mechanosensitive alterations of action potentials and membrane currents in healthy and diseased cardiomyocytes: cardiac tissue and isolated cell. In: Andre K, Irina K (eds) Mechanosensitivity in cells and tissues 3. Mechanosensitivity of the heart. Springer, New York, NY, pp 185–238

Mascareno E, El-Shafei M, Maulik N, Sato M, Guo Y, Das DK, Siddiqui MAQ (2001) JAK/STAT signaling is associated with cardiac dysfunction during ischemia and reperfusion. Circulation 104:325–329

McKenzie AN, Zurawski G (1995) Interleukin-13: characterization and biologic properties. Cancer Treat Res 80:367–378

Minty A, Chalon P, Derocq JM, Dumont X, Guillemot JC, Kaghad M, Labit C, Leplatois P, Liauzun P, Miloux B (1993) Interleukin-13 is a new human lymphokine regulating inflammatory and immune responses. Nature 362(6417):248–250

Nishimura Y, Inoue T, Nitto T, Morooka T, Node K (2007) Increased interleukin-13 levels in patients with chronic heart failure. Int J Cardiol 131(3):421–423

Nishimura Y, Inoue T, Nitto T, Morooka T, Node K (2008) Mechanical stretch and angiotensin II increase interleukin-13 production and interleukin-13 receptor α2 expression in rat neonatal cardiomyocytes. Circ J 72:647–653

Ohtsuka T, Inoue K, Hara Y, Morioka N, Ohshima K, Suzuki J, Ogimoto A, Shigematsu Y, Higaki J (2005) Serum markers of angiogenesis and myocardial ultrasonic tissue characterization in patients with dilated cardiomyopathy. Eur J Heart Fail 7(4):689–695

Pan J, Fukuda K, Saito M, Matsuzaki J, Kodama H, Sano M, Takahashi T, Kato T, Ogawa S (1999) Mechanical stretch activates the JAK/STAT pathway in rat cardiomyocytes. Circ Res 84:1127–1136

Prabhu SD (2004) Cytokine-induced modulation of cardiac function. Circ Res 95:1140–1153

Pulkki KJ (1997) Cytokines and cardiomyocyte death. Ann Med 29(4):339–343

Sadoshima J, Izumo S (1997a) The cellular and molecular response of cardiac myocytes to mechanical stress. Annu Rev Physiol 59:551–571

Sadoshima J, Izumo S (1997b) Tyrosine kinases mediation of c-fos expression by cell swelling in cardiac myocytes. Heart Vessels (Suppl 12):194–197

Welham MJ, Learmonth L, Bone H, Schrader JW (1995) Interleukin-13 signal transduction in lymphohemopoietic cells. Similarities and differences in signal transduction with interleukin-4 and insulin. J Biol Chem 270:12286–12296

Wills-Karp M (2001) IL-12/IL-13 axis in allergic asthma. J Allergy Clin Immunol 107:9–18

Wynn TA (2003) IL-13 effector functions. Annu Rev Immunol 21:425–456

Chapter 7
Mechanical Stretch and Cytokine Synthesis in Pulmonary Endothelial Cells

Satoru Ito and Yoshinori Hasegawa

Abstract The respiratory system is dynamically exposed to physical forces derived from tidal breathing and blood flow. These physical forces are important for maintaining homeostasis of the respiratory system and normal lung development in vivo. In contrast, excessive mechanical forces lead to the pathogenesis of pulmonary diseases. Pulmonary endothelial cells serve as a barrier that regulates fluid balance and inflammatory cell accumulation in the lung. The functions of pulmonary endothelial cells are significantly affected by mechanical stimuli such as shear stress and mechanical stretch. For example, pulmonary endothelial cells produce cytokines/chemokines and nitric oxide in response to mechanical stimuli. The mechanotransduction pathways for mechanical force transmission into the intracellular signals are through mechanosensitive ion channels, integrins, actin cytoskeleton, and activation of mitogen-activated protein-kinases. Alterations in pulmonary endothelial cell properties due to excessive mechanical stress are suggested to play an important role in the disease progression of lung injury and inflammation. This chapter focuses on recent evidence regarding regulation of mechanical stress-induced cytokine/chemokine production by pulmonary endothelial cells and the role of pulmonary endothelial cells in the pathogenesis of pulmonary diseases related to mechanical stress.

Keywords ARDS · Mechanotransduction · Stretch · p38 · Ventilator-induced lung injury · Biotrauma · IL-8 · Pulmonary microvascular endothelial cells

Abbreviations

ALI	acute lung injury
ARDS	acute respiratory distress syndrome
$[Ca^{2+}]_I$	intracellular Ca^{2+} concentration
COPD	chronic obstructive pulmonary disease
DAD	diffuse alveolar damage
EC	endothelial cell

S. Ito (✉)
Department of Respiratory Medicine, Nagoya University Graduate School of Medicine, Showa-ku, Nagoya 466-8550, Japan
e-mail: itori@med.nagoya-u.ac.jp

A. Kamkin, I. Kiseleva (eds.), *Mechanical Stretch and Cytokines*, Mechanosensitivity in Cells and Tissues 5, DOI 10.1007/978-94-007-2004-6_7,

ECM	extracellular matrix
ELISA	enzyme-linked immunosorbent assay
eNOS	endothelial cell nitric oxide synthase
EPC	endothelial progenitor cell
ERK	extracellular signal-regulated kinase
GM-CSF	granulocyte/macrophage colony stimulating factor
HUVEC	human umbilical vein endothelial cell
IL	interleukin
JNK	c-Jun NH2-terminal kinase
LPS	lipopolysaccharide
LVRS	lung volume reduction surgery
MAP-kinase	mitogen-activated protein-kinase
MCP-1	monocyte chemoattractant protein-1
NO	nitric oxide
PECAM-1	platelet/endothelial cell adhesion molecule-1
TNF-α	tumor necrosis factor-α
TRP	transient receptor potential
VEGF	vascular endothelial growth factor
VILI	ventilator-induced lung injury

7.1 Introduction

The pulmonary endothelial cells (ECs) consist of three diverse types: artery, capillary (microvascular), and vein ECs. These pulmonary ECs have a variety of functions including regulation of vascular smooth muscle tone, host-defense response, angiogenesis, prevention of thrombosis, tissue fluid homeostasis, and inflammatory responses in the lung (Nilius and Droogmans, 2001; Mehta and Malik, 2006). Like all other cell types, pulmonary ECs are able to sense chemical ligands as well as mechanical stimuli such as shear stress and mechanical stretch (Chien, 2007; Ando and Yamamoto, 2009). The pulmonary EC functions regulated by physical forces involve endothelial permeability, cell orientation, cell differentiation, nitric oxide (NO) production, and cytokine/chemokine release (Rubanyi et al., 1986; Awolesi et al., 1995; Davies, 1995; Uematsu et al., 1995; Ando and Yamamoto, 2009).

Mechanical stress is important for maintaining homeostasis of the respiratory system as well as normal lung development in vivo. Nevertheless, excessive physical forces are known to contribute to the pathogenesis of pulmonary diseases in certain conditions. One example of a pulmonary disease caused by mechanical stress is ventilator-induced lung injury (VILI) (Uhlig and Uhlig, 2004; Vlahakis and Hubmayr, 2005; Dos Santos and Slutsky, 2006). Mechanical ventilation is often used to maintain adequate oxygenation in patients with acute lung injury (ALI) and acute respiratory distress syndrome (ARDS), which is characterized by acute respiratory failure that results from non-cardiac pulmonary edema and systemic inflammation (ARDSnet, 2000). However, the mechanically ventilated lung may be exposed to excessive cyclic mechanical strain that causes further lung damage and alveolar-capillary barrier dysfunction (West and Mathieu-Costello, 1992; Artigas et al., 1998; Dreyfuss and Saumon, 1998; Uhlig and Uhlig, 2004; Vlahakis and Hubmayr, 2005; Dos Santos and Slutsky, 2006).

Pulmonary microvascular ECs are considered one of the primary targets of stretch during mechanical ventilation because they form the alveolar-capillary barrier with alveolar epithelial cells (Haseneen et al., 2003; Marini et al., 2003; West, 2003). The pulmonary microvascular ECs localized at the alveolar wall have the vital function of regulating the tissue fluid balance in the lung (Mehta and Malik, 2006). When pulmonary microvascular ECs are exposed to repeated cyclic stretch in vitro, they synthesize inflammatory cytokines and chemokines including interleukin (IL)-6, IL-8, and the CC-chemokine monocyte chemoattractant protein-1 (MCP-1), all of which are involved in the pathogenesis of ALI/ARDS and VILI (Miller et al., 1992; Dreyfuss and Saumon, 1998; Ranieri et al., 1999; Oudin and Pugin, 2002; Puneet et al., 2005; Iwaki et al., 2009). These findings highlight the importance of cytokine/chemokine production by pulmonary microvascular ECs in biotrauma as a mechanism of VILI (Dos Santos and Slutsky, 2006).

7.2 Endothelial Cells

7.2.1 Heterogeneity of Pulmonary Endothelial Cells

Vascular ECs are defined by the ability to take up LDL cholesterol and to express a variety of endothelial markers, including VIII/von Willebrand factor, CD31 (platelet/endothelial cell adhesion molecule-1; PECAM-1), vascular endothelial cadherin (VE-cadherin), and endothelial cell nitric oxide synthase (eNOS). However, pulmonary ECs exhibit diversity in structure and functions among artery, capillary (microvascular), and vein ECs (King et al., 2004; Adkison et al., 2006; Stevens et al., 2008). It is known that macrovascular and micovascular ECs can be distinguished by their distinctive lectin bindings (Gebb and Stevens, 2004; King et al., 2004). Microvascular ECs uniquely bind *Griffonia simplicifolia*, a lectin that specifically interacts with α-galactose (Magee et al., 1994; Gebb and Stevens, 2004; King et al., 2004). In contrast, macrovascular ECs such as pulmonary artery ECs bind *Helix pomatia*, a lectin that interacts with α- and β-*N*-acetyl-galactosamine (Gebb and Stevens, 2004; King et al., 2004).

Functional differences between pulmonary artery ECs and pulmonary microvascular ECs have been demonstrated (King et al., 2004). Pulmonary microvascular ECs possess a more restrictive barrier than pulmonary artery ECs (Parker and Yoshikawa, 2002). A previous in vitro study demonstrated that when confluent monolayers of ECs are stimulated by thapsigargin, an inhibitor of sarcoendoplasmic reticulum Ca^{2+}-ATPase, which elicits store-operated Ca^{2+} entry, the permeability of FITC-labeled dextran was increased in pulmonary artery ECs but not in pulmonary microvascular ECs (Gebb and Stevens, 2004). Pulmonary microvascular ECs also possess a greater proliferative ability than pulmonary arterial ECs. King et al. (2004) reported that pulmonary microvascular ECs grow approximately two times faster in response to 10% fetal bovine serum than pulmonary artery ECs in rats. It is known that adult (postnatal) endothelium does not actively proliferate in the intact mature blood vessel. However, injured or senescent ECs must be continually replenished to maintain homeostasis of the vascular system (Aird, 2007a, b). In general, microvascular ECs have an intrinsic ability to proliferate at high rates, which is essential for vascular homeostasis and repair (Carmeliet, 2003).

In adults, ECs differentiate from endothelial progenitor cells (EPCs) which have been isolated from the bone marrow, blood stream, and intima or adventitia of vascular walls (Asahara et al., 1997; Reyes et al., 2002; Carmeliet, 2003). Asahara et al. (1999) found circulating EPCs expressing both endothelial surface markers and leukocyte markers (CD14 and CD45). Alvarez et al. (2008) demonstrated that pulmonary microvascular ECs obtained from rat lungs express both endothelial markers (CD31, CD144, eNOS, and von Willebrand factor) and progenitor cell antigens (CD34 and CD309) but not the leukocyte marker CD45. In that report, these EPC-like "resident pulmonary microvascular EPCs" exhibited highly proliferative behavior and formed vascular networks in an in vitro Matrigel assay (Alvarez et al., 2008). Thus, it is likely that pulmonary microvascular ECs are enriched with EPCs that display vasculogenic ability while maintaining functional microvascular specificity.

7.2.2 Role of Mechanical Stress in Endothelial Cell Functions

Each EC senses both biochemical (e.g. cytokines, chemokines, growth factors, hormones, and lipid mediators) and biophysical signals (e.g. fluid shear stress, cyclic stretch, and blood pressure) from the surrounding microenvironment. Many functional properties of ECs are dynamically regulated by these mechanical forces in both physiological and pathological conditions (Davies, 1995; Chien, 2007). The effects of physical forces on EC functions have been investigated intensively using a device that can apply stretch or fluid shear stress to cultured ECs or isolated vascular tissues. The EC properties affected by physical forces involve morphology, vascular tone, anti-thrombotic activity, cell differentiation, mediator release, and gene expression (Ando and Yamamoto, 2009). The best known EC function regulated by fluid shear stress and cyclic stretch is NO production via eNOS activation, which causes vascular dilation (Rubanyi et al., 1986; Awolesi et al., 1995; Uematsu et al., 1995). The endogenous NO production by pulmonary ECs is important for maintaining the relatively lower vascular resistance in the lung. An in situ study by Kuebler et al. clearly demonstrated that NO production is regulated by shear stress and cyclic stretch in pulmonary microvascular ECs of the isolated lung from rats and mice (Kuebler et al., 2003). Loss of shear stress due to ischemia also up-regulates

NO production by altering intracellular Ca^{2+} signaling in pulmonary microvascular ECs isolated from rats (Al-Mehdi et al., 2000).

Mechanical stimulation of ECs alters the secretion of another vasodilator, prostacyclin, and the vasoconstrictor endothelin-1 (Frangos et al., 1985; Kuchan and Frangos, 1993). cDNA microarray analysis demonstrated that gene expression patterns of heat shock 10-kDa protein-1 (HSP-10), thrombomodulin, and catalase are regulated by cyclic stretch in human umbilical vein ECs (HUVECs) (Frye et al., 2005). Similarly, gene expression profiling by cDNA microarray analysis is significantly affected by fluid shear stress in human aortic ECs (Brooks et al., 2002).

7.2.3 Mechanical Stretch Induces Cytokine Production by Pulmonary Endothelial Cells

The microvasculature of the lung is involved in the recruitment of inflammatory cells into the inter-alveolar space and is therefore crucial in inflammatory lung diseases. It is well established that vascular ECs produce a variety of cytokines and chemokines such as IL-1, IL-6, IL-8, IL-18, tumor necrosis factor (TNF)-α, MCP-1, and granulocyte/macrophage colony stimulating factor (GM-CSF) in response to both chemical and physical stimuli (Wung et al., 1996; Okada et al., 1998; Kawai et al., 2002; Kobayashi et al., 2003; Minami and Aird, 2005; Sasamoto et al., 2005; Ando and Yamamoto, 2009). Human pulmonary microvascular ECs also produce IL-8, MCP-1, and GM-CSF in response to TNF-α, lipopolysaccharide (LPS), or IL-1β (Beck et al., 1999; Burg et al., 2002). These inflammatory cytokines and chemokines play important roles in the pathogenesis of ALI/ARDS (Puneet et al., 2005).

The pulmonary microvasculature is exposed to cyclic stretch during the tidal breathing necessary for gas exchange. Moreover, cyclic stretch is the predominant mechanical stress for pulmonary microvascular ECs in the mechanically ventilated lung. Experimental studies using cultured cells have demonstrated that mechanical stretch induces synthesis of cytokines and chemokines in pulmonary microvascular ECs. When human pulmonary microvascular ECs cultured on fibronectin-coated elastic silicone membranes were stretched uniaxially (20% strain, 50 cycles/min) using a cell stretch apparatus in vitro, the concentrations of IL-6, IL-8 and MCP-1 in the cell culture medium as assessed by enzyme-linked immunosorbent assay (ELISA) were significantly increased (Iwaki et al., 2009). In contrast, IL-1β production was not observed in either the static or stretched condition in human pulmonary microvascular ECs. The increases in IL-8 mRNA expression and protein synthesis were enhanced by application of 20% stretch in a time-dependent manner. When human pulmonary microvascular ECs were exposed to cyclic stretch with 5 or 20% amplitude for 12 h, the IL-8 production was significantly higher in the cells exposed to the larger (20%) strain than the lower (5%) strain (Iwaki et al., 2009). Consistent with these results, Pinhu et al. (2008) reported that cyclic stretch (30% elongation) induced IL-8 mRNA expression in human pulmonary microvascular ECs. These in vitro findings in pulmonary microvascular ECs indicate that large stretching of the lung enhanced pulmonary inflammation via cytokine/chemokine production in mechanically ventilated patients with ALI/ARDS in vivo (Dreyfuss and Saumon, 1998; Ranieri et al., 1999; Li et al., 2009).

Fluid shear stress is another essential mechanical force for maintaining vasculature homeostasis (Davies, 1995; Ando and Yamamoto, 2009). ECs are not only able to sense the fluid shear stress but also to discriminate among distinct patterns of flow. It has been shown that fluid shear stress increases production of MCP-1 and GM-CSF in HUVECs (Shyy et al., 1994; Kosaki et al., 1998) and IL-1 and IL-6 in bovine aortic ECs (Sterpetti et al., 1993). In bovine pulmonary microvascular ECs, a high pulsatile flow induced mRNA expression of MCP-1, adhesion molecules (ICAM-1, E-selectin), vascular endothelial growth factor (VEGF) and its receptor Flt-1 (Li et al., 2009). Therefore, altered EC responses to fluid shear stress may be relate to the pathophysiology of certain pulmonary diseases such as ALI/ARDS, pulmonary embolism, and pulmonary hypertension.

7.2.4 Role of p38 MAP-Kinase in Stretch-Induced IL-8 Production in Pulmonary Microvascular Endothelial Cells

The mechanisms of mechanical force sensing (mechanosensing) and conversion of these forces into intracellular biochemical signals (mechanotransduction) involve membrane extracellular matrix (ECM) receptors (e.g. integrins), focal adhesions, actin cytoskeleton, mechano-sensitive ion channels, GTP-binding proteins, reactive oxygen species, cyclooxygenase, and many protein kinases (Hamill and Martinac, 2001; Ali et al., 2006; Ito et al., 2006a, 2008, 2010; Chien, 2007; Copland and Post, 2007; Cohen et al., 2010;). Although the specific mechanosensor or mechanoreceptor in ECs has not been identified yet, the ECM-integrins-cytoskeleton pathway is considered to act as a mechanosensing complex (Ingber, 1993; Ali and Schumacker, 2002).

The family of mitogen-activated protein-kinases (MAP-kinases) involves p38 MAP-kinase, extracellular signal-regulated kinase (ERK)1/2, and c-Jun NH2-terminal kinase (JNK). In general, IL-8 production is regulated by activation of the MAP-kinase family (Hoffmann et al., 2002). Hashimoto et al. (1999) demonstrated that activation of p38 MAP-kinase regulates TNF-α- and IL-1α-induced IL-8 production in human pulmonary artery ECs. The MAP-kinases are representative protein kinases involved in mechanotransduction processes in ECs (Tseng et al., 1995; Kito et al., 2000; Oudin and Pugin, 2002; Chien, 2007; Iwaki et al., 2009). The activation of p38 MAP-kinase regulates the IL-8 synthesis induced by cyclic stretch in human pulmonary microvascular ECs (Iwaki et al., 2009). In human pulmonary microvascular ECs, IL-8 synthesis induced by cyclic stretch (20%, 12 h) was significantly blocked by the p38 inhibitor SB203580, whereas neither the ERK1/2 inhibitor U0126 nor JNK inhibitor SP600125 affected it (Iwaki et al., 2009). Moreover, 20% cyclic stretch transiently upregulated phosphorylation of p38 MAP-kinase, demonstrating that p38 activation is essential to initiate IL-8 synthesis following cyclic stretching in human pulmonary microvascular ECs (Fig. 7.1). In contrast, neither JNK nor ERK1/2 is involved in these mechanisms.

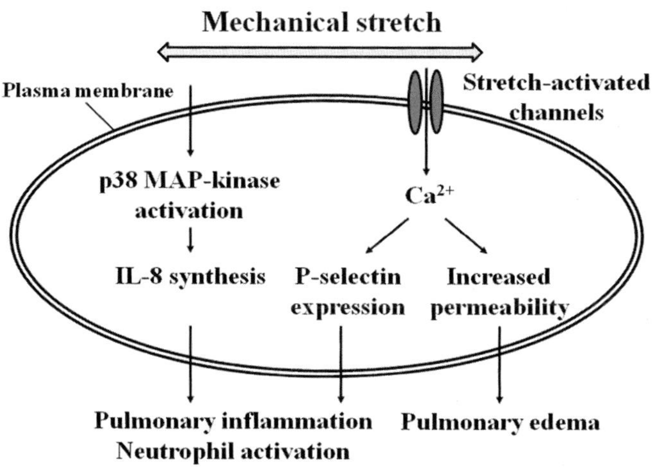

Fig. 7.1 Mechanotransduction pathways activated by mechanical stretch in pulmonary microvascular endothelial cells. Activation of p38 MAP-kinase is involved in IL-8 synthesis. It is considered that Ca^{2+} influx via stretch-activated channels induces P-selectin expression and increases endothelial permeability

It is well known that transcription of the IL-8 gene is regulated by nuclear factor (NF)-κB (Blackwell and Christman, 1997; Hoffmann et al., 2002). Mechanical stretch of human pulmonary microvascular ECs induces translocation of p65/RelA into the nucleus (Iwaki et al., 2009), which indicates NF-κB activation (Blackwell and Christman, 1997). However, the IL-8 production induced by cyclic stretch is not inhibited by NF-κB inhibitors, either SN50 or ammonium pyrrolidinedithiocarbamate. Because IL-8 gene expression is regulated by multiple transcription factors (Hoffmann et al., 2002), transcription factors other than NF-κB may be involved in the stretch-induced IL-8 gene expression in human pulmonary microvascular ECs. Additionally, it is possible that activation of p38 MAP-kinase regulates IL-8 production via stabilization of IL-8 mRNA rather than inducing mRNA synthesis (Hoffmann et al., 2002).

7.2.5 Stretch-Activated Calcium Channels in Pulmonary Endothelial Cells

An increase in intracellular Ca^{2+} concentration ($[Ca^{2+}]_i$) is essential for a number of important EC functions such as NO production and vascular permeability to maintain normal vascular homeostasis (Nilius and Droogmans, 2001; Wu et al., 2005). On the other hand, an excessive $[Ca^{2+}]_i$ elevation induced by ionomycin or thrombin promotes EC permeability and edema formation in the lung (Garcia et al., 1997; Mehta and Malik, 2006). As in other mammalian cell types, pulmonary ECs possess mechanosensitive ion channels including Ca^{2+}-permeable stretch-activated cation channels (Naruse et al., 1998; Nilius and Droogmans, 2001; Boriek and Kumar, 2008; Hayakawa et al., 2008; Ito et al., 2010). Uni-axial stretching of human pulmonary microvascular ECs in vitro increases $[Ca^{2+}]_i$, which are inhibited by the extracellular Ca^{2+}-free condition or gadolinium ion (Gd^{3+}), a potent inhibitor of stretch-activated channels (Ito et al., 2010). This Ca^{2+} influx is activated by a large ($\geq$20%) strain and is increased in a strain-dependent manner (Ito et al., 2010). "Part I. Mechanosensitive Channels, Mechanosensitivity in Cells and Tissues Vol. 1, edited by Kamkin and Kiseleva" provides a more detailed description of mechanosensitive ion channels. Regulation of stretch-activated Ca^{2+} channels by the actin cytoskeleton has been reported as one of the mechanisms of ECs (Hayakawa et al., 2008; Ito et al., 2010). The stretch-activated Ca^{2+} influx was inhibited by actin disruption with cytochalasin D and enhanced by increased actin polymerization with jasplakinolide or sphingosine-1-phosphate (Ito et al., 2010). On the other hand, the increases in $[Ca^{2+}]_i$ due to Ca^{2+} influx and release induced by thapsigargin or thrombin were preserved in human pulmonary microvascular ECs (Ito et al., 2010). Therefore, stretch-activated Ca^{2+} channels and other Ca^{2+} mobilizing pathways, specifically receptor-operated channels and store-operated channels (Nilius and Droogmans, 2001; Yao and Garland, 2005), are differently regulated by the actin cytoskeleton in human pulmonary microvascular ECs.

In spite of the crucial role of $[Ca^{2+}]_i$ elevation in the neuropeptide-induced IL-8 production in the NCM460 colon epithelial cell line (Zhao et al., 2001), increased $[Ca^{2+}]_i$ via the stretch-activated channel is not involved in the mechanisms of stretch-induced IL-8 production in human pulmonary microvascular ECs (Fig. 7.1). The IL-8 production induced by cyclic stretch (20%, 12 h) is not inhibited by the blockade of Ca^{2+} influx pathways with the Ca^{2+} chelator EGTA or Gd^{3+} in human pulmonary microvascular ECs (Iwaki et al., 2009).

There has been some evidence for a contribution by increases in $[Ca^{2+}]_i$ to lung injury induced by physical forces. An ex vivo study demonstrated that elevation of hydrostatic pressure to 20 cmH_2O increased endothelial $[Ca^{2+}]_i$ with exocytosis of the leukocyte adhesion receptor P-selectin in a single lung venular capillary isolated from the rat (Kuebler et al., 1999). Endothelial expression of P-selectin increases the number of rolling leukocytes on the vessel surface, indicating that P-selectin expression is related to the pathogenesis of lung injury (Mulligan et al., 1992). Parker et al. (1998) demonstrated that Gd^{3+} prevented the increase in vascular permeability induced by high airway pressure ventilation in isolated rat lung. Furthermore, both in vivo and ex vivo studies in rodents have shown that inhibition of TRPV-4, a subfamily of transient receptor potential (TRP) cation channels, reduced the lung injury induced by mechanical ventilation (Alvarez et al., 2006; Hamanaka et al., 2007; Jian et al., 2008). These findings suggest that excessive $[Ca^{2+}]_i$ elevation via stretch-activated channels, possibly TRPV-4, in response to physical forces may contribute to pulmonary edema and inflammation as the pathogenesis of physical force-induced lung injury (Kuebler et al., 1999; Hamanaka et al., 2007). Nevertheless, the specific TRP genes for putative stretch-activated Ca^{2+} channels have not been identified in ECs yet. Collectively, $[Ca^{2+}]_i$ elevation via stretch-activated Ca^{2+} channels is responsible for endothelial permeability and P-selectin expression as a mechanotransduction signal but is not involved in IL-8 production in pulmonary ECs (Fig. 7.1).

7.2.6 Induction of Endothelial Cell Alignment by Cyclic Stretch

Cyclic stretch and fluid shear stress modulate the cell morphology of ECs (Naruse et al., 1998; Ando and Yamamoto, 2009). In general, longitudinal axes of cultured EC monolayers are distributed randomly in vitro. In contrast, ECs, including human pulmonary microvascular ECs, subjected to cyclic stretch are elongated, and their long axes are aligned perpendicularly to the stretch axis and accompanied by actin polymerization (Naruse et al., 1998; Kaunas et al., 2005; Iwaki et al., 2009). There is no significant orientation of stress fibers following uni-axial stretching of less than 3% in amplitude in bovine aortic ECs (Kaunas et al., 2005). Both in vivo and in vitro studies have demonstrated that the fluid shear stress induces EC alignment parallel to the direction of fluid flow with actin remodeling (Langille and Adamson, 1981; Ando and Yamamoto, 2009). Taken together, reorganization and remodeling of the actin cytoskeleton is important for determining EC alignment in response to cyclic stretch and shear stress.

Several pathways are proposed for the mechanotransduction responsible for cell alignment induced by cyclic stretch or shear stress. In bovine pulmonary arterial ECs, p38 MAP-kinase regulated shear stress-induced cell alignment but not cyclic stretch-induced cell reorientation (Kito et al., 2000; Azuma et al., 2001). Similarly, the stretch-induced cell reorientation was not affected by p38 MAP-kinase inhibition in human pulmonary microvascular ECs either (Iwaki et al., 2009). Ca^{2+} influx via the stretch-activated cation channels is involved in the cyclic stretch-induced cell alignment of ECs (Naruse et al., 1998; Thodeti et al., 2009). Thodeti et al. (2009) demonstrated that inhibition of TRPV-4 using a specific small interfering RNA (siRNA) and the pharmacological inhibitor ruthenium red suppressed the cell alignment induced by cyclic stretch in bovine adrenal cortex capillary ECs. Another study reported that fluid shear stress-mediated cell alignment was regulated by Ca^{2+} influx via P2X4, a subtype of ATP-operated cation channel P2X purinoceptors, in HUVECs (Yamamoto et al., 2000). Although the role of stretch-induced reorientation of pulmonary ECs in the pathophysiology of pulmonary diseases remains unclear, the capacity of ECs to realign in response to the direction of physical forces is considered to contribute to vascular homeostasis, wound repair, angiogenesis, and pathogenesis (Chien, 2007), hence uncovering the molecular mechanisms underlying reorientation should contribute to a better understanding and improving lung repair during mechanical ventilation considered next.

7.3 Ventilator-Induced Lung Injury

7.3.1 Mechanical Ventilation in Acute Lung Injury and ARDS

ALI and ARDS are characterized by acute respiratory failure that results from non-cardiac pulmonary edema and systemic inflammation caused by direct (e.g., pneumonia or acid aspiration) or indirect pulmonary injury (e.g., sepsis, pancreatitis, or severe trauma) (Ashbaugh et al., 1967; Tate and Repine, 1983; Bernard et al., 1994; ARDSnet, 2000; Matthay et al., 2003). Despite improvements in the management and treatment of ARDS, mortality rates in patients with ARDS remain high (approximately 40%) due to severe respiratory failure and multiple organ failure (Kollef and Schuster, 1995; Slutsky and Tremblay, 1998). To date, there is no universally effective therapy for this disease. Mechanical ventilation is often used to maintain adequate oxygenation and to reduce respiratory muscle efforts in patients with ALI/ARDS. However, the mechanically ventilated lung is exposed to excessive cyclic mechanical strain and positive pressure, which causes further lung damage called VILI (Dreyfuss and Saumon, 1998; Gattinoni et al., 2003; Pinhu et al., 2003; Uhlig and Uhlig, 2004; Vlahakis and Hubmayr, 2005; Oeckler and Hubmayr, 2007; Pelosi and Rocco, 2008). Because VILI is the major critical complication in mechanically ventilated patients with ALI/ARDS, lung protective ventilation strategies that reduce mechanical stress are necessary to reduce mortality rates (Suki et al., 1998; ARDSnet, 2000; Gattinoni et al., 2003; Peek et al., 2009; Sud et al., 2010a, b).

7.3.2 Mechanical Stress as Pathogenesis of Ventilator-Induced Lung Injury

Several mechanisms of VILI have been proposed. Mechanical ventilation at high peak pressures causes lung damage (barotrauma) characterized by air leaks into extra-alveolar spaces due to disruption of the airspace wall (Denney and Glas, 1964). Alveolar overdistension at large tidal volume also causes lung injury (volutrauma) (Dreyfuss et al., 1988; Pinhu et al., 2003). In ALI/ARDS, the loss of integrity of the alveolar-capillary membrane leads to exudation of protein-rich fluid into the airspaces, followed by hyaline membrane deposition and surfactant destruction (Dos Santos and Slutsky, 2006). Moreover, neutrophil recruitment to the airspace causes a severe inflammatory response (Dos Santos and Slutsky, 2000). As a result, the number of normal alveolar units is reduced, and the distribution of injured lung tissue becomes heterogeneous. The regular tidal volume is directed into smaller normal lung regions which results in excessive stretch in those regions. Furthermore, injured alveoli that remain collapsed are exposed to repetitive cyclic strain as well as shear stress as the lung units collapse and reopen during mechanical ventilation. The repeated recruitment and derecruitment of the collapsed lung cause further lung damage called atelectrauma (Gattinoni et al., 1987; Pinhu et al., 2003; Dos Santos and Slutsky, 2006). The mechanism by which cellular damage occurs during atelectrauma is related to the mechanical stress-induced separation of the plasma membrane from the cytoskeleton which offers a simple clinical solution by strengthening the adhesion between the membrane and the cytoskeleton using hypertonic treatment of the lung (Oeckler et al., 2010).

In the three mechanisms described above (barotraumas, volutrauma, and atelectrauma), the lung is directly injured by physical forces (i.e. transpulmonary pressure, mechanical stretch, and shear stress). In contrast to these pathways, mechanical ventilation injures the lung via activating cellular responses (biotrauma). During mechanical ventilation, various kinds of pulmonary cells, alveolar and bronchial epithelial cells, vascular ECs, airway smooth muscle cells, fibroblasts, and macrophages, are subjected to mechanical forces (Marini et al., 2003; Vlahakis and Hubmayr, 2005). If excessive, mechanical stress directly induces cellular stress failure of the epithelial cells and capillary ECs (West and Mathieu-Costello, 1992; Vlahakis and Hubmayr, 2005), which results in pulmonary edema and inflammation. Furthermore, the biophysical forces alter normal cellular functions in the lung (Iwaki et al., 2009; Ito et al., 2010), which leads to inflammatory mediator production, abnormal tissue remodeling and repair, and apoptosis (Slutsky and Tremblay, 1998; Vlahakis and Hubmayr, 2005; Dos Santos and Slutsky, 2006).

7.3.3 Role of Pulmonary Endothelial Cells in Pathophysiology of Ventilator-Induced Lung Injury

Because pulmonary microvascular ECs form the alveolar-capillary barrier together with alveolar epithelial cells, dysfunction of the pulmonary EC barrier plays a key role in the pathophysiology of VILI (Matthay et al., 2003). Pulmonary microvascular ECs are one of the primary targets of mechanical stretch during mechanical ventilation (West and Mathieu-Costello, 1992; Haseneen et al., 2003; Marini et al., 2003). An experimental study demonstrated that endothelial permeability in response to high peak inflation pressure as assessed by filtration coefficients (K_f), is greater in the microvascular region than in the pulmonary artery lesion in isolated rat lungs (Parker and Yoshikawa, 2002).

There are several mechanisms by which mechanical force causes endothelial barrier disruption and increases endothelial permeability in the lung. The first mechanism is direct injury to the pulmonary ECs by physical forces. Mechanical stress is able to directly induce stress failure of the alveolar-endothelial barrier during mechanical ventilation (West and Mathieu-Costello, 1992; Wirtz and Dobbs, 2000; Vlahakis and Hubmayr, 2005). At a cellular level, mechanical stress could cause stress failure of pulmonary microvascular ECs, leading to necrosis and apoptosis (Vlahakis and Hubmayr, 2005; Adkison et al., 2006). Raaz et al. found that cyclic stretch (30% increase in membrane surface area, cycling frequency 40/min) employed for 24 h induced apoptosis with a decrease in Akt phosphorylation in human pulmonary microvascular ECs (Raaz et al., 2009). The second mechanism is alterations in the cytoskeletal structure without ultra-structural cell damage. Changes in the integrity of the endothelial barrier and gap formation between neighboring ECs are caused by mechanical stretch and shear stress via cytoskeletal rearrangement and generation of tensile forces within the cell (Birukov et al., 2002, 2003). This mechanism involves cellular contraction via small GTPase RhoA activation followed by phosphorylation of myosin light chain. The third mechanism is the effects of mechanical forces on EC functions as described in Sections 7.2.2 and 7.2.3. Pulmonary microvascular ECs produce a variety of mediators, cytokines, and chemokines in response to physical forces. Several in vitro studies have demonstrated that cyclic stretch of human pulmonary microvascular ECs induces IL-6, IL-8, MCP-1, and matrix metalloproteinase-1 and -2, all of which may contribute to the pathogenesis of VILI (Haseneen et al., 2003; Pinhu et al., 2008; Iwaki et al., 2009). These studies suggest that pulmonary ECs enhance inflammation of the lung via cytokine/chemokine production in response to large stretching in mechanically ventilated patients with ALI/ARDS which may ultimately lead to multiple organ failure.

How alveoli and the microvasculature react to distension during ventilation in vivo is not fully understood because the distribution of edema and inflammation is spatially and temporally heterogeneous in the lung with ARDS (Dreyfuss and Saumon, 1998; Uhlig and Uhlig, 2004). When the lung volume is increased by high tidal ventilation, alveoli and pulmonary microvascular ECs are expected to be stretched by 20% or more (Tschumperlin and Margulies, 1999). In cultured alveolar epithelial cells, a 25% increase in cell surface area corresponding to a 8–12% linear distension correlates with physiological levels of mechanical cyclic strain in vivo (Tschumperlin et al., 2000). By contrast, cyclic stretch with a 37–50% increase in cell surface area corresponding to a 17–22% linear distension is relevant to pathophysiological conditions induced by mechanical ventilation. Birukov et al. (2003) elongated human pulmonary artery ECs 5 and 18% using a cell stretch device and found that the larger mechanical stretch reduced the endothelial barrier function. It was reported that 17–18.5% elongation of human pulmonary microvascular ECs induced matrix metalloproteinase-2 (Haseneen et al., 2003). Pinhu et al. reported that cyclic stretch (30% elongation) induced IL-8 mRNA expression in human pulmonary microvascular ECs. Taken together, mechanical strain of pulmonary microvascular ECs beyond the physiological levels alters cellular properties, which may lead to progression of VILI in mechanically ventilated patients with ALI/ARDS.

7.3.4 Neutrophil Activation in Ventilator-Induced Lung Injury

Neutrophil recruitment to the airspace plays a critical role in the inflammatory response of ALI/ARDS and VILI (Fowler et al., 1987; Ware and Matthay, 2000; Matthay and Zimmerman, 2005; Puneet et al., 2005). Although neutrophil activation protects against pulmonary infection by bacteria, its excessive activation may have harmful effects on the lungs. Neutrophils are major effector cells in the generation of the tissue injury characteristic of VILI (Dreyfuss and Saumon, 1998; Dos Santos and Slutsky, 2000). Both in vitro and in vivo studies have provided evidence that neutrophil oxidants and proteases can injure cells forming the alveolar-capillary membrane (Cochrane et al., 1983a). Neutrophil elastase is a potent proteolytic enzyme targeting the extracellular matrix (ECM) components elastin, fibrinogen, proteoglycan, and collagen. Furthermore, neutrophil elastase induces pulmonary EC damage, which causes pulmonary edema (Lee et al., 1981; Cochrane et al., 1983b; Kaynar et al., 2008).

Normally, more than 90% of the cells in the air space are alveolar macrophages, less than 10% are lymphocytes, and only 1–2% are neutrophils (Puneet et al., 2005). Neutrophils predominate in the pulmonary edema fluid and bronchoalveolar lavage fluid of patients with ARDS (Pittet et al., 1997), even in the absence of pulmonary infection (Idell and Cohen, 1985). Mechanical ventilation leads to release of mediators and cytokines/chemokines that prime neutrophils. Tsuno et al. observed interstitial lymphoid infiltration and alveolar macrophage and neutrophil infiltration were in the lungs of piglets mechanically ventilated at high tidal volume for 22 h (Tsuno et al., 1991). To elucidate how mechanical stress to the lung induces neutrophil activation is important for better understanding the mechanisms of VILI.

7.3.5 Roles of IL-8 in Ventilator-Induced Lung Injury

IL-8 is the major chemoattractant CXC chemokine for neutrophils (Donnelly et al., 1993; Goodman et al., 1996; Puneet et al., 2005). Therefore, release of IL-8 is considered to play an important role in the inflammatory response and progression of VILI (Oudin and Pugin, 2002). In 1987, IL-8 was first cloned as a neutrophil chemotactic factor from LPS-stimulated human mononuclear cell culture supernatants Yoshimura et al., 1987). In the lung, various types of cells such as leukocytes, macrophages, alveolar epithelial cells, and ECs produce IL-8 in response to mechanical stretch or pro-inflammatory cytokines (Pugin et al., 1998; Beck et al., 1999; Hashimoto et al., 1999; Vlahakis et al., 1999; Burg et al., 2002; Mukaida, 2003; Iwaki et al., 2009). Again, mechanical stretch-induced IL-8 release by pulmonary microvascular ECs would be involved in the mechanisms of neutrophil activation in VILI.

7.4 Pathogenesis of COPD

7.4.1 Mechanical Damage of the Parenchyma in Pulmonary Emphysema

Another mechanical stress-related pulmonary disorder is chronic obstructive pulmonary disease (COPD), which is characterized by airflow limitation that is not fully reversible and is associated with an abnormal inflammatory response in the small airways and alveoli (Cosio et al., 2009). The primary risk factor for this disease is long-term exposure to cigarette smoke (Rabe et al., 2007). Pulmonary emphysema defined by destruction of lung parenchyma distal to the terminal bronchioles is the major pathologic feature of COPD. An early study demonstrated that one-way valves placed in the airways caused mechanical hyperinflation and produced emphysema-like changes in the remaining tissues (Harris and Chillingworth, 1919). Therefore, besides various biological factors such as inflammation (Barnes, 2000), mechanical force has long been considered a potential etiologic factor in the development and progression of emphysema (West, 1971; Suki et al., 2003; Ito et al., 2005, 2006).

Previous clinical evidence reveals that mechanical forces accelerate the progression of emphysema in patients with severe emphysema undergoing lung volume reduction surgery (LVRS). LVRS involves surgical resection and resizing of the hyper-expanded lung to fit the chest wall (Brantigan and Mueller, 1957; Cooper et al., 1995; Ingenito et al., 1998; Meyers and Patterson, 2003). Because the remaining lung tissue of these patients is stretched to fill the thoracic cavity, mechanical stress on the lung increases significantly. Indeed, it has been reported that patients who undergo LVRS experience a deterioration in lung function over time, which is accelerated compared with their rate of lung function decline before surgery (Gelb et al., 2001). Thus, the increased mechanical forces on the weak fiber network of emphysematous lung parenchyma that follow LVRS may be responsible for the accelerated rate of mechanical damage to the alveolar walls (Suki et al., 2003). Nevertheless, little is known about how the altered mechanical forces affect cellular signaling in the lung.

7.4.2 Role of Endothelial Cells in Pathogenesis of COPD

Dysfunctions of pulmonary microvascular ECs have been implicated in the pathogenesis of COPD (Kasahara et al., 2001; Nana-Sinkam et al., 2007). Indeed, cigarette smoking impairs endothelial functions and integrity. Inhibition of VEGF and VEGF receptor significantly contributes to alveolar cell death and emphysematous change (Kasahara et al., 2001). Moreover, circulating EPCs are decreased in patients with COPD (Palange et al., 2006). Taken together, both the lung endothelial disorder and impairment of microvasculature regeneration are considered to be involved in the pathogenesis of COPD. Interestingly, statins, which improve the integrity of the microvasculature, have inhibitory effects on the development of emphysema in rodent models of COPD (Lee et al., 2005; Takahashi et al., 2008). Clinical evidence suggests that statin treatment may reduce mortality from COPD (Dobler et al., 2009). While the mechanism is unclear, pulmonary microvascular ECs may be a therapeutic target of this disease which is important since currently no medication is known to stop the progression of emphysema.

7.5 Role of Mechanical Stress in Normal Lung Physiology and Development

The respiratory system composed of the lung, airways, and chest wall, is a dynamic organ because it is exposed to various kinds of physical forces derived from tidal breathing, blood flow, and surface tension (Tschumperlin and Drazen, 2006). It is well documented that these mechanical forces are important for fetal lung maturation, vasculogenesis, development of the respiratory system, and maintenance of lung homeostasis and structure in vivo (Riley et al., 1990; Harding and Hooper, 1996; Torday et al., 1998; Liu and Post, 2000; Wirtz and Dobbs, 2000; Sanchez-Esteban et al., 2001).

Many important cellular functions are directly affected by mechanical stimuli in the physiological condition. An important example is secretion of pulmonary surfactant, a complex mixture of phospholipids and apoproteins, that lines the air-liquid interface throughout the lung, lowering surface tension, and contributing to alveolar stability (Weaver and Whitsett, 1991). Pulmonary surfactant synthesis and secretion by alveolar epithelial type II cells is regulated by mechanical stretch (Chander and Fisher, 1990; Wirtz and Dobbs, 1990; Arold et al., 2009). The biomechanical properties of the connective tissue network composed of cells and ECM including collagen, elastic fibers (mostly elastin), and proteoglycans determine the physiology of the lung parenchyma (Fredberg and Stamenovic, 1989; Cavalcante et al., 2005; Suki et al., 2005; Pelosi and Rocco, 2008). Moreover, mechanical interaction between cells and the ECM via integrin adhesion receptors have major regulatory effects on cellular physiology (Geiger and Bershadsky, 2002; Moore et al., 2010).

Mechanical forces affect enzymatic activity in living tissues (Khan and Sheetz, 1997; Zhang et al., 2009). Jesudason et al. (2010) demonstrated that physiologically relevant macroscopic mechanical forces modified enzyme activity of the lung parenchyma in mice. Using fluorescently conjugated porcine pancreatic elastase and fluorescent recovery after a photobleach (FRAP) technique, they found that elastase activity was significantly stronger in stretched than in unstretched (relaxed) alveolar walls. These findings suggest an important role for the coupling between mechanical forces and enzyme activity in ECM breakdown and remodeling in normal lung physiology as well as in certain disease conditions such as pulmonary emphysema and fibrosis.

7.6 Conclusion and Perspectives

Although mechanical stress is important for maintaining normal lung homeostasis and development, lack or excess of physiological levels of mechanical forces contribute to the pathogenesis and progression of pulmonary diseases such as VILI and COPD. The pulmonary microvascular ECs forming the alveolar-capillary barrier with alveolar epithelial cells play an important role in regulating vascular permeability and pulmonary inflammation. When the lung is exposed to large cyclic stretching during mechanical ventilation, the pulmonary microvasculature may induce production of inflammatory cytokines/chemokines and mediators, leading to increased severity of VILI. Understanding how mechanical stretch alters the functions and morphology of pulmonary microvascular ECs may lead to novel treatment strategies for physical force-related pulmonary diseases.

Acknowledgements This work was supported by Grants-in-Aid from the Ministry of Education, Culture, Sports, Science, and Technology of Japan (#19689017 and #22890837 to S.I.). We thank Dr. Béla Suki, Department of Biomedical Engineering, Boston University, for the helpful comments. We also thank Ms. Katherine Ono for providing language help.

References

Adkison JB, Miller GT, Weber DS, Miyahara T, Ballard ST, Frost JR, Parker JC (2006) Differential responses of pulmonary endothelial phenotypes to cyclical stretch. Microvasc Res 71:175–184

Aird WC (2007a) Phenotypic heterogeneity of the endothelium: I. Structure, function, and mechanisms. Circ Res 100:158–173

Aird WC (2007b) Phenotypic heterogeneity of the endothelium: II. Representative vascular beds. Circ Res 100:174–190

Al-Mehdi AB, Song C, Tozawa K, Fisher AB (2000) Ca2+- and phosphatidylinositol 3-kinase-dependent nitric oxide generation in lung endothelial cells in situ with ischemia. J Biol Chem 275:39807–39810

Ali MH, Schumacker PT (2002) Endothelial responses to mechanical stress: where is the mechanosensor? Crit Care Med 30:S198–S206

Ali MH, Mungai PT, Schumacker PT (2006) Stretch-induced phosphorylation of focal adhesion kinase in endothelial cells: role of mitochondrial oxidants. Am J Physiol Lung Cell Mol Physiol 291:L38–L45

Alvarez DF, King JA, Weber D, Addison E, Liedtke W, Townsley MI (2006) Transient receptor potential vanilloid 4-mediated disruption of the alveolar septal barrier: a novel mechanism of acute lung injury. Circ Res 99:988–995

Alvarez DF, Huang L, King JA, Elzarrad MK, Yoder MC, Stevens T (2008) Lung microvascular endothelium is enriched with progenitor cells that exhibit vasculogenic capacity. Am J Physiol Lung Cell Mol Physiol 294:L419–L430

Ando J, Yamamoto K (2009) Vascular mechanobiology: endothelial cell responses to fluid shear stress. Circ J 73:1983–1992

ARDSnet (2000) Ventilation with lower tidal volumes as compared with traditional tidal volumes for acute lung injury and the acute respiratory distress syndrome. N Engl J Med 342: 1301–1308

Arold SP, Bartolak-Suki E, Suki B (2009) Variable stretch pattern enhances surfactant secretion in alveolar type II cells in culture. Am J Physiol Lung Cell Mol Physiol 296:L574–L581

Artigas A, Bernard GR, Carlet J, Dreyfuss D, Gattinoni L, Hudson L, Lamy M, Marini JJ, Matthay MA, Pinsky MR, Spragg R, Suter PM (1998) The American-European Consensus Conference on ARDS, part 2: ventilatory, pharmacologic, supportive therapy, study design strategies, and issues related to recovery and remodeling. Acute respiratory distress syndrome. Am J Respir Crit Care Med 157:1332–1347

Asahara T, Murohara T, Sullivan A, Silver M, Van Der Zee R, Li T, Witzenbichler B, Schatteman G, Isner JM (1997) Isolation of putative progenitor endothelial cells for angiogenesis. Science 275:964–967

Asahara T, Masuda H, Takahashi T, Kalka C, Pastore C, Silver M, Kearne M, Magner M, Isner JM (1999) Bone marrow origin of endothelial progenitor cells responsible for postnatal vasculogenesis in physiological and pathological neovascularization. Circ Res 85:221–228

Ashbaugh DG, Bigelow DB, Petty TL, Levine BE (1967) Acute respiratory distress in adults. Lancet 2:319–323

Awolesi MA, Sessa WC, Sumpio BE (1995) Cyclic strain upregulates nitric oxide synthase in cultured bovine aortic endothelial cells. J Clin Invest 96:1449–1454

Azuma N, Akasaka N, Kito H, Ikeda M, Gahtan V, Sasajima T, Sumpio BE (2001) Role of p38 MAP kinase in endothelial cell alignment induced by fluid shear stress. Am J Physiol Heart Circ Physiol 280:H189–H197

Barnes PJ (2000) Chronic obstructive pulmonary disease. N Engl J Med 343:269–280

Beck GC, Yard BA, Breedijk AJ, Van Ackern K, Van Der Woude FJ (1999) Release of CXC-chemokines by human lung microvascular endothelial cells (LMVEC) compared with macrovascular umbilical vein endothelial cells. Clin Exp Immunol 118:298–303

Bernard GR, Artigas A, Brigham KL, Carlet J, Falke K, Hudson L, Lamy M, Legall JR, Morris A, Spragg R (1994) The American-European Consensus Conference on ARDS. Definitions,

mechanisms, relevant outcomes, and clinical trial coordination. Am J Respir Crit Care Med 149:818–824
Birukov KG, Birukova AA, Dudek SM, Verin AD, Crow MT, Zhan X, Depaola N, Garcia JG (2002) Shear stress-mediated cytoskeletal remodeling and cortactin translocation in pulmonary endothelial cells. Am J Respir Cell Mol Biol 26:453–464
Birukov KG, Jacobson JR, Flores AA, Ye SQ, Birukova AA, Verin AD, Garcia JG (2003) Magnitude-dependent regulation of pulmonary endothelial cell barrier function by cyclic stretch. Am J Physiol Lung Cell Mol Physiol 285:L785–L797
Blackwell TS, Christman JW (1997) The role of nuclear factor-kappa B in cytokine gene regulation. Am J Respir Cell Mol Biol 17:3–9
Boriek AM, Kumar A (2008) Regulation of intracellular signal transduction pathways by mechanosensitive ion channels. In: Kamkin A, Kiseleva I (eds) Mechanosensitivity in cells and tissues No 1. Mechanosensitive ion channels. Springer, New York, NY, pp 303–327
Brantigan OC, Mueller E (1957) Surgical treatment of pulmonary emphysema. Am Surg 23:789–804
Brooks AR, Lelkes PI, Rubanyi GM (2002) Gene expression profiling of human aortic endothelial cells exposed to disturbed flow and steady laminar flow. Physiol Genomics 9:27–41
Burg J, Krump-Konvalinkova V, Bittinger F, Kirkpatrick CJ (2002) GM-CSF expression by human lung microvascular endothelial cells: in vitro and in vivo findings. Am J Physiol Lung Cell Mol Physiol 283:L460–L467
Carmeliet P (2003) Angiogenesis in health and disease. Nat Med 9:653–660
Cavalcante FS, Ito S, Brewer K, Sakai H, Alencar AM, Almeida MP, Andrade JS Jr, Majumdar A, Ingenito EP, Suki B (2005) Mechanical interactions between collagen and proteoglycans: implications for the stability of lung tissue. J Appl Physiol 98:672–679
Chander A, Fisher AB (1990) Regulation of lung surfactant secretion. Am J Physiol 258: L241–L253
Chien S (2007) Mechanotransduction and endothelial cell homeostasis: the wisdom of the cell. Am J Physiol Heart Circ Physiol 292:H1209–H1224
Cochrane CG, Spragg R, Revak SD (1983a) Pathogenesis of the adult respiratory distress syndrome. Evidence of oxidant activity in bronchoalveolar lavage fluid. J Clin Invest 71: 754–761
Cochrane CG, Spragg RG, Revak SD, Cohen AB, Mcguire WW (1983b) The presence of neutrophil elastase and evidence of oxidation activity in bronchoalveolar lavage fluid of patients with adult respiratory distress syndrome. Am Rev Respir Dis 127:S25–S27
Cohen TS, Gray Lawrence G, Khasgiwala A, Margulies SS (2010) MAPK activation modulates permeability of isolated rat alveolar epithelial cell monolayers following cyclic stretch. PLoS One 5:e10385
Cooper JD, Trulock EP, Triantafillou AN, Patterson GA, Pohl MS, Deloney PA, Sundaresan RS, Roper CL (1995) Bilateral pneumectomy (volume reduction) for chronic obstructive pulmonary disease. J Thorac Cardiovasc Surg 109:106–116; discussion 116–109
Copland IB, Post M (2007) Stretch-activated signaling pathways responsible for early response gene expression in fetal lung epithelial cells. J Cell Physiol 210:133–143
Cosio MG, Saetta M, Agusti A (2009) Immunologic aspects of chronic obstructive pulmonary disease. N Engl J Med 360:2445–2454
Davies PF (1995) Flow-mediated endothelial mechanotransduction. Physiol Rev 75:519–560
Denney MK, Glas WW (1964) Experimental studies in barotrauma. J Trauma 4:791–796
Dobler CC, Wong KK, Marks GB (2009) Associations between statins and COPD: a systematic review. BMC Pulm Med 9:32
Donnelly SC, Strieter RM, Kunkel SL, Walz A, Robertson CR, Carter DC, Grant IS, Pollok AJ, Haslett C (1993) Interleukin-8 and development of adult respiratory distress syndrome in at-risk patient groups. Lancet 341:643–647
Dos Santos CC, Slutsky AS (2000) Invited review: mechanisms of ventilator-induced lung injury: a perspective. J Appl Physiol 89:1645–1655

Dos Santos CC, Slutsky AS (2006) The contribution of biophysical lung injury to the development of biotrauma. Annu Rev Physiol 68:585–618

Dreyfuss D, Saumon G (1998) Ventilator-induced lung injury: lessons from experimental studies. Am J Respir Crit Care Med 157:294–323

Dreyfuss D, Soler P, Basset G, Saumon G (1988) High inflation pressure pulmonary edema. Respective effects of high airway pressure, high tidal volume, and positive end-expiratory pressure. Am Rev Respir Dis 137:1159–1164

Fowler AA, Hyers TM, Fisher BJ, Bechard DE, Centor RM, Webster RO (1987) The adult respiratory distress syndrome. Cell populations and soluble mediators in the air spaces of patients at high risk. Am Rev Respir Dis 136:1225–1231

Frangos JA, Eskin SG, Mcintire LV, Ives CL (1985) Flow effects on prostacyclin production by cultured human endothelial cells. Science 227:1477–1479

Fredberg JJ, Stamenovic D (1989) On the imperfect elasticity of lung tissue. J Appl Physiol 67:2408–2419

Frye SR, Yee A, Eskin SG, Guerra R, Cong X, Mcintire LV (2005) cDNA microarray analysis of endothelial cells subjected to cyclic mechanical strain: importance of motion control. Physiol Genomics 21:124–130

Garcia JG, Schaphorst KL, Shi S, Verin AD, Hart CM, Callahan KS, Patterson CE (1997) Mechanisms of ionomycin-induced endothelial cell barrier dysfunction. Am J Physiol 273:L172–L184

Gattinoni L, Pesenti A, Avalli L, Rossi F, Bombino M (1987) Pressure-volume curve of total respiratory system in acute respiratory failure. Computed tomographic scan study. Am Rev Respir Dis 136:730–736

Gattinoni L, Carlesso E, Cadringher P, Valenza F, Vagginelli F, Chiumello D (2003) Physical and biological triggers of ventilator-induced lung injury and its prevention. Eur Respir J Suppl 47:15s–25s

Gebb S, Stevens T (2004) On lung endothelial cell heterogeneity. Microvasc Res 68:1–12

Geiger B, Bershadsky A (2002) Exploring the neighborhood: adhesion-coupled cell mechanosensors. Cell 110:139–142

Gelb AF, Mckenna RJ Jr, Brenner M, Epstein JD, Zamel N (2001) Lung function 5 yr after lung volume reduction surgery for emphysema. Am J Respir Crit Care Med 163:1562–1566

Goodman RB, Strieter RM, Martin DP, Steinberg KP, Milberg JA, Maunder RJ, Kunkel SL, Walz A, Hudson LD, Martin TR (1996) Inflammatory cytokines in patients with persistence of the acute respiratory distress syndrome. Am J Respir Crit Care Med 154:602–611

Hamanaka K, Jian MY, Weber DS, Alvarez DF, Townsley MI, Al-Mehdi AB, King JA, Liedtke W, Parker JC (2007) TRPV4 initiates the acute calcium-dependent permeability increase during ventilator-induced lung injury in isolated mouse lungs. Am J Physiol Lung Cell Mol Physiol 293:L923–L932

Hamill OP, Martinac B (2001) Molecular basis of mechanotransduction in living cells. Physiol Rev 81:685–740

Harding R, Hooper SB (1996) Regulation of lung expansion and lung growth before birth. J Appl Physiol 81:209–224

Harris WH, Chillingworth FP (1919) The experimental production in dogs of emphysema with associated asthmatic syndrome by means of an intratracheal ball valve. J Exp Med 30:75–85

Haseneen NA, Vaday GG, Zucker S, Foda HD (2003) Mechanical stretch induces MMP-2 release and activation in lung endothelium: role of EMMPRIN. Am J Physiol Lung Cell Mol Physiol 284:L541–L547

Hashimoto S, Matsumoto K, Gon Y, Maruoka S, Takeshita I, Hayashi S, Koura T, Kujime K, Horie T (1999) p38 mitogen-activated protein kinase regulates IL-8 expression in human pulmonary vascular endothelial cells. Eur Respir J 13:1357–1364

Hayakawa K, Tatsumi H, Sokabe M (2008) Actin stress fibers transmit and focus force to activate mechanosensitive channels. J Cell Sci 121:496–503

Hoffmann E, Dittrich-Breiholz O, Holtmann H, Kracht M (2002) Multiple control of interleukin-8 gene expression. J Leukoc Biol 72:847–855

Idell S, Cohen AB (1985) Bronchoalveolar lavage in patients with the adult respiratory distress syndrome. Clin Chest Med 6:459–471

Ingber DE (1993) Cellular tensegrity: defining new rules of biological design that govern the cytoskeleton. J Cell Sci 104(Pt 3):613–627

Ingenito EP, Evans RB, Loring SH, Kaczka DW, Rodenhouse JD, Body SC, Sugarbaker DJ, Mentzer SJ, Decamp MM, Reilly JJ Jr (1998) Relation between preoperative inspiratory lung resistance and the outcome of lung-volume-reduction surgery for emphysema. N Engl J Med 338:1181–1185

Ito S, Ingenito EP, Brewer KK, Black LD, Parameswaran H, Lutchen KR, Suki B (2005) Mechanics, nonlinearity, and failure strength of lung tissue in a mouse model of emphysema: possible role of collagen remodeling. J Appl Physiol 98:503–511

Ito S, Bartolak-Suki E, Shipley JM, Parameswaran H, Majumdar A, Suki B (2006a) Early emphysema in the tight skin and pallid mice: roles of microfibril-associated glycoproteins, collagen, and mechanical forces. Am J Respir Cell Mol Biol 34:688–694

Ito S, Kume H, Oguma T, Ito Y, Kondo M, Shimokata K, Suki B, Naruse K (2006b) Roles of stretch-activated cation channel and Rho-kinase in the spontaneous contraction of airway smooth muscle. Eur J Pharmacol 552:135–142

Ito S, Kume H, Naruse K, Kondo M, Takeda N, Iwata S, Hasegawa Y, Sokabe M (2008) A novel Ca^{2+} influx pathway activated by mechanical stretch in human airway smooth muscle cells. Am J Respir Cell Mol Biol 38:407–413

Ito S, Suki B, Kume H, Numaguchi Y, Ishii M, Iwaki M, Kondo M, Naruse K, Hasegawa Y, Sokabe M (2010) Actin cytoskeleton regulates stretch-activated Ca2+ influx in human pulmonary microvascular endothelial cells. Am J Respir Cell Mol Biol 43:26–34

Iwaki M, Ito S, Morioka M, Iwata S, Numaguchi Y, Ishii M, Kondo M, Kume H, Naruse K, Sokabe M, Hasegawa Y (2009) Mechanical stretch enhances IL-8 production in pulmonary microvascular endothelial cells. Biochem Biophys Res Commun 389:531–536

Jesudason R, Sato S, Parameswaran H, Araujo AD, Majumdar A, Allen PG, Bartolak-Suki E, Suki B (2010) Mechanical forces regulate elastase activity and binding site availability in lung elastin. Biophys J 99:3076–3083

Jian MY, King JA, Al-Mehdi AB, Liedtke W, Townsley MI (2008) High vascular pressure-induced lung injury requires P450 epoxygenase-dependent activation of TRPV4. Am J Respir Cell Mol Biol 38:386–392

Kasahara Y, Tuder RM, Cool CD, Lynch DA, Flores SC, Voelkel NF (2001) Endothelial cell death and decreased expression of vascular endothelial growth factor and vascular endothelial growth factor receptor 2 in emphysema. Am J Respir Crit Care Med 163:737–744

Kaunas R, Nguyen P, Usami S, Chien S (2005) Cooperative effects of Rho and mechanical stretch on stress fiber organization. Proc Natl Acad Sci USA 102:15895–15900

Kawai M, Naruse K, Komatsu S, Kobayashi S, Nagino M, Nimura Y, Sokabe M (2002) Mechanical stress-dependent secretion of interleukin 6 by endothelial cells after portal vein embolization: clinical and experimental studies. J Hepatol 37:240–246

Kaynar AM, Houghton AM, Lum EH, Pitt BR, Shapiro SD (2008) Neutrophil elastase is needed for neutrophil emigration into lungs in ventilator-induced lung injury. Am J Respir Cell Mol Biol 39:53–60

Khan S, Sheetz MP (1997) Force effects on biochemical kinetics. Annu Rev Biochem 66:785–805

King J, Hamil T, Creighton J, Wu S, Bhat P, Mcdonald F, Stevens T (2004) Structural and functional characteristics of lung macro- and microvascular endothelial cell phenotypes. Microvasc Res 67:139–151

Kito H, Chen EL, Wang X, Ikeda M, Azuma N, Nakajima N, Gahtan V, Sumpio BE (2000) Role of mitogen-activated protein kinases in pulmonary endothelial cells exposed to cyclic strain. J Appl Physiol 89:2391–2400

Kobayashi S, Nagino M, Komatsu S, Naruse K, Nimura Y, Nakanishi M, Sokabe M (2003) Stretch-induced IL-6 secretion from endothelial cells requires NF-kappaB activation. Biochem Biophys Res Commun 308:306–312

Kollef MH, Schuster DP (1995) The acute respiratory distress syndrome. N Engl J Med 332:27–37
Kosaki K, Ando J, Korenaga R, Kurokawa T, Kamiya A (1998) Fluid shear stress increases the production of granulocyte-macrophage colony-stimulating factor by endothelial cells via mRNA stabilization. Circ Res 82:794–802
Kuchan MJ, Frangos JA (1993) Shear stress regulates endothelin-1 release via protein kinase C and cGMP in cultured endothelial cells. Am J Physiol 264:H150–H156
Kuebler WM, Ying X, Singh B, Issekutz AC, Bhattacharya J (1999) Pressure is proinflammatory in lung venular capillaries. J Clin Invest 104:495–502
Kuebler WM, Uhlig U, Goldmann T, Schael G, Kerem A, Exner K, Martin C, Vollmer E, Uhlig S (2003) Stretch activates nitric oxide production in pulmonary vascular endothelial cells in situ. Am J Respir Crit Care Med 168:1391–1398
Langille BL, Adamson SL (1981) Relationship between blood flow direction and endothelial cell orientation at arterial branch sites in rabbits and mice. Circ Res 48:481–488
Lee CT, Fein AM, Lippmann M, Holtzman H, Kimbel P, Weinbaum G (1981) Elastolytic activity in pulmonary lavage fluid from patients with adult respiratory-distress syndrome. N Engl J Med 304:192–196
Lee JH, Lee DS, Kim EK, Choe KH, Oh YM, Shim TS, Kim SE, Lee YS, Lee SD (2005) Simvastatin inhibits cigarette smoking-induced emphysema and pulmonary hypertension in rat lungs. Am J Respir Crit Care Med 172:987–993
Li M, Scott DE, Shandas R, Stenmark KR, Tan W (2009) High pulsatility flow induces adhesion molecule and cytokine mRNA expression in distal pulmonary artery endothelial cells. Ann Biomed Eng 37:1082–1092
Liu M, Post M (2000) Invited review: mechanochemical signal transduction in the fetal lung. J Appl Physiol 89:2078–2084
Magee JC, Stone AE, Oldham KT, Guice KS (1994) Isolation, culture, and characterization of rat lung microvascular endothelial cells. Am J Physiol 267:L433–L441
Marini JJ, Hotchkiss JR, Broccard AF (2003) Bench-to-bedside review: microvascular and airspace linkage in ventilator-induced lung injury. Crit Care 7:435–444
Matthay MA, Zimmerman GA (2005) Acute lung injury and the acute respiratory distress syndrome: four decades of inquiry into pathogenesis and rational management. Am J Respir Cell Mol Biol 33:319–327
Matthay MA, Zimmerman GA, Esmon C, Bhattacharya J, Coller B, Doerschuk CM, Floros J, Gimbrone MA Jr, Hoffman E, Hubmayr RD, Leppert M, Matalon S, Munford R, Parsons P, Slutsky AS, Tracey KJ, Ward P, Gail DB, Harabin AL (2003) Future research directions in acute lung injury: summary of a National Heart, Lung, and Blood Institute working group. Am J Respir Crit Care Med 167:1027–1035
Mehta D, Malik AB (2006) Signaling mechanisms regulating endothelial permeability. Physiol Rev 86:279–367
Meyers BF, Patterson GA (2003) Chronic obstructive pulmonary disease. 10: Bullectomy, lung volume reduction surgery, and transplantation for patients with chronic obstructive pulmonary disease. Thorax 58:634–638
Miller EJ, Cohen AB, Nagao S, Griffith D, Maunder RJ, Martin TR, Weiner-Kronish JP, Sticherling M, Christophers E, Matthay MA (1992) Elevated levels of NAP-1/interleukin-8 are present in the airspaces of patients with the adult respiratory distress syndrome and are associated with increased mortality. Am Rev Respir Dis 146:427–432
Minami T, Aird WC (2005) Endothelial cell gene regulation. Trends Cardiovasc Med 15:174–184
Moore SW, Roca-Cusachs P, Sheetz MP (2010) Stretchy proteins on stretchy substrates: the important elements of integrin-mediated rigidity sensing. Dev Cell 19:194–206
Mukaida N (2003) Pathophysiological roles of interleukin-8/CXCL8 in pulmonary diseases. Am J Physiol Lung Cell Mol Physiol 284:L566–L577
Mulligan MS, Polley MJ, Bayer RJ, Nunn MF, Paulson JC, Ward PA (1992) Neutrophil-dependent acute lung injury. Requirement for P-selectin (GMP-140. J Clin Invest 90: 1600–1607

Nana-Sinkam SP, Lee JD, Sotto-Santiago S, Stearman RS, Keith RL, Choudhury Q, Cool C, Parr J, Moore MD, Bull TM, Voelkel NF, Geraci MW (2007) Prostacyclin prevents pulmonary endothelial cell apoptosis induced by cigarette smoke. Am J Respir Crit Care Med 175:676–685

Naruse K, Yamada T, Sokabe M (1998) Involvement of SA channels in orienting response of cultured endothelial cells to cyclic stretch. Am J Physiol 274:H1532–H1538

Nilius B, Droogmans G (2001) Ion channels and their functional role in vascular endothelium. Physiol Rev 81:1415–1459

Oeckler RA, Hubmayr RD (2007) Ventilator-associated lung injury: a search for better therapeutic targets. Eur Respir J 30:1216–1226

Oeckler RA, Lee WY, Park MG, Kofler O, Rasmussen DL, Lee HB, Belete H, Walters BJ, Stroetz RW, Hubmayr RD (2010) Determinants of plasma membrane wounding by deforming stress. Am J Physiol Lung Cell Mol Physiol 299:L826–L833

Okada M, Matsumori A, Ono K, Furukawa Y, Shioi T, Iwasaki A, Matsushima K, Sasayama S (1998) Cyclic stretch upregulates production of interleukin-8 and monocyte chemotactic and activating factor/monocyte chemoattractant protein-1 in human endothelial cells. Arterioscler Thromb Vasc Biol 18:894–901

Oudin S, Pugin J (2002) Role of MAP kinase activation in interleukin-8 production by human BEAS-2B bronchial epithelial cells submitted to cyclic stretch. Am J Respir Cell Mol Biol 27:107–114

Palange P, Testa U, Huertas A, Calabro L, Antonucci R, Petrucci E, Pelosi E, Pasquini L, Satta A, Morici G, Vignola MA, Bonsignore MR (2006) Circulating haemopoietic and endothelial progenitor cells are decreased in COPD. Eur Respir J 27:529–541

Parker JC, Yoshikawa S (2002) Vascular segmental permeabilities at high peak inflation pressure in isolated rat lungs. Am J Physiol Lung Cell Mol Physiol 283:L1203–L1209

Parker JC, Ivey CL, Tucker JA (1998) Gadolinium prevents high airway pressure-induced permeability increases in isolated rat lungs. J Appl Physiol 84:1113–1118

Peek GJ, Mugford M, Tiruvoipati R, Wilson A, Allen E, Thalanany MM, Hibbert CL, Truesdale A, Clemens F, Cooper N, Firmin RK, Elbourne D (2009) Efficacy and economic assessment of conventional ventilatory support versus extracorporeal membrane oxygenation for severe adult respiratory failure (CESAR): a multicentre randomised controlled trial. Lancet 374:1351–1363

Pelosi P, Rocco PR (2008) Effects of mechanical ventilation on the extracellular matrix. Intensive Care Med 34:631–639

Pinhu L, Whitehead T, Evans T, Griffiths M (2003) Ventilator-associated lung injury. Lancet 361:332–340

Pinhu L, Park JE, Yao W, Griffiths MJ (2008) Reference gene selection for real-time polymerase chain reaction in human lung cells subjected to cyclic mechanical strain. Respirology 13: 990–999

Pittet JF, Mackersie RC, Martin TR, Matthay MA (1997) Biological markers of acute lung injury: prognostic and pathogenetic significance. Am J Respir Crit Care Med 155:1187–1205

Pugin J, Dunn I, Jolliet P, Tassaux D, Magnenat JL, Nicod LP, Chevrolet JC (1998) Activation of human macrophages by mechanical ventilation in vitro. Am J Physiol 275:L1040–L1050

Puneet P, Moochhala S, Bhatia M (2005) Chemokines in acute respiratory distress syndrome. Am J Physiol Lung Cell Mol Physiol 288:L3–L15

Raaz U, Kuhn H, Wirtz H, Hammerschmidt S (2009) Rapamycin reduces high-amplitude, mechanical stretch-induced apoptosis in pulmonary microvascular endothelial cells. Microvasc Res 77:297–303

Rabe KF, Hurd S, Anzueto A, Barnes PJ, Buist SA, Calverley P, Fukuchi Y, Jenkins C, Rodriguez-Roisin R, Van Weel C, Zielinski J (2007) Global strategy for the diagnosis, management, and prevention of chronic obstructive pulmonary disease: GOLD executive summary. Am J Respir Crit Care Med 176:532–555

Ranieri VM, Suter PM, Tortorella C, De Tullio R, Dayer JM, Brienza A, Bruno F, Slutsky AS (1999) Effect of mechanical ventilation on inflammatory mediators in patients with acute respiratory distress syndrome: a randomized controlled trial. JAMA 282:54–61

Reyes M, Dudek A, Jahagirdar B, Koodie L, Marker PH, Verfaillie CM (2002) Origin of endothelial progenitors in human postnatal bone marrow. J Clin Invest 109:337–346

Riley DJ, Rannels DE, Low RB, Jensen L, Jacobs TP (1990) NHLBI Workshop Summary. Effect of physical forces on lung structure, function, and metabolism. Am Rev Respir Dis 142:910–914

Rubanyi GM, Romero JC, Vanhoutte PM (1986) Flow-induced release of endothelium-derived relaxing factor. Am J Physiol 250:H1145–H1149

Sanchez-Esteban J, Cicchiello LA, Wang Y, Tsai SW, Williams LK, Torday JS, Rubin LP (2001) Mechanical stretch promotes alveolar epithelial type II cell differentiation. J Appl Physiol 91:589–595

Sasamoto A, Nagino M, Kobayashi S, Naruse K, Nimura Y, Sokabe M (2005) Mechanotransduction by integrin is essential for IL-6 secretion from endothelial cells in response to uniaxial continuous stretch. Am J Physiol Cell Physiol 288:C1012–C1022

Shyy YJ, Hsieh HJ, Usami S, Chien S (1994) Fluid shear stress induces a biphasic response of human monocyte chemotactic protein 1 gene expression in vascular endothelium. Proc Natl Acad Sci USA 91:4678–4682

Slutsky AS, Tremblay LN (1998) Multiple system organ failure. Is mechanical ventilation a contributing factor? Am J Respir Crit Care Med 157:1721–1725

Sterpetti AV, Cucina A, Morena AR, Di Donna S, D'angelo LS, Cavalarro A, Stipa S (1993) Shear stress increases the release of interleukin-1 and interleukin-6 by aortic endothelial cells. Surgery 114:911–914

Stevens T, Phan S, Frid MG, Alvarez D, Herzog E, Stenmark KR (2008) Lung vascular cell heterogeneity: endothelium, smooth muscle, and fibroblasts. Proc Am Thorac Soc 5:783–791

Sud S, Friedrich JO, Taccone P, Polli F, Adhikari NK, Latini R, Pesenti A, Guerin C, Mancebo J, Curley MA, Fernandez R, Chan MC, Beuret P, Voggenreiter G, Sud M, Tognoni G, Gattinoni L (2010a) Prone ventilation reduces mortality in patients with acute respiratory failure and severe hypoxemia: systematic review and meta-analysis. Intensive Care Med 36:585–599

Sud S, Sud M, Friedrich JO, Meade MO, Ferguson ND, Wunsch H, Adhikari NK (2010b) High frequency oscillation in patients with acute lung injury and acute respiratory distress syndrome (ARDS): systematic review and meta-analysis. BMJ 340:c2327

Suki B, Alencar AM, Sujeer MK, Lutchen KR, Collins JJ, Andrade JS Jr, Ingenito EP, Zapperi S, Stanley HE (1998) Life-support system benefits from noise. Nature 393:127–128

Suki B, Lutchen KR, Ingenito EP (2003) On the progressive nature of emphysema: roles of proteases, inflammation, and mechanical forces. Am J Respir Crit Care Med 168:516–521

Suki B, Ito S, Stamenovic D, Lutchen KR, Ingenito EP (2005) Biomechanics of the lung parenchyma: critical roles of collagen and mechanical forces. J Appl Physiol 98:1892–1899

Takahashi S, Nakamura H, Seki M, Shiraishi Y, Yamamoto M, Furuuchi M, Nakajima T, Tsujimura S, Shirahata T, Nakamura M, Minematsu N, Yamasaki M, Tateno H, Ishizaka A (2008) Reversal of elastase-induced pulmonary emphysema and promotion of alveolar epithelial cell proliferation by simvastatin in mice. Am J Physiol Lung Cell Mol Physiol 294:L882–L890

Tate RM, Repine JE (1983) Neutrophils and the adult respiratory distress syndrome. Am Rev Respir Dis 128:552–559

Thodeti CK, Matthews B, Ravi A, Mammoto A, Ghosh K, Bracha AL, Ingber DE (2009) TRPV4 channels mediate cyclic strain-induced endothelial cell reorientation through integrin-to-integrin signaling. Circ Res 104:1123–1130

Torday JS, Sanchez-Esteban J, Rubin LP (1998) Paracrine mediators of mechanotransduction in lung development. Am J Med Sci 316:205–208

Tschumperlin DJ, Drazen JM (2006) Chronic effects of mechanical force on airways. Annu Rev Physiol 68:563–583

Tschumperlin DJ, Margulies SS (1999) Alveolar epithelial surface area-volume relationship in isolated rat lungs. J Appl Physiol 86:2026–2033

Tschumperlin DJ, Oswari J, Margulies AS (2000) Deformation-induced injury of alveolar epithelial cells. Effect of frequency, duration, and amplitude. Am J Respir Crit Care Med 162:357–362

Tseng H, Peterson TE, Berk BC (1995) Fluid shear stress stimulates mitogen-activated protein kinase in endothelial cells. Circ Res 77:869–878

Tsuno K, Miura K, Takeya M, Kolobow T, Morioka T (1991) Histopathologic pulmonary changes from mechanical ventilation at high peak airway pressures. Am Rev Respir Dis 143:1115–1120

Uematsu M, Ohara Y, Navas JP, Nishida K, Murphy TJ, Alexander RW, Nerem RM, Harrison DG (1995) Regulation of endothelial cell nitric oxide synthase mRNA expression by shear stress. Am J Physiol 269:C1371–C1378

Uhlig S, Uhlig U (2004) Pharmacological interventions in ventilator-induced lung injury. Trends Pharmacol Sci 25:592–600

Vlahakis NE, Hubmayr RD (2005) Cellular stress failure in ventilator-injured lungs. Am J Respir Crit Care Med 171:1328–1342

Vlahakis NE, Schroeder MA, Limper AH, Hubmayr RD (1999) Stretch induces cytokine release by alveolar epithelial cells in vitro. Am J Physiol 277:L167–L173

Ware LB, Matthay MA (2000) The acute respiratory distress syndrome. N Engl J Med 342:1334–1349

Weaver TE, Whitsett JA (1991) Function and regulation of expression of pulmonary surfactant-associated proteins. Biochem J 273(Pt 2):249–264

West JB (1971) Distribution of mechanical stress in the lung, a possible factor in localisation of pulmonary disease. Lancet 1:839–841

West JB (2003) Thoughts on the pulmonary blood-gas barrier. Am J Physiol Lung Cell Mol Physiol 285:L501–L513

West JB, Mathieu-Costello O (1992) Stress failure of pulmonary capillaries: role in lung and heart disease. Lancet 340:762–767

Wirtz HR, Dobbs LG (1990) Calcium mobilization and exocytosis after one mechanical stretch of lung epithelial cells. Science 250:1266–1269

Wirtz HR, Dobbs LG (2000) The effects of mechanical forces on lung functions. Respir Physiol 119:1–17

Wu S, Cioffi EA, Alvarez D, Sayner SL, Chen H, Cioffi DL, King J, Creighton JR, Townsley M, Goodman SR, Stevens T (2005) Essential role of a Ca^{2+}-selective, store-operated current (ISOC) in endothelial cell permeability: determinants of the vascular leak site. Circ Res 96: 856–863

Wung BS, Cheng JJ, Chao YJ, Lin J, Shyy YJ, Wang DL (1996) Cyclical strain increases monocyte chemotactic protein-1 secretion in human endothelial cells. Am J Physiol 270:H1462–H1468

Yamamoto K, Korenaga R, Kamiya A, Ando J (2000) Fluid shear stress activates Ca^{2+} influx into human endothelial cells via P2X4 purinoceptors. Circ Res 87:385–391

Yao X, Garland CJ (2005) Recent developments in vascular endothelial cell transient receptor potential channels. Circ Res 97:853–863

Yoshimura T, Matsushima K, Tanaka S, Robinson EA, Appella E, Oppenheim JJ, Leonard EJ (1987) Purification of a human monocyte-derived neutrophil chemotactic factor that has peptide sequence similarity to other host defense cytokines. Proc Natl Acad Sci USA 84:9233–9237

Zhang X, Halvorsen K, Zhang CZ, Wong WP, Springer TA (2009) Mechanoenzymatic cleavage of the ultralarge vascular protein von Willebrand factor. Science 324:1330–1334

Zhao D, Keates AC, Kuhnt-Moore S, Moyer MP, Kelly CP, Pothoulakis C (2001) Signal transduction pathways mediating neurotensin-stimulated interleukin-8 expression in human colonocytes. J Biol Chem 276:44464–44471

Chapter 8
Implications of Cytokines in Cochlear Pathophysiology

Keiji Tabuchi and Akira Hara

Abstract The cochlea is a mechanosensitive organ that perceives sounds. Hair cells with mechanosensitive transducer channels are the sensory cells of hearing. The stria vascularis generates specific electrochemical gradients of potassium and an endocochlear potential that are needed for the mechanoelectrical transduction of hair cells. Spiral ganglion neurons form the VIIIth cranial nerve and conduct action potentials. So far, little is known about the effects of cytokines/chemokines on the cochlear physiology in auditory perception. Recent reports have clarified the involvements of cytokines in the death of cochlear cells in various cochlear injuries. On the other hand, neurotrophic factors play key roles in the development and maintenance of spiral ganglion neurons. This review summarizes what is currently known about the involvement of cytokines in cochlear pathophysiology.

Keywords Cochlea · Cytokine · Hair cell · Spiral ganglion · Stria vascularis

8.1 Introduction

There are three steps in an acoustical situation: generation of sound by a source, propagation of sound from the source to a receiver, and reception of the sound. Sound generation usually originates from mechanical vibration, and sound is transferred by the movement of air molecules in the direction of propagation (Yates, 1995; Hartmann, 1995). The movement of air molecules in the external auditory canal causes movement of the tympanic membrane, which, in turn, vibrates the middle ear ossicles and cochlea. The cochlea is a specialized organ for the mechanotransduction of auditory stimuli, and the sense of hearing relies on the fine structures of the cochlea. In this review, we first describe the cochlear structures and auditory transduction.

Cytokines are produced by a variety of cells and are used in the cell-cell communication at immediate and intermediate ranges. They act by binding to their distinct

K. Tabuchi (✉)
Department of Otolaryngology, Graduate School of Comprehensive Human Sciences, University of Tsukuba, Tsukuba 305-8575, Japan
e-mail: ktabuchi@md.tsukuba.ac.jp

A. Kamkin, I. Kiseleva (eds.), *Mechanical Stretch and Cytokines*,
Mechanosensitivity in Cells and Tissues 5, DOI 10.1007/978-94-007-2004-6_8,

receptors and subsequently activate intracellular signaling cascades. It has been demonstrated that inflammatory cytokines play a key role in the death of hair cells in various cochlear injuries as well as cellular stress responses. This review focuses on mechanisms of hair cell death induced by inflammatory cytokines. Based on their anti-inflammatory and anti-apoptotic actions, glucocorticoids have been widely used in daily clinics in the treatment of sensorineural hearing loss of acute onset. The protective actions of glucocorticoids against cytokine/chemokine-induced cochlear injuries are also reviewed.

Growth factors are essential for the development and maintenance of organs. The effects of growth factors on hair cell development are still unclear, but recent findings have demonstrated that neurotrophin 3 (NT3) and brain-derived neurotrophic factor (BDNF) are of particular importance in the development of cochlear neuronal cells. Therefore, the effects of these neurotrophic factors on auditory afferent neurons are also reviewed in addition to the pathological actions of inflammatory cytokines in the cochlea.

8.2 Overview of the Cochlear Structure and Transduction of Acoustic Stimuli

The cochlea adopts a characteristic spiral form for sound reception. In humans, the total length of the cochlea is about 35 mm, coiled into 2 1/2-2 3/4 turns. It is divided into three compartments: the scala vestibuli, scala media, and scala tympani. The scala vestibuli is contiguous with the scala tympani at the top of the cochlea (helicotrema), and they are filled with perilymph. The scala media is filled with endolymph. Perilymph is high in sodium and low in potassium, but endolymph shows unique characteristics as extracellular fluid. Namely, endolymph is high in potassium and low in sodium, and has an extracellular potential (termed the endocochlear potential (EP)) of approximately +80 mV compared to perilymph. The stria vascularis plays a critical role in EP generation by recycling potassium in the cochlea.

The scala media and scala tympani are separated by the basilar membrane, on which is the organ of Corti. Acoustical stimuli move the tympanic membrane and then the stapes footplate. The movement of the stapes footplate causes fluctuation in the volume of the scala vestibuli. This volume fluctuation induces vertical displacement of the basilar membrane and generates a travelling wave.

Displacement of the basilar membrane causes oscillation of the organ of Corti. The organ of Corti has two types of cochlear sensory cells: inner hair cells (IHCs) and outer hair cells (OHCs). IHCs are flask-shaped and true afferent receptor cells. On the other hand, OHCs are cylindrically shaped cells that form three rows. OHCs contribute to the active process, also being known as cochlear amplifier. OHCs add energy to the traveling wave of the basilar membrane to enhance its movement. Two mechanisms have been proposed for cochlear amplification in mammals: somatic electromotility, and active hair-bundle motility (Gillespie and

Muller, 2009). Somatic electromotility is caused by voltage-dependent changes in the confirmation of molecules of prestin expressed in the lateral wall of OHCs (Dallos, 2008; Dallos et al., 2008).

Cochlear hair cells face endolymph only in their apical membrane and hair bundles. Other parts of hair cells (basolateral parts of hair cells) face perilymph. Transduction channels exist in the tip region of hair bundles. They are mechanically-sensitive nonselective cation channels with a high permeability to calcium (Corey and Hudspeth, 1979; Ohmori, 1985; Jaramillo and Hudspeth, 1991). Travelling waves cause deflection of the stereocilia (the hair of hair cells) toward the tallest stereocilia, which leads to the shearing of shorter stereocilia. This deflection of the stereocilia opens the transduction channeles, results in an influx of cations, and depolarizes hair cells. Because endolymph is high in potassium, potassium entery into hair cells results in hair cell depolarization. EP is the driving force for the influx of potassium from the endolymphatic space into hair cells, contributing to the perception of auditory stimuli. Although the mechanotransduction of hair cells has been studied in considerable detail, the molecules that form the transduction channels have not been clarified (Tabuchi et al., 2005; Tabuchi and Hara, 2009; Gillespie and Muller, 2009).

8.3 Production of Proinflammatory Cytokines in the Cochlea

Recent studies have demonstrated that proinflammatory cytokines are secreted by cells in the modiolus and organ of Corti as well as fibrocytes in the lateral wall (Kahn et al., 2010). However, it has been suggested that proinflammatory cytokines are produced in the cochlea mainly by stimulated fibrocytes in the spiral ligament and by the infiltrating leukocytes (Yoshida et al., 1999; Satoh et al., 2002). The treatment of cultured primary spiral ligament fibrocytes with recombinant interleukin-1β (IL-1β stimulates the release of IL-6 and tumor necrosis factor a (TNFα). In addition, other proinflammatory cytokines/chemokines like monocyte chemoattractant protein-1, macrophage inflammatory protein-2, and soluble adhesion molecules are also produced by the IL-1β stimulated fibrocytes (Ichimiya et al., 2000; Yoshida et al., 1999). The secretion of these cytokines/chemokines and other mediators induces inflammatory cell movement, which would prolong inflammatory responses in the cochlea. Since an increase of proinflammatory cytokines in the cochlea plays a key role in cochlear infiltration leukocytes, generation of proinflammatory cytokines by fibrocytes in the lateral wall is an important step in cochlear inflammatory responses.

The causes and mechanisms of proinflammatory cytokine generation in the cochlear fibrocytes have not been clarified. However, it has been suggested that reactive oxygen species generated in various cochlear injuries are one of the main causative factors for the generation of proinflammatory cytokines. Based on a culture experiment using HEI-OC1 cells, cisplatin treatment induces the activation of NADPH oxidase, which, in turn, causes the overproduction of reactive oxygen species (Kim et al., 2010). Reactive oxygen species generated by cisplatin exposure increase the synthesis of proinflammatory cytokines, including TNFα, IL-1β, and IL6, though the activation of MEK1/extracellular signal-regulated kinase (ERK) (Kim et al., 2010).

8.4 Involvement of Cytokines in Cochlear Hair Cell Death in Cochlear Injuries

Various forms of trauma cause damage to hair cells. Recent studies have shown that hair cell death occurs through apoptosis, which leads to permanent hearing loss. Apoptosis mainly occurs through the sequential actions of caspases. Two upstream initiator caspases, caspase-8 and caspase-9, and their associated extrinsic and intrinsic pathways are well known. The extrinsic pathway begins with the activation of cell surface death receptors. Activation of these death receptors cleaves caspase-8, which, in turn, activates downstream effector caspases such as caspase-3. In contrast, the intrinsic pathway is initiated by changes in the mitochondrial membrane permeability. Cytochrome c released from mitochondria forms a protein complex with Apaf 1. This, in turn, activates caspase-9 and the downstream effector caspases. Activated effector caspases then cleave critical intracellular proteins to induce the final stages of cell death. The current literature, using cell lines or explants from neonatal rodents, supports the intrinsic apoptotic pathway as the major pathway of hair cell deth induced by aminoglycosides in the cochlea (Tabuchi et al., 2007; Ding et al., 2009). However, recent findings have suggested that the extrinsic as well as intrinsic pathways are activated in the course of hair cell death in various cochlear injuries (Nicotera et al., 2003; Ding et al., 2007).

It has been demonstrated that the inner ear functions, hearing and balance functions, are influenced by immune responses inside the inner ear (Rahman et al., 2001). Although immune responses are necessary to protect the inner ear against infectious labyrinthitis, they also damage the inner ear, leading to cochlear degeneration and permanent hearing loss (Ryan et al., 2002; Satoh et al., 2003). Proinflammatory cytokines including TNFα, IL-1β, and IL-6 are expressed after ototoxic insults, such as the excessive administration of cisplatin (So et al., 2007), hypoxia (Kahn et al., 2010), vibration (Zou et al., 2005), and acoustic overstimulation (Fujioka et al., 2006).

Mitogen-activated protein kinases (MAPKs) are serine-threonine kinases that mediate intracellular signaling associated with a variety of cellular activities including cell proliferation, differentiation, survival, death, and transformation. The mammalian MAPK family consists of ERK, p38 MAPK, and c-Jun NH_2-terminal kinase (JNK), also known as stress-activated protein kinase (SAPK). The JNK and p38 signaling pathways are activated by pro-inflammatory cytokines such as TNFa and IL-1β in addition to cellular stresses (Kim and Choi, 2009). The activation of MAPKs is demonstrated after the IL-1β treatment of cochlear culture cells (Nam, 2006). These pathways are activated in various cochlear injuries, and reportedly involved in hair cell death through apoptotic mechanisms. Several reports indicated that the JNK pathway was activated during the course of the aminoglycoside-induced apoptosis of hair cells, and that aminoglycoside-induced hearing loss was alleviated by JNK inhibitors (Murai et al., 2008; Ylikoski et al., 2002; Wang et al., 2003; Nakamagoe et al., 2010). Recent reports have indicated that the JNK pathway is also involved in hair cell death in acoustic injury and age-related hearing loss (Wang et al., 2003; Sha et al., 2009). The excessive administration of gentamicin, acoustic overstimulation, and aging are also known to activate p38 MAPK (Sha et al., 2009; Tabuchi et al., 2010; Wei et al., 2005). Upregulation of the mRNA level of p38 MAPK was reported in acute phases of cases with parmanent threshold shifts in acoustic injury (Meltser et al., 2010).

It has been demonstrated that the cochlea can be protected against various cochlear injuries by the inhibition of proinflammatory cytokines and MAPK pathways, at least in part (Abi-Hachem et al., 2010; Dinh and Van De Water, 2009). Cochlear damage induced by cisplatin is attenuated by the downregulation of proinflammatory cytokines (So et al., 2008). The blockade of IL-6 signaling using anti-IL-6 receptor antibody (MR16-1) suppressed the cochlear inflammatory response and improved hearing impairment in the noise-damaged cochlea (Wakabayashi et al., 2010).

Proinflammatory cytokines can induce hair cell death via MAPK pathways (Abi-Hachem et al., 2010; Dinh and Van De Water, 2009). In addition to directly inducing hair cell death, proinflammatory cytokines/chemokines can damage the cochlea by the recruitment of inflammatory cells. TNFa infused into the scala tympani of the cochlea resulted in the infiltration of leukocytes within the scala tympani (Keithley et al., 2008). Gene array analysis showed increased expressions of cytokines/chemokines after acoustic overexposure (Tornabene et al., 2006). The recruitment of leukocytes and macrophages to the cochlea was observed in acoustic injury (Hirose et al., 2005; Miyao et al., 2008). These leukocytes further damage the cochlea by the production of inflammatory cytokines and reactive oxygen species.

8.5 Involvement of Cytokines in Degeneration of Spiral Ganglion Neurons

The role of proinflammatory cytokines in the injury of spiral ganglion neurons is currently unclear. However, recent findings have suggested that same deteriorative actions of proinflammatory cytokines as observed in hair cells may exist in cochlear neuronal injury. Wei et al. (2010) demonstrated that salicylate induces apoptotic genes, particularly among members of the TNF family, including Tnfsf10, Lta (TNF ligands), and Fas (TNF receptor), in the culture explant of spiral ganglion cells. As salicylate induces the soma shrinkage of cultured cells and causes the apoptosis of these cells, their findings suggests that TNF family substances are involved in the salicylate-induced apoptosis of spiral ganglion cells. Furthermore, blockade of the IL-6 signaling pathway decreased the infiltration of activated macrophages around spiral ganglion cells and suppressed the loss of spiral ganglion neurons in cochlear acoustic injury (Wakabayashi et al., 2010).

8.6 Neurite Outgrowth with Growth Factors

Generally, developing neurons depend on trophic factors for continued survival, growth, and differentiation. Members of the family of neurotrophins, especially neurotrophin 3 (NT3) and brain-derived neurotrophic factor (BDNF), and their high-affinity receptors, TrkB and TrkC, respectively, reportedly play an important role in the development of auditory afferent neurons. They are also involved in the maintenance of their innervations to hair cells. The neurotrophins are first expressed in the otocyst around the time afferent sensory neurons become postmitotic, and they are critical for the survival of spiral ganglion neurons. The distribution of neurotrophins varies along a base-to-apex gradient in the cochlea. NT-3 knockout mice show a complete loss of spiral ganglion neurons in the basal turn of the cochlea, but only a partial loss in middle and apical turns at the embryonic stage (Fritzsch et al., 1997). The distribution pattern of NT3 reportedly changes during development. Namely, trophic effects of NT3 are more evident in the basal turn in the embryonic stage, as described above. However, the expression of NT3 decreases from the apical to basal cochlear turn in the post-natal stage, and this expression pattern persists into adulthood (Sugawara et al., 2007). BDNF, on the other hand, supports the sensory neurons of the apex and middle turn of the cochlea and of the semicircular canals.

In addition to the roles in the development of spiral ganglion neurons, NT-3 and BDNF exhibited neuroprotective and neurotrophic effects in the presence of cochlear injuries. Namely, they protect spiral ganglion neurons from cochlear injuries. They significantly enhance the survival of spiral ganglion neurons and promote neurite outgrowth (Staecker et al., 1996; Altschuler et al., 1999). The overexpression of BDNF significantly increased the survival rate of spiral ganglion neurons in the deafened ear (Miller et al., 1997). The pharmacological application of BDNF and NT-3 also increases the survival rate of spiral ganglion neurons in culture, and protects spiral ganglion neurons from a variety of ototoxic drugs (Zheng and Gao, 1996; Gao, 1999; McGuinness and Shepherd, 2005). The most prominent effect on survival was seen when the concentration of both neurotrophins was 5 ng/ml (Mou et al., 1998). Recently, Sun and Salvi (2009) demonstrated that BDNF and NT3 modulate neurotransmitter receptor expressions on developing spiral ganglion neurons. BDNF and NT3 enhanced the expression of gamma-aminobutyric acid (GABA) receptors and suppressed glycine receptor expression on spiral ganglion neurons in mouse pups. In addition, it has been demonstrated that NT3 is capable of altering the firing characteristics of spiral ganglion neurons (Zhou et al., 2005).

Hair cells and supporting cells are major sources of NT3 in the cochlea, and, thus, the loss of inner hair cells may lead to the degeneration of nerve fibers from the sensory epithelium, and eventually to the degeneration of spiral ganglion neurons (Nadol, 1997; Sugawara et al., 2007). Mammalian auditory hair cells and spiral ganglion neurons are not considered to spontaneously regenerate after cell death. Cochlear implants are useful to restore the auditory function in patients with severe or profound hearing loss of hair cell origin. Namely, in the absence of hair cells, cochlear implant electrodes can stimulate spiral ganglion neurons. From this point of view, the effects of BDNF and NT3 to enhance the survival of spiral ganglion neurons may be important (Shibata et al., 2010). Furthermore, elevated levels of BDNF and NT3 not only increased the survival of spiral ganglion neurons, but also induced the sprouting of peripheral nerve fibers. Shibata et al. (2010) reported that forced expression of the BDNF gene in epithelial or mesothelial cells that remain in the deaf ear induced the robust regrowth of nerve fibers towards the cells that secrete the neurotrophin.

8.7 The Lateral Wall and Cytokines/Chemokines

Cochlear acoustic injury and the ototoxicity of aminoglycosides and cisplatin predominantly damage hair cells. In several diseases, however, it has been suggested that the cochlear lateral wall including stria vascularis is mainly damaged and cochlear inflammatory responses may be involved in the generation of injury of the cochlear lateral wall. Ichimiya et al. (1999) showed that immunostaining for connexin 26 was decreased in the spiral ligament, accompanied by marked fibrinogen staining in this region, after the administration of viable *Streptococcus pneumoniae* into the middle ear cavity in mice. Connexin 26 is critical for potassium recycling and EP generation in the cochlea. Pendred syndrome is an autosomal-recessive disorder characterized by deafness and goiter. This disorder is caused by a mutation of *Slc26a4*, which codes for pendrin, an anion exchanger. In *Slc26a4*-knockout mice, marginal cells of the stria vascularis are disorganized (Jabba et al., 2006). In this knockout mouse, macrophages infiltrate the cochlea, and macrophage infiltration is observed only in the stria vascularis (Jabba et al., 2006). Based on these findings, Jabba et al. (2006) suggested that inflammatory actions in the stria vascularis might be the main cause of this disease.

8.8 Protective Effects of Glucocorticoids Against Cochlear Injury

Glucocorticoids have been used for the treatment of inner ear disorders such as idiopathic sudden sensorineural hearing loss, acoustic injury, Meniere's disease, and immune-mediated hearing loss for decades. Most clinicians accept glucocorticoids as the treatment of choice for sensorineural hearing loss of sudden-onset, although their precise protective mechanism has not yet been clarified. Animal studies have demonstrated that glucocorticoids exhibit protective effects against various cochlear injuries. Prednisolone and methylprednisolone ameliorated ischemia-reperfusion injury of the cochlea in the guinea pig and chinchilla (Tabuchi et al., 2003). Dexamethasone, prednisolone, and methylprednisolone were reported to decrease the threshold shift of the hearing level and prevent hair cell death induced by acoustic overexposure (Tabuchi et al., 2006; Hirose et al., 2007). Glucocorticoids also protect the cochlea from ototoxic agents. Glucocorticoids reportedly protect hair cells against the ototoxicity of aminoglycosides (Himeno et al., 2002). Generally, glucocorticoids inhibit the release of proinflammatory cytokines and stimulate the production of anti-inflammatory cytokines. It has been demonstrated that glucocorticoids inhibit the synthesis and release of inflammatory cytokines in the cochlea (Maeda et al., 2005). In addition, Dinh et al. (2008a, b) suggested that TNFa may increase Bax expression, change the Bax/Bcl ratio, and initiate the apoptosis of hair cells, and that dexamethasone may decrease BAX expression induced by TNFa in hair cells and protect such cells against TNFa-induced apoptosis. Kalinec et al. (2009) recently showed that glucocorticoids activate a myosin IIC-mediated mechanism that drives annexin A1 from the lipid droplets to the apical region of Hensen cells, where it is released into the external milieu. Their findings suggest that annexin A1 may be one of the major mediators of the anti-inflammatory effects of glucocorticoids in the cochlea.

8.9 Conclusion and Perspectives

Recent findings have clarified that inflammatory cytokines are involved in the activation of MAPKs and hair cell death in various cochlear diseases such as acoustic injury and aminoglycoside ototoxicity. The activation of MAPK cascades contributes to disease progression through inducing the apoptosis of auditory hair cells. In addition to the actions of inflammatory cytokines directly inducing hair cell death, they are also involved in the infiltration of inflammatory cells into the cochlea. These cells enhance the production of cytokines and generate reactive oxygen species. The deteriorating roles of inflammatory cytokines are also suspected in injuries of the spiral ganglion neurons and cochlear lateral wall in some situations. Glucocorticoids have been used in the treatment of cochlear diseases. The anti-inflammatory actions of glucocorticoids partially serve to preserve the cochlear function.

Neurotrophic factors play critical roles in the development and survival of spiral ganglion neurons. Advances in cochlear implant surgery have made the issue of cochlear neuronal survival more prominent.

The cochlea is a mechanosensory organ that facilitates hearing. Our understanding of the effects of cytokines will offer new insights regarding cochlear pathophysiology and the treatment of sensorineural hearing loss.

Acknowledgement This work was supported by a Grant-in-aid for Scientific Research ((C) 20591969) from the Ministry of Education, Culture, Sports, Science, and Technology of Japan.

References

Abi-Hachem RN, Zine A, Van De V, De Water TR (2010) The injured cochlea as a target for inflammatory processes, initiation of cell death pathways and application of related otoprotective strategies. Recent Pat Drug Discov 5:147–163

Altschuler RA, Cho Y, Ylikoski J, Pirvola U, Magal E, Miller JM (1999) Rescue and regrowth of sensory nerves following deafferentation by neurotrophic factors. Ann N Y Acad Sci 884:305–311

Corey DP, Hudspeth AJ (1979) Ionic basis of the receptor potential in a vertebrate hair cell. Nature 281:675–677

Dallos P (2008) Cochear amplification, outer hair cells and prestin. Curr Opin Neurobiol 18:370–376

Dallos P, Wu X, Cheatham MA, Gao J, Zheng J, Anderson CT, Jia S, Cheng WH, Sengupta S, He DZ, Zuo J (2008) Prestin-based outer hair cell motility is necessary for mammalian cochlear amplification. Neuron 58:299–301

Ding D, Jiang H, Wang P, Salvi R (2007) Cell death after co-administration of cisplatin and ethacrynic acid. Hear Res 226:129–139

Ding D, Jiang H, Salvi RJ (2009) Mechanisms of rapid sensory hair-cell death following co-administration of gentamicin and ethacrynic acid. Hear Res 259:16–23

Dinh CT, Van De Water TR (2009) Blocking pro-cell-death pathways to conserve hearing. Audiol Neurootol 14:383–392

Dinh CT, Haake S, Chen S, Hoang K, Nong E, Eshraghi AA, Balkany TJ, Van De Water TR (2008a) Dexamethasone protects organ of Corti explants against tumor necrosis factor -alpha-induced loss of auditory hair cells and alters the expression levels of apoptosis-related genes. Neuroscience 157:405–413

Dinh CT, Hoang K, Haake S, Chen S, Angeli S, Nong E, Eshraghi AA, Balkany TJ, Van De Water TR (2008b) Biopolymer-released dexamethasone prevents tumor necrosis factor alpha-induced loss of auditory hair cells in vitro: implications toward the development of a drug-eluting cochlear implant electrode array. Otol Neurotol 29:1012–1019

Fritzsch B, Silos-Santiago I, Bianchi LM, Farinas I (1997) The role of neurotrophic factors in regulating the development of inner ear innervations. Trends Neurosci 20:159–164

Fujioka M, Kanzaki S, Okano HJ, Masuda M, Ogawa K, Okano H (2006) Proinflammatory cytokines expression in noise-induced damaged cochlea. J Neurosci Res 83:575–583

Gao WQ (1999) Role of neurotrophins and lectins in prevention of ototoxicity. Ann N Y Acad Sci 884:312–327

Gillespie PG, Muller U (2009) Mechanotransduction by hair cells: models, molecules, and mechanisms. Cell 139:33–44

Hartmann WM (1995) The physical description of signals. In: Moore BCJ (ed) Hearing. Academic, San Diego, CA, pp 1–40

Himeno C, Komeda M, Izumikawa M, Takemura K, Yagi M, Wiping Y, Doi T, Kuriyama H, Miller JM, Yamashita T (2002) Intra-cochlear administration of dexamethasone attenuates aminoglycoside ototoxicity in the guinea pig. Hear Res 167:61–70

Hirose K, Discolo CM, Keasler JR, Ransohoff R (2005) Mononuclear phagocytes migrate into the murine cochlea after acoustic trauma. J Comp Neurol 22:180–194

Hirose Y, Tabuchi K, Oikawa K, Murashita H, Sakai S, Hara A (2007) The effects of the glucocorticoid receptor antagonist RU486 and phospholipase A2 inhbitor quinacrine on acoustic injury of the mouse cochlea. Neurosci Lett 413:63–67

Ichimiya I, Suzuki M, Hirano T, Mogi G (1999) The influence of pneumococcal otitis media on the cochlear lateral wall. Hear Res 131:128–134

Ichimiya I, Yoshida K, Hirano T, Suzuki M, Mogi G (2000) Significance of spiral ligament fibrocytes with cochlear inflammation. Int J Pediatr Otorhinolaryngol 56:45–51

Jabba SV, Oelke A, Singh R, Maganti RJ, Fleming S, Wall SM, Everett LA, Green ED, Wangemann P (2006) Macrophage invasion contributes to degeneration of stria vascularis in Pendred syndrome mouse model. BMC Med 4:37

Jaramillo F, Hudspeth AJ (1991) Localization of the hair cell–s transduction channels at the hair bundle–s top by iontophoretic application of a channel blocker. Neuron 7:409–420

Kahn M, Szczepek AJ, Haupt H, Olze H, Mazurek B (2010) Expression of the proinflammtory cytokines in cochlear explants cultures: influence of normoxia and hypoxia. Neurosci Lett 479:249–252

Kalinec F, Webster P, Maricle A, Guerrero D, Chakravarti DN, Chakravarti B, Gellibolian R, Kalinec G (2009) Glucocorticoid-stimulated, transcription-independent release of annexin A1 by cochlear Hensen cells. Br J Pharmacol 158:1820–1834

Keithley EM, Wang X, Barkdull GC (2008) Tumor necrosis alpha can induce recruitment of inflammatory cells to the cochlea. Otol Neurotol 29:854–859

Kim EK, Choi E-J (2009) Pathological roles of MAPK signaling pathways in human diseases. Biochim Biophys Acta 1802:396–405

Kim HJ, Lee JH, Kim SJ, Oh GS, Moon HD, Kwon KB, Park C, Park BH, Lee HK, Chung SY, Park R, So HS (2010) Roles of NADPH oxidase in cisplatin-induced reactive oxygen species generation and ototoxicity. J Neurosci 30:3933–3946

Maeda K, Yoshida K, Ichimiya I, Suzuki M (2005) Dexamethasone inhibits tumor necrosis factor-alpha-induced cytokine secretion from spiral ligament fibrocytes. Hear Res 202:154–160

McGuinness SL, Shepherd RK (2005) Exogenous BDNF rescues rat spiral ganglion neurons in vivo. Otol Neurotol 26:1064–1072

Meltser I, Tahera Y, Canlon B (2010) Differential activation of mitogen-activated protein kinases and brain-derived neurotrophic factor after temporary or permanent damage to a sensory system. Neuroscience 165:1439–1446

Miller JM, Chi DH, O–Keeffe LJ, Kruzska P, Raphael Y, Altschuler RA (1997) Neurotrophins can enhance spiral ganglion cell survival after inner ear hair cell loss. Int J Dev Neurosci 15:631–643

Miyao M, Firestein GS, Keithley EM (2008) Acoustic trauma augments the cochlear immune response to antigen. Laryngoscope 118:1–8

Mou K, Adamson CL, Davis RL (1998) Time-dependence and cell-type specificity of synergistic neurotrophin actions on spiral ganglion neurons. J Comp Neurol 402:129–139

Murai N, Kirkegaard M, Jariebark LL, Risling M, Sneson A, Ulfendahl M (2008) Activation of JNK in the inner ear following impulse noise exposure. J Neurotrauma 25:72–77

Nadol JB Jr (1997) Patterns of neural degeneration in the human cochlea and auditory nerve: implications for cochlear implantation. Otolaryngol Head Neck Surg 117:220–228

Nakamagoe M, Tabuchi K, Uemaetomari I, Nishimura B, Hara A (2010) Estradiol protects the cochlea against gentamicin ototoxicity through inhibition of JNK pathway. Hear Res 261:67–74

Nam SI (2006) Interleukin-1beta up-regulates inducible nitric oxide by way of phosphoinositide 3-kinase-dependent in a cochlear cell model. Laryngoscope 116:2166–2170

Nicotera TM, Hu BH, Henderson D (2003) The caspase pathway in noise-induced apoptosis of the chinchilla cochlea. J Assoc Res Otolaryngol 4:466–477

Ohmori H (1985) Mechano-electrical transduction currents in isolated vestibular hair cells of the chick. J Physiol (London) 359:189–217

Rahman MU, Poe DS, Choi HK (2001) Autoimmune vestibule-cochlear disorders. Curr Opin Rheumatol 13:184–189

Ryan AF, Harris JP, Keithley EM (2002) Immune-mediated hearing loss: basic mechanisms and options for therapy. Acta Otolayrngol Suppl 548:38–43

Satoh H, Firestein GS, Billings PB, Harris JP, Keithley EM (2002) Tumor necrosis factor-alpha, an initiator, and etanercept, an inhibitor of cochlear inflammation. Laryngoscope 112:1627–1634

Satoh H, Firestein GS, Billings PB, Harris JP, Keithley EM (2003) Proinflammatory cytokined expression in the endolymphatic sac during inner ear inflammation. J Assoc Res Otorlayngol 4:139–147

Sha S-H, Chen F-Q, Schacht J (2009) Activation of cell death pathways in the inner ear of the aging CBA/J mouse. Hear Res 254:92–99

Shibata SB, Cortez SR, Beyer LA, Wiler JA, Di Polo A, Pfingst BE, Raphael Y (2010) Transgenic BDNF induces nerve fiber regrowth into the auditory epithelium in deaf cochleae. Exp Neurol 223:464–472

So H, Kim H, Lee JH, Park C, Kim Y, Kim E, Kim JK, Yun KJ, Lee KM, Lee HY, Moon SK, Lim DJ, Park R (2007) Cisplatin ototoxicity of auditory cells requires secretions of proinflammatory cytokines via activation of ERK and NF-kappaB. J Assoc Res Otolaryngol 8:338–355

So H, Kim H, Kim Y, Kim E, Pae HO, Chung HT, Kim HJ, Kwon KB, Lee KM, Lee HY, Moon SK, Park R (2008) Evidence that cisplatin-induced auditory damage is attenuated

by downregulation of pro-inflammatory cytokines via Nrf2/Ho-1. J Assoc Res Otolaryngol 8:338–355
Staecker H, Kopke R, Malgange B, Lefebvre P, Van de Water TR (1996) NT-3 and/or BDNF therapy prevents loss of auditory neurons following loss of hair cells. NeuroReport 7:889–894
Sugawara M, Muritie JC, Stankovic KM, Lierman MC, Corfas G (2007) Dyanmic patterns of neurotrophin 3 expression in the postnatal mouse inner ear. J Com Neurol 501:30–37
Sun W, Salvi RJ (2009) Brain derived neurotrophic factor and neurotrophic factor 3 modulate neurotransmitter receptor expressions on developing spiral ganglion neurons. Neuroscience 164:1854–1866
Tabuchi K, Hara A (2009) Mechanosensitivity of the cochlea. In: Kamkin A, Kiseleva I (Eds) Mechanosensitivity of the nervous system. Springer, New York, NY, pp 107–114
Tabuchi K, Oikawa K, Uemaetomari I, Tsuji S, Wada T, Hara A (2003) Glucocorticoids and dehydroepiandrosterone sulfate ameliorate ischemia-induced injury of the cochlea. Hear Res 180:51–60
Tabuchi K, Suzuki M, Mizuno A, Hara A (2005) Hearing impairment in TRPV4 knockout mice. Neurosci Lett 382:304–308
Tabuchi K, Murashita H, Sakai S, Hoshino T, Uemaetomari I, Hara A (2006) Therapeutic time window of methylprednisolone in acoustic injury. Otol Neurotol 27:1176–1179
Tabuchi K, Park K, Chavez E, Ryan AF (2007) Role of inhibitor of apoptosis protein in gentamicin-induced cochlear hair cell damage. Neuroscience 149:213–222
Tabuchi K, Oikawa K, Hoshino T, Nishimura B, Hayashi K, Yanagawa T, Warabi E, Ishii T, Tanaka S, Hara A (2010) Cochlear protection from acoustic injury by inhibitors of p38 mitogen-activated protein kinase and sequestosome 1 stress protein. Neuroscience 166:665–670
Tornabene SV, Sato K, Pham L, Billings P, Keithley EM (2006) Immune cell recruitment following acoustic trauma. Hear Res 222:115–124
Wakabayashi K, Fujioka M, Kanzaki S, Okano HJ, Shibata S, Yamashita D, Masuda M, Mihara M, Ohsugi Y, Ogawa K, Okano H (2010) Blockade of interleukin-6 signaling suppressed cochlear inflammatory response and improved hearing impairment in nois-damaged mice cochlea. Neurosci Res 66:345–352
Wang J, Van De Water TR, Bonny C, de Ribaupierre F, Puel JL, Zine A (2003) A peptide inhibitor of c-Jun N-terminal kinase protects against both aminoglycoside and acoustic trauma-induced auditory hair cell death and hearing loss. J Neurosci 23:8596–8607
Wei X, Zhao L, Liu J, Dodel RC, Farlow MR, Du Y (2005) Minocycline prevents gentamicin-induced ototoxicity by inhibiting p38 MAP kinase phosphorylation and caspase 3 activation. Neuroscience 131:513–521
Wei L, Ding D, Salvi R (2010) Salicylate-induced degeneration of the cochlea spiral ganglion neurons-apoptosis signaling. Neuroscience 168:288–299
Yates GK (1995) Cochlear structure and function. In: Moore BCJ (ed) Hearing. Academic, San Diego, CA, pp 41–74
Ylikoski J, Xing-Qun L, Virkkala J, Pirvola U (2002) Blockade of c-Jun N-terminal kinase pathway attenuates gentamicin-induced cochlear and vestibular hair cell death. Hear Res 163:71–81
Yoshida K, Ichimiya I, Suzuki M, Mogi G (1999) Effect of proinflammatory cytokines on cultured spiral ligament fibrocytes. Hear Res 137:155–159
Zheng JL, Gao WQ (1996) Differential damage to auditory neurons and hair cells by ototoxins and neuroprotection by specific neurotrophins in rat cochlear organotypic cultures. Eur J Neurosci 8:1897–1905
Zhou Z, Liu Q, Davis RL (2005) Complex regulation of spiral ganglion neuron firing patterns by neurotrophin-3. J Neurosci 25:7558–7566
Zou J, Pyykko I, Sutinen P, Toppila E (2005) Vibration induced hearing loss in guinea pig cochlea: expression of TNF-alpha and VEGF. Hear Res 202:13–20

Chapter 9
Primary Cilia are Mechanosensory Organelles with Chemosensory Roles

Brian S. Muntean, Xingjian Jin, and Surya M. Nauli

Abstract In the previous volume, we discussed the roles of primary cilia as mechanosensory organelles. Primary cilia have also been proposed to be chemosensory organelles. As a mechanosensory organelle, primary cilium could initiate the release of various cytokines-like substances, involving nitric oxide and purinergic agonists. As a chemosensory organelle, primary cilium also contains various receptors responding to hedgehog and Wnt signaling systems. In this section, we will describe cilia as newly recognized communication devices in response to agonist to regulate cell cycle and cellular development.

Keywords Centriole · Centrosome · Basal body · Primary cilium · Shear stress · Signal transduction

9.1 Introduction

The cilium family consists of motile and non-motile (primary) cilia. One difference between the two is their physiological tissue expression. Motile cilia are seen in places such as the respiratory system. They have the ability to beat in a particular motion that helps with the passage of fluid over cells. When present, motile cilia are seen on a cell in multiple copies. In contrast, the primary cilium does not possess similar motor functions and exists as a single copy on the cell.

The structure of a motile cilium mimics that of a flagellum (Fig. 9.1). The core axoneme structure is a circular arrangement of nine outer microtubule doublets (A+B) along with two inner microtubule singlets. The primary non-motile cilium has the outer nine microtubule doublets but lacks the inner pair of microtubules and thus is designated as a "9+0" axoneme (versus "9+2" for the former, motile cilia). In either case, the cilium is surrounded by the plasma membrane. Motile and non-motile cilia may look quite similar, but as this basic structural information eludes, they have drastically different functions.

S.M. Nauli (✉)
Department of Pharmacology, The University of Toledo, Toledo, OH 43614, USA
e-mail: Surya.Nauli@UToledo.Edu

A. Kamkin, I. Kiseleva (eds.), *Mechanical Stretch and Cytokines*, Mechanosensitivity in Cells and Tissues 5, DOI 10.1007/978-94-007-2004-6_9,

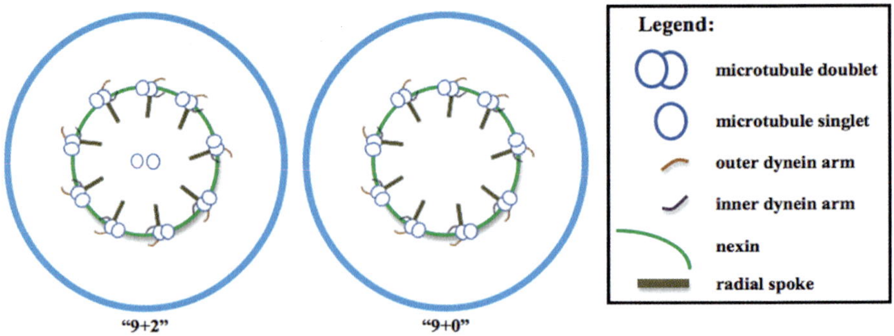

Fig. 9.1 Cross sections of motile and non-motile cilium. Although there is some exception, "9+2" cilia are usually referred as motile cilia. In general, the non-motile cilia have "9+0" microtubules arrangement. A structural nexin protein connects the outer microtubule doublets to form a ring in both motile and non-motile cilium. The radial spokes are oriented inward from the outer microtubule doublets

A structural nexin protein connects the outer microtubule doublets to form a ring in both motile and nonmotile cilium (Fig. 9.1). The motile cilium is then complemented with radial spokes oriented inward from the outer microtubule doublets. It also features two dynein arms that are present per A-tubule of each outer microtubule doublet. They are oriented to approach the adjacent B-tubule of the next microtubule doublet. These arms slide along the axoneme, providing motile cilia its ability to beat.

A primary cilium is a small hair-like projection extended from the basal body or mother centriole (Fig. 9.2). This region, where the triplet microtubules of the basal body give rise to the doublet microtubules of the cilium, is referred to as the transition zone (Rohatgi and Snell). A series of membrane proteins at the transition zone, termed the ciliary necklace, help distinguish the ciliary membrane from the cell's plasma membrane (Gilula and Satir, 1972). Ciliary pockets can also be found on each side of the primary cilium. These are invaginations into the cell membrane adjacent to the ciliary necklace common to many species (Sorokin, 1962; Gilula and Satir, 1972; Moser et al., 2009). This aperture creates a semi-enclosed area at the transition zone (designated as the ciliary sheath) that is thought to help restrict protein and lipid entry into the cilium and is formed during ciliogenesis (Rohatgi and Snell, 2010).

A primary cilium is generally present on most cell types in eukaryotes, and it is notably absent in plants and fungi. Henneguy and Lenhossék first described the cilium in the late 1800s. For many years following, the primary cilium was considered an evolutionary relic. However, the past couple of decades of research have radically changed the scientific perspective on cilia biology. The cellular functions of this mysterious organelle have recently begun to unfold. The primary cilium, a specialized compartment potentially housing thousands of proteins, is now recognized for its central role in various signaling pathways and as a cellular sensory organelle. These new discoveries allow us to appreciate its significant physiological importance. Accordingly, disruptions or mutations in ciliary components gives rise

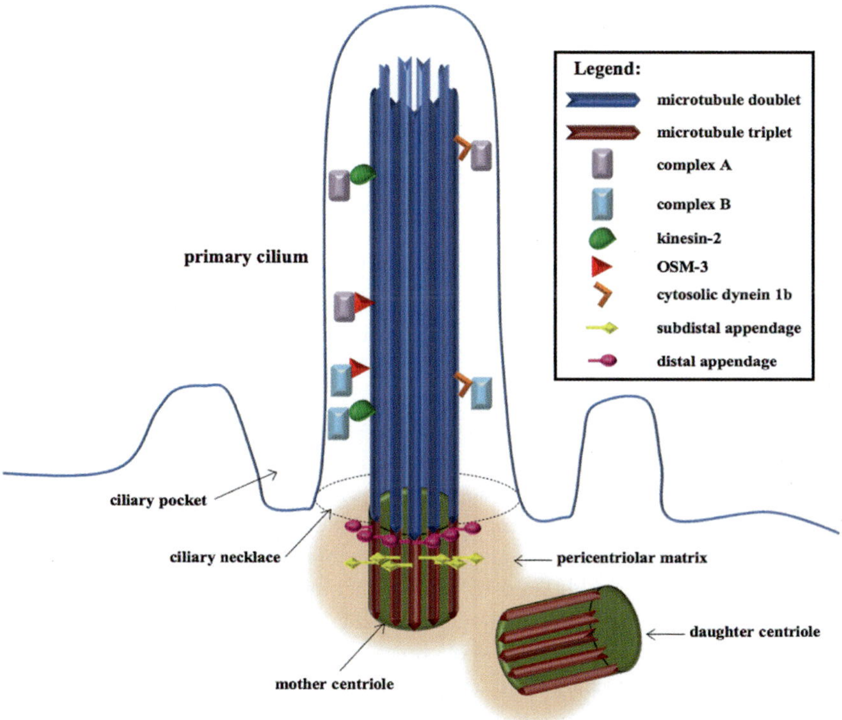

Fig. 9.2 Non-motile primary cilium. A primary cilium is extended from the mother centriole, complemented by distal and subdistal appendages. Ciliary pockets are found on each side of the cilium, and ciliary necklace distinguishes the ciliary membrane from the cell's plasma membrane. The IFT cargo consists of two distinct particles referred to as complex A and complex B, involved in retrograde and anterograde transports, respectively. This system is critical for assembling the primary cilium by adding new axonemal subunits. Kinesin-2 and OSM-3 are the molecular motors involved in anterograde movement, while cytoplasmic dynein 1b is used for recycling IFT components by retrograde movement

to an astounding array of diseases encompassing various tissue types. A few of these cilium-related disorders include polycystic diseases of the liver and kidney, Bardet-Biedl syndrome, Joubert syndrome, Meckel-Gruber syndrome, nephronophthisis, and defects in left-right asymmetry. As they serve as unique sensory organelles (Nauli and Zhou, 2004; Kolb and Nauli, 2008; Ratnam and Nauli, 2010), the remainder of this text will focus entirely on the primary non-motile cilium.

9.2 Primary Cilia-Induced Nitric Oxide and ATP Release

The primary cilium acts as a mechanosensory organelle that responds to external stimulation through polycystin-1 and polycystin-2. In response to the flow of fluid, polycystin-1 becomes activated and its extracellular domain is thought to stimulate polycystin-2 (McGrath et al., 2003; Chauvet et al., 2004; Masyuk et al., 2006; Nauli et al., 2006, 2008; Xiao et al., 2006; Xu et al., 2007, 2009; AbouAlaiwi et al., 2009b; Hou et al., 2009). This leads to a calcium influx through the polycystin-2 channel (Abdul-Majeed and Nauli, 2010). Prominent examples of this are seen in vascular endothelial and renal epithelial cells.

In vascular endothelial cells, subtle flow changes sensed by the primary cilium induce nitric oxide (NO) release (Nauli et al., 2008; AbouAlaiwi et al., 2009b). In response to fluid shear stress, endothelial cells have the ability to generate NO through a complex biochemical cascade involving calcium, calmodulin, Akt/PKB, and protein kinase C. In the isolated artery, NO production is further enhanced by a pressurized biomechanical force that involves purinergic receptor activation. Primary cilium and its constituent sensory proteins, polycystin-1 and polycystin-2, play a crucial role in transmitting extracellular shear stress to intracellular NO biosynthesis.

In renal epithelial cells, subtle flow changes sensed by the primary cilium induce ATP release, which further amplifies the epithelial response to cilia activation (Hovater et al., 2008; Praetorius and Leipziger, 2009; Xu et al., 2009). Loss of apical cilia on renal epithelial cells impairs ATP secretion across the apical cell surface and ATP-dependent calcium signals. The flow-sensitive ATP release and paracrine receptor activation are modulated by ecto-nucleotidase activity and abrogated by purinergic receptor inhibition or extracellular ATP hydrolysis. The purinergic receptor is also expressed in epithelial cilia of cholangiocytes (Masyuk et al., 2008). Thus, primary cilia have been recognized as both mechano- and sensory organelles.

9.3 Chemosensory Signaling Through Primary Cilia

Aside from purinergic system, roles of primary cilia in signaling of Hedgehog and Wnt have been very well studied. Thus, we will take a look at these signaling systems in this section.

9.3.1 Hedgehog

The Hedgehog (Hh) signal transduction pathway is dependent upon the Patched 1 (Ptch1) and Smoothened (Smo) transmembrane receptors (Fig. 9.3). When the Hh protein morphogen is absent, inactive Smo is localized on the membrane of intracellular vesicles whereas Ptch1 is displayed on the plasma membrane of the cell. Under these circumstances, the regulatory protein glioma-associated oncogene (Gli1 as an activator; Gli2 and Gli3 as latent repressors) is phosphorylated by protein kinase A (PKA), casein kinase 1α (CK1α), and glycogen synthase kinase 3β (GSK3). Gli is also ubiquitinated and then processed by proteasomes. The digested form of Gli passes into the nucleus and represses Hh genes to regulate the signal. The hydrolysis of Gli is reliant on the scaffold protein Costal2 (Cos2) that is not only the docking partner of the three kinases but also connects Gli to microtubules, thus regulating its cellular location.

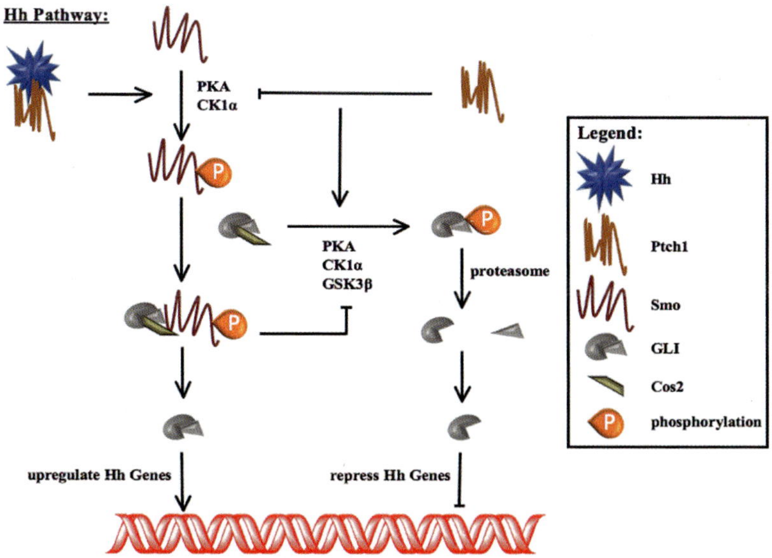

Fig. 9.3 Hedgehog signaling. Hedgehog (Hh) binds to Patched-1 (Ptch1), leading to the phosphorylation of Smoothened (Smo) by protein kinase A (PKA) and casein kinase 1α (CK1α). The adaptor protein Cos2 binds to GLI and is recruited to the activated Smo. GLI then translocates to the nucleus to upregulate Hh target genes. In the absence of Hh, Smo phosphorylation is inhibited. This prevents Cos2 from binding Smo and allows Cos2 to dock PKA, CK1α, and glycogen synthase kinase 3α (GSK3). These kinases phosphorylate and tag GLI for degradation. The hydrolyzed GLI product represses transcription of Hh target genes

The three Hh protein homologs are Sonic hedgehog (Shh), Indian hedgehog (Ihh), and Desert hedgehog (Dhh). Hh proteins have a C-terminal cysteine that is attached to a cholesterol moiety, and the amino terminus is palmitoylated (Bumcrot et al., 1995; Saqui-Salces and Merchant, 2010). These hydrophobic accessories make Hh less soluble, however, they also enable the formation of covalently cross-linked oligomers that are extremely stable. This increases Hh's biological activity and promotes long range signaling, despite the hydrophobicity (Marti et al., 1995; Porter et al., 1995; Dierker et al., 2009). Dispatched transmembrane receptor and a disintegrin and metalloprotease 17 (ADAM17) are needed for cellular secretion of Hh (Dierker et al., 2009).

When present, Hh binds to Ptch1, leading to the internalization and lysosomal degradation of Patched receptors (Fig. 9.3). This allows Smo to be phosphorylated by PKA and CK1α, thus allowing movement to the cell's surface where its vesicle fuses with the plasma membrane. This activated Smoothened initiates the downstream cascade of the hedgehog pathway, beginning with the recruitment of Cos2 and its attached Gli. The binding to Smoothened prevents Cos2 from docking any kinases. This prevents Gli from being hydrolyzed and allows Gli to translocate to the nucleus intact, whereupon it upregulates Hh gene transcription.

Hh signaling is essential for embryonic development. The concentration gradient of Hh controls segment patterning and cell fate and regulates cell growth and proliferation (Jia and Jiang, 2006). Although Hh signaling is drastically reduced in adult tissues, it is still needed to maintain post-natal tissue, as its overexpression or other alterations in the pathway are implicated in various cancers (Johnson et al., 1996; Dahmane et al., 1997; Taylor et al., 2002; Evangelista et al., 2006; Caro and Low, 2010).

The primary cilium has been found to be essential for proper Hh signaling. The intraflagellar transport (IFT) components IFT172, IFT88/Polaris, and Kif3A act downstream of Ptch1 (Huangfu et al., 2003). IFT172 and, surprisingly, both retrograde and anterograde motors (cDHC1b/DHC2 and Kif3A) are required for Gli processing. IFT172 also acts downstream of Smo (Huangfu and Anderson, 2005). When IFT proteins are eliminated, Hh cannot stimulate Gli activation (Liu et al., 2005). It is interesting that IFT components appear to play a role in both positive and negative feedback of the Hh response.

In addition, many other players in the Hh pathway have been geographically linked to the primary cilium. The Gli proteins (Gli1, Gli2, and Gli3) have been localized to the base of the cilium through immunostaining (Haycraft et al., 2005). Fused (Fu) and its nemesis Suppressor of Fused (SuFu) are other important regulators of Hh in vertebrate systems. Fu is a kinase that promotes Gli2/Gli3 processing and ultimate upregulation of Hh signaling. SuFu rivals this activity by keeping Gli2/Gli3 in the cytosol.

Smo has been localized to the primary cilium in the presence of Hh. A hydrophobic and basic residue found on the carboxy terminal adjacent to the seventh transmembrane segment of Smo allows this translocation (Corbit et al., 2005). This sequence is conserved among other seven-pass transmembrane ciliary proteins, including neuronal somatostatin receptor 3 and serotonin receptor 6 (Handel et al., 1999; Brailov et al., 2000). Mutations of these residues prevent Smo from traveling to the ciliary membrane regardless of Hh's presence. In this event, Smo localizes to the plasma membrane instead (Corbit et al., 2005). Ptch1 is also found on cilia. In the absence of Hh, Ptch1 regulates Smo activity by preventing Smo from translocating to the cilia. Prior to Smo translocation, Ptch1 is observed to leave the cilium (Corbit et al., 2005; Rohatgi et al., 2007). An increasing number of recent investigations are exploring the small molecule regulation of this receptor translocation (Stanton and Peng, 2010).

The evidence of IFT's importance toward Hh signal transduction suggests that actual IFT proteins could be involved in the Hh pathway in some unknown manner. This may explain the role of Hh signaling through the cilium. Another theory is that the confinement of Hh components to a small organelle, such as the membrane enclosed primary cilium, will increase their local concentration and yield a more efficient response (Jia and Jiang, 2006). In fact, there may be a synergistic effect between the two ideas. Nonetheless, cilia are absolutely necessary for vertebrate Hh signaling but are surprisingly dispensable for invertebrates.

9.3.2 Wnt

Sharing similarities with the Hh pathway, Wnt signaling is another cilia-dependent pathway in development (Figs. 9.4 and 9.5). The Wnt protein family consists of 19 secreted glycoproteins. Wnt ligands activate seven-pass transmembrane G-protein coupled receptors in the Frizzled (Fz) family. There are various intracellular processes regulated by Wnt signaling. There are two distinct Wnt pathways: the canonical (Wnt-β-catenin) and non-canonical (β-catenin independent) pathways (Abdul-Majeed and Nauli, 2011). Of these, the Wnt-β-catenin dependent pathway will be discussed first.

When Wnt ligands are absent, glycogen synthase kinase 3 (GSK3), CK1α, axin, and adenomatous polyposis coli (APC) form a complex in the cytosol (Fig. 9.4). Axin binds GSK3 and CK1α. β-catenin then joins by binding Axin and APC. β-catenin is phosphorylated first by CK1α and then by GSK3. This second phosphorylation by GSK3 marks β-catenin for ubiquitylation by SCF E3 ligases and proteasomal degradation (Kimelman and Xu, 2006). Hence, these four proteins are referred to as a degradation or destruction complex and maintain very low cellular levels of β-catenin. Under these conditions, Groucho binds to DNA regulatory

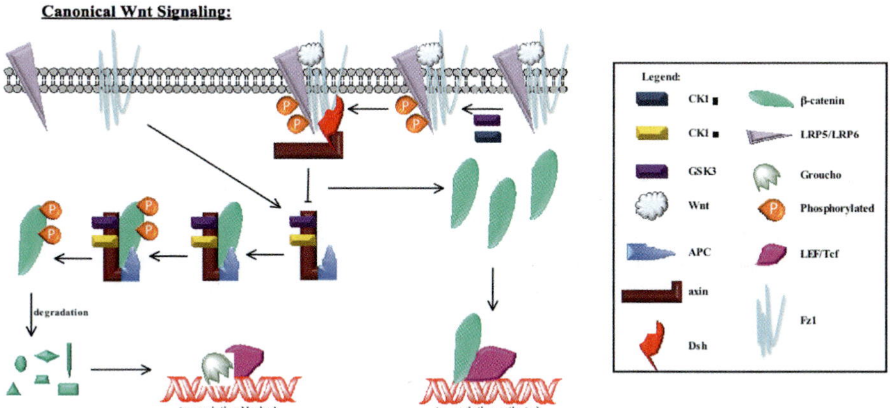

Fig. 9.4 Canonical Wnt signaling. GSK3, CK1α, Axin, and APC form a cytosolic degradation complex in the absence of Wnt ligands. β-catenin binds to this complex through interactions with axin and APC and is sequentially phosphorylated by CK1α and GSK3. This marks β-catenin for ubiquitylation and proteasomal degradation. Under these conditions, Groucho prevents transcription of Wnt target genes by binding to the regulatory proteins LEF and TCF. Conversely, the binding of Wnt to Fz and LRP tethers the cytosolic domains of the receptors together, allowing GSK3 and CK1α to then phosphorylate LRP. Dsh binds to Axin and is recruited to Fz at the plasma membrane. By reducing the cellular pool of Axin, the formation of the degradation complex is prevented. This leads to a buildup of β-catenin which then translocates to the nucleus, displaces Groucho from LEF and TCF, and activates the transcription of Wnt target genes such as c-Myc, cyclin-D1, fibronectin, and connexin 43

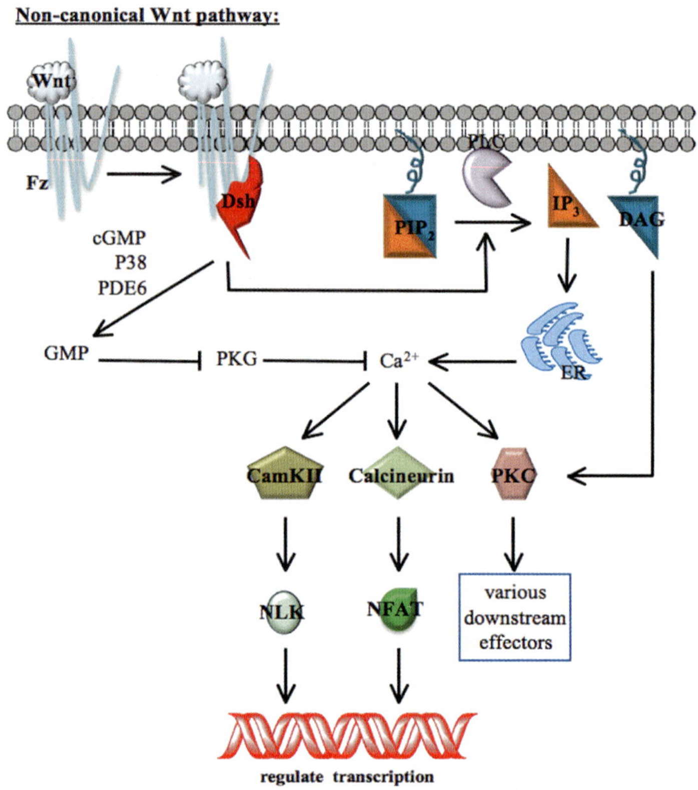

Fig. 9.5 Non-canonical Wnt signaling. Wnt binding to Frizzled (Fz) in the absence of low-density lipoprotein receptor-related proteins (LRP) brings Dishevelled (Dsh) to the plasma membrane. As seen with traditional G-protein coupled receptors, phospholipase C (PLC) is activated and cleaves phosphatidylinositol 4,5-bisphosphate (PIP2) into inositol 1,4,5-trisphosphate (IP3) and diacylglycerol (DAG). IP3 activates calcium channels on the endosplasmic reticulum, and DAG activates calcium-dependent protein kinase (PKC). Additionally, the Wnt/Fz/Dsh complex causes p38 kinase (p38) and phosphodiesterase 6 (PDE6) to hydrolyze cyclic GMP (cGMP) which then inactivates cGMP-dependent protein kinase (PKG), leading to an increase in intracellular calcium. The rise in calcium is followed by the activation of PKC, calcium/calmodulin-dependent kinase II (CamKII), and calcineurin. PKC is a regulator of various downstream processes. CamKII activates Nemo-like kinase (NLK), and calcineurin activates nuclear factor of activated T cells (NFAT). NLK and NFAT translocate to the nucleus to regulate gene expression

proteins lymphoid enhancer-binding factor (LEF) and T cell factor (TCF) to prevent transcription of Wnt target genes such as c-Myc, cyclin-D1, fibronectin, and connexin 43.

When Wnt is present (Fig. 9.5), it binds to Fz receptors, such as Fz1, and low-density lipoprotein receptor-related proteins (LRP5 or LRP6) (Yang-Snyder et al., 1996). This tethers the two transmembrane receptors, allowing GSK3 and casein kinase 1γ (CK1γ) to phosphorylate the cytoplasmic tails of LRP5 or LRP6. Dishevelled (Dvl) then brings Axin to LRP5/LRP6 and Fz at the cell surface. This disperses the destruction complex proteins, which then prevents the degradation of β-catenin. The increased levels of β-catenin are able to enter the nucleus, displace Groucho from LEF/TCF, and activate gene transcription.

An alternative LRP5/LRP6 independent pathway exists by Wnt interacting with only Fz receptors (commonly referred to as the non-canonical Wnt or β-catenin independent pathway). There are various ways to activate this pathway, and the molecular details are not as clear. In the most studied mechanism, Fz2 is activated by Wnt5a in the absence of LRP co-receptors, and Dvl associates with the complex at the cell membrane (Sheldahl et al., 1999). As seen in traditional GPCRs, this binding activates phospholipase C (PLC). PLC then cleaves PIP_2 into IP_3 and DAG. IP_3 opens calcium channels on the endoplasmic reticulum to release calcium into the cytosol (Slusarski et al., 1997a). The Wnt/Fz/Dvl complex also causes p38 kinase and phosphodiesterase 6 to hydrolyze cyclic GMP (Ahumada et al., 2002; Wang et al., 2004; Ma and Wang, 2007). This effectively inactivates protein kinase G, which is not only critical to the pathway but also leads to a further increase of intracellular calcium (Ma and Wang, 2006).

This build up of cellular calcium is followed by protein kinase C (PKC) activation (Slusarski et al., 1997b; Ahumada et al., 2002), calcium/calmodulin-dependent kinase II (CamKII) (Kuhl et al., 2000), and calcineurin (Saneyoshi et al., 2002). PKC can have various downstream effectors and is also activated by the previously synthesized DAG. CamKII functions to activate Nemo-like kinase (NLK) while calcineurin activates nuclear factor of activated T cells (NFAT) (Ishitani et al., 2003; Dejmek et al., 2006). NLK and NFAT translocate to the nucleus where they regulate gene expression (Ishitani et al., 1999; Saneyoshi et al., 2002).

9.4 Centrosome as a Cellular Division Marker

The centrosome is a cytoplasmic organelle consisting of a pair of centrioles encircled by an electron dense mass of material termed the pericentriolar matrix (PCM). A single centrosome is generally present in each cell. The centrioles are composed of cylindrical units that lie next to each other. Each one is approximately 200 nm wide and 400 nm in length. The centrioles differ in that the older "mother" centriole was constructed at least two cell cycles ago and is complemented by distal and subdistal appendages. The juxtaposed "daughter" centriole was constructed in the previous cell cycle and does not bear appendage structures (Loncarek and Khodjakov, 2009).

Within the PCM, there is a "proteinous cloud" at the proximal end and appendage structures at the distal end of the mother centriole (Paintrand et al., 1992). The proximal end is composed of nine microtubule triplets. The distal domain of the centriole differs, as it is constructed of nine microtubule doublets in addition to projecting nine distal and subdistal appendages toward the plasma membrane perpendicular from the centriole. The distal appendages (sometimes referred to as transitional fibers) are extremities found at the extreme distal end of the centriole. The subdistal appendages, proximal to the distal appendages, are filaments whose tips fuse with microtubules (Paintrand et al., 1992; Chretien et al., 1997; Kenney et al., 1997; Ibrahim et al., 2009).

9.4.1 Pericentriolar Matrix and Microtubules

The PCM provides the centrosome its ability to serve as the microtubule organizing center (MTOC) of the cell (Kellogg et al., 1994). A unique form γ-tubulin is found exclusively in the PCM where γ-tubulin ring complexes (γ-TuRC) are formed (Zheng et al., 1995). These complexes nucleate microtubule growth in such a fashion that the minus end of the microtubule is attached to the centrosome, leaving the plus end directed toward the plasma membrane. In addition, γ-TuRC are thought to influence the lattice structure of microtubules. The microtubules nucleated from γ-TuRC consist of 13 protofilaments that are able to assemble at concentrations lower than those required for spontaneous microtubule nucleation; spontaneous microtubule nucleation usually consists of 14 protofilaments (Evans et al., 1985).

There is an uneven association of microtubules between the centrioles during G1 phase. The mother centriole is responsible for the array of microtubules seen from the MTOC, while the microtubules originating from the daughter are quickly released into the cytoplasm. The daughter centriole is also oriented away from the mother centriole's radial array of microtubules (Piel et al., 2000, 2001). While there have been many proteins identified within the PCM, the general structure of this fibrous mass is still unclear. Electron microscopy has revealed staggering differences in the PCM surrounding the mother and daughter centrioles.

The intensity of the PCM varies throughout the cell cycle and is inconsistent across different cell types (Rusan and Rogers, 2009). In G1 and S phase, electron micrographs indicate that the PCM levels are at their lowest (Vorobjev and Chentsov Yu, 1982). Similarly, γ-tubulin and γ-TuRC are also at their lowest concentration during G1 and S phase (Khodjakov and Rieder, 1999; Piehl et al., 2004). As the cell arrives at G2, centrosome duplication is complete and PCM/γ-tubulin levels of each centrosome are similar to those in G1/S. The centrosomes then mature throughout mitosis, which is hallmarked by increased PCM/γ-tubulin concentrations as well as rates of microtubule nucleation (Snyder and McIntosh, 1975; Kuriyama and Borisy, 1981; Vorobjev and Chentsov Yu, 1982; Khodjakov and Rieder, 1999; Piehl et al., 2004). These levels peak during prophase and metaphase, which allows the mitotic spindle to form. The PCM, γ-tubulin, and microtubule nucleation sharply decrease after mitosis is complete.

9.4.2 Pericentriolar Matrix Composition

Several proteins have been linked to these structures in the PCM. Ninein has been localized to the subdistal appendages and shown to bind minus-end microtubules. As microtubules are released from γ-tubulin at the centrosome, ninein is proposed to be involved in the translocation and anchoring of the microtubules to apical surfaces in the cell (Moudjou et al., 1996; Mogensen et al., 2000; Piel et al., 2000; Ou et al., 2002).

Centrosomal protein 170 (CEP170) localizes to subdistal appendages of mature centrioles during interphase. CEP170 assists in maintaining microtubule organization and, although expressed throughout the cell cycle, is phosphorylated by Polo-like kinase 1 (Plk1) during mitosis. The result of this interaction makes CEP170 capable of serving as a marker for centriole maturation (Guarguaglini et al., 2005).

ε-tubulin is another centrosomal protein that is unique to the subdistal appendages. It has also been co-localized with ninein and is required for both centriole duplication and microtubule organization (Dupuis-Williams et al., 2002; Chang et al., 2003).

Cenexin, previously referred to as Outer Dense Fiber 2 (ODF2), is a scaffold protein found at the subdistal/distal appendages of the mature centriole (Nakagawa et al., 2001). It has been shown that cenexin is necessary for appendage formation and ciliogenesis (Ishikawa et al., 2005). As its localization at the appendages depends on the cell cycle, cenexin is able to serve as a marker for centriole maturation (Lange and Gull, 1995).

hCenexin1, a splice variant of cenexin/ODF2, also localizes to the centrosome and has specialized functions throughout the cell cycle. hCenexin1 is responsible for recruiting ninein to the centrosome during interphase and ciliogenesis during G0/G1. hCenexin1 is phosphorylated by Cdc2 in late G2/early M phase to form a complex with Plk-1 which can in turn recruit and dock pericentrin and γ-tubulin to the centrosome (Soung et al., 2006, 2009). It has been proposed that the ODF2, originally discovered in rat sperm tail cytoskeleton, but previously characterized as a centrosomal protein, has likely been hCenexin1 all along (Brohmann et al., 1997; Turner et al., 1997).

Centrosomal P4.1-associated protein (CPAP; human analog to SAS-4) has been localized with γ-tubulin in the centrsome (Hung et al., 2000). The cellular CPAP concentration gradient is regulated by the cell cycle; it increases in S phase and decreases after the completion of mitosis (Tang et al., 2009). CPAP's functional role is to interact with γ-tubulin and recruit tubulin to promote procentriole growth (Hung et al., 2000, 2004; Hsu et al., 2008). A cellular deficiency in CPAP inhibits procentriole formation, whereas overexpressed CPAP emanates an overly long procentriole (Kohlmaier et al., 2009; Schmidt et al., 2009; Tang et al., 2009). CPAP has recently been shown to form a homodimer that prevents centrosome separation before G2/M phase (Zhao et al., 2010).

CP110, or CEP110, serves as a cap to limit procentriole growth and antagonizes the function of CPAP (Schmidt et al., 2009). CP110 is essential for centriole duplication and is stabilized by CEP97 (Spektor et al., 2007). Cellular levels of CP110 are dependent on the cell cycle and are highest during phase G1/S (Chen et al., 2002). CP110 is a substrate for CDK-cyclin phosphorylation, and siRNA knockdown can block centriole duplication (Chen et al., 2002; Kleylein-Sohn et al., 2007).

CDK5RAP2 (Cdk5 regulatory subunit associated protein or CEP215) is found in the PCM and localized with γ-TuRC throughout the cell cycle (Andersen et al., 2003). It is required for γ-TuRC attachment to the centrosome. CDK5RAP2 overexpression or depletion corresponded in the appropriate over-recruitment of centrosomal proteins or unorganization of microtubule arrays (Fong et al., 2008). CDK5RAP2 also appears to be essential for centriole cohesion during interphase. This is unique in a sense that CDK5RAP2 does not associate with known centriole fiber proteins such as rootletin, C-NAP1, or CEP86. However, CDK5RAP2 is speculated to couple with pericentrin to carry out its cohesive role (Graser et al., 2007). Another role of CDK5RAP2 is to localize dynein to the centrosome (Lee and Rhee, 2010).

9.4.3 Centriole Replication

The centrosome, which is composed of a mother centriole and a daughter centriole, is replicated once per cell cycle in a process described as the centriolar cycle (Fig. 9.6). The mother centriole can be distinguished by the presence of two appendages. In G1 phase, the two centrioles are held together through cohesion fibers. Rootletin and C-Nap1 both interact to help maintain this linkage. Rootletin overexpression results in greater fiber formation, whereas siRNA depletion of either rootletin or C-NAP1 causes the centrosomes to bifurcate (Fry et al., 1998; Mayor et al., 2000; Bahe et al., 2005). CEP68, while not shown to bind directly to either rootletin or C-NAP1, also appears to play a role in centriole cohesion (Graser et al., 2007).

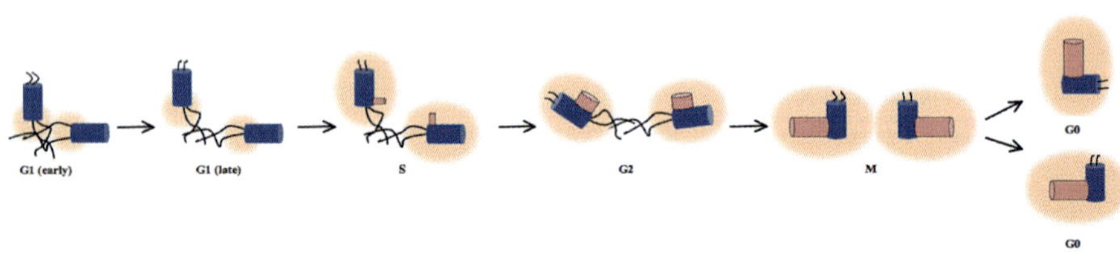

Fig. 9.6 Centriole cycle. In early G1, the perpendicular centrioles are held together through cohesion fibers maintained by rootletin, C-Nap1, and CEP68. As the replication begins in late G1, the centrioles begin to split and separate. This coincides with elevated levels of Cdk2. During S phase of cell cycle, a procentriole structure begins to form at the proximal ends of the parent centrioles. In G2 phase, the procentriole continues to elongate throughout the cell cycle. This complex is referred to as "engaged", because all of the structures are held relatively close together, commonly called the diplosome. During mitotic phase, procentriole growth is complete. The diplosome becomes disengaged, and the two centrosomes are localized at opposite poles in the cell. Following mitosis, the two daughter cells will enter G0 phase. Each of the daughter cells has one centrosome. Each of the centrosomes consists of a pair of centrioles. The centriole that was present at least two cell cycles ago is denoted as mother centriole, whereas the centriole that was constructed in the previous cell cycle is termed daughter centriole

Centriole replication begins at G1/S phase (Fig. 9.6). As the levels of Cdk2 begin to rise during this stage of the cell cycle, the two mature centrioles disorientate from each other and lose their orthogonal arrangement (Crasta and Surana, 2006). This allows a procentriole structure to begin to elongate at a right angle on the proximal end of each of the existing centrioles (Vorobjev and Chentsov Yu, 1982; Chretien et al., 1997). Cdk2-cyclin E and Cdk2-cyclin A have been reported to be involved in centriole duplication (Meraldi et al., 1999). In fact, cyclin E has been localized to γ-tubulin in the centrsome and shown to enhance S phase entry when in excess (Matsumoto and Maller, 2004).

The procentriole grows to reach the full length of a mature centriole sometime during mitosis (Fig. 9.6). The completion time of procentriole growth varies across species (Vorobjev and Chentsov Yu, 1982). The mature centriole and elongated daughter centriole remain at right angles with respect to one another. The complex of all four centrioles remains "engaged" and is designated as the "diplosome" (Hinchcliffe and Sluder, 2001; Loncarek and Khodjakov, 2009). This diplosome is lost in anaphase when the spindle poles are assembled in the cell, and thus the centrosome becomes "disengaged". During disengagement, the distance between centrosomes increases and the perpendicular mother/daughter centriole geometry dissipates. By the end of mitosis (usually around anaphase), four equally sized centrioles are present. Following telophase and cytokinesis, each daughter cell receives one equivalent centrosome, which consists of a mother centriole that was carried over from the existing cell and a daughter centriole that was newly constructed during the cell cycle.

9.4.4 Key Proteins in the Centriolar Cycle

Separase, the protease associated with chromosome segregation, which is activated by anaphase-promoting complex (APC/C), is thought to be required for centriolar disengagement. Procentriole growth can only occur after the centrosome splits in G1/S phase. Therefore, it is proposed that the centriole engagement/disengagement prevents overduplication of centrioles in normal cells. This arrangement of events can occur because procentriole growth in G1/S phase and disengagement by separase in anaphase occur at distant time points in the cell cycle, ultimately limiting centrosome replication to take place only once per cell division (Tsou and Stearns, 2006). Centriole reduplication has been shown in CHO cells arrested in S phase (upon hydroxyurea treatment) after laser ablation to physically remove daughter centrioles (Loncarek et al., 2008). It seems logical that the close proximity of juxtaposed centrioles provides a steric hindrance that blocks further procentriole formation.

There is, however, criticism of separase's involvement in this process. These arguments stem from the observation that centrioles can disengage when separase is not expected to be active (Loncarek and Khodjakov, 2009). For example, daughter centrioles are seen to form and then disengage during interphase. They then travel to the cell surface to facilitate ciliogenesis (Dawe et al., 2007). In addition, centrosomal disengagement during mitotic arrest has been observed (Keryer et al., 1984; Sluder and Begg, 1985; Loncarek and Khodjakov, 2009).

Nucleophosmin (NPM/B23) is bound to centrosomes and dissociates upon phosphorylation by Cdk2-cyclin E during G1/S phase. This event is required for centriole duplication (as NPM/B23 antibodies prevent centriole replication) and allows splitting of the mature centrioles. Following mitosis, unphosphorylated NPM/B23 exits the nucleus and attaches to the centrosome to regulate the process (Okuda et al., 2000).

Mps1 protein kinase, localized to the centrosome when phosphorylated, is another substrate for Cdk2 and is thought to play a role in centrosome duplication (Fisk et al., 2003). It has been deemed necessary for spindle assembly checkpoint (Stucke et al., 2002). Mps1 degradation is prevented by Cdk2 and may be necessary for centriole duplication (Kasbek et al., 2007), although this report conflicts with a previous attempt to link Mps1 with centriole duplication (Stucke et al., 2002).

9.4.5 Procentriole Formation

The procentriole formation varies slightly depending on the species. In *C. elegans*, the procentriole begins as a small central tube, roughly 60 nm long and perpendicular to the existing centriole (Pelletier et al., 2006). The tube will grow into a full-length centriole, and microtubules are added to its surface in a precise manner to achieve the nine-fold symmetrical pattern observed in mature centrioles (Kemp et al., 2004; Pelletier et al., 2004, 2006; Strnad and Gonczy, 2008). This process requires several centrosomal proteins. SPD-2, analogous to CEP192, recruits ZYG-1, analogous to Plk4/SAK, to the centrosome (Andersen et al., 2003; Pelletier et al., 2004). ZYG-1/Plk4/SAK is then able to draft SAS-5, followed by SAS-6/hsSAS-6, to the centrosome. Only after these events occur can central tube development continue (Bettencourt-Dias et al., 2005; Kleylein-Sohn et al., 2007). Next, SAS-4/CPAP/DSAS-4 is required at the centrosome in order for microtubular attachment to occur (Leidel et al., 2005; Peel et al., 2007; Zhu et al., 2008). It is noteworthy that an SAS-5 functional analogue has not been recognized in other species, and the Drosophila SPD-2 analogue (DSpd2) is not required and has comparatively nonspecific centriolar localization during somatic cell centriole duplication (Dix and Raff, 2007).

In other organisms, including *Drosophila*, *Chlamydomonas reinhardtii*, and numerous vertebrates, the central tube development occurs as a cartwheel structure (Fig. 9.7). The cartwheel does not have attached microtubules, yet it portrays centriolar symmetry due to the extension of nine spoke-like structures emerging from a central hub. The ends of the spokes are capped by pinhead looking structures that are later attached to triplet microtubules (Cavalier-Smith, 1974; O'Toole et al., 2003). Bld10p (CEP135 in humans; Bld10 in drosophila) is one constituent of these pinhead caps (Matsuura et al., 2004). By exploiting mutated versions of Bld10p, it has been deduced that both the cartwheel structure and Bld10p are central to developing the nine-microtubule triplet symmetry and obtaining the correct

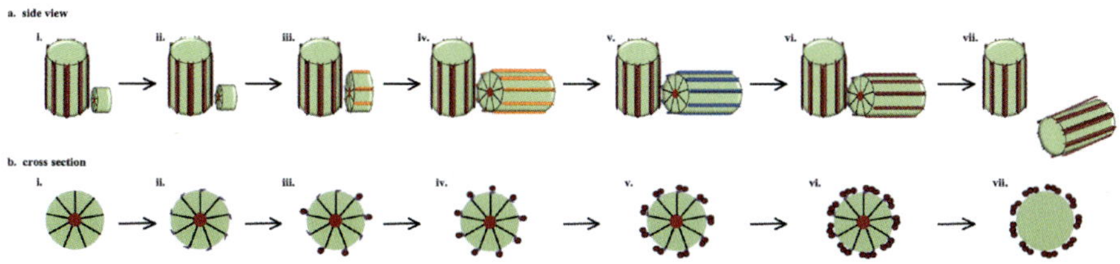

Fig. 9.7 Procentriole formation in vertebrates. A procentriole is an immature form of centriole seen in an early replication stage. Various replication stages of a procentriole are shown from the side view (**a**) and the cross section (**b**). **i.** A procentriole begins with a cartwheel structure that is formed at a right angle to the proximal end of the mature centriole. These spokes share the same nine-fold symmetry as observed in mature centrioles. **ii.** The procentriole consists of a central hub that radially emanates nine spoke like structures. A pinhead shaped figure caps the tip of each spoke. **iii.** As the procentriole enlarges, a microtubule singlet (shown in *orange*) is attached to each of the pinheads. **iv.** The procentriole core continues to grow, and each pinhead radiates a strand of microtubules. **v.** A microtubule doublet (shown in *blue*) is formed at the pinhead by the addition of new microtubules. **vi.** Microtubules are continually added until the microtubule triplet (shown in *red*) configuration is achieved. **vii.** As a mature centrosome, each centriole is an equivalently sized barrel displaying the nine-fold microtubule triplet. The cartwheel template used in procentriole formation has not been observed in a mature centrosome

length of mature centrioles. Furthermore, the Bld10p mutants gave rise to centrioles with shortened spokes and abnormal symmetry (Hiraki et al., 2007). In paramecium, Bld10p is needed to build the cartwheel (as depletion effects cartwheel formation) but does not appear to be necessary for the symmetry. Instead, Sas6p ensures that the proper centriolar symmetry is achieved (Jerka-Dziadosz et al., 2010; Kitagawa et al., 2011).

As mentioned previously for centriole replication, this process requires centrosomal proteins such as CPAP, CP110, and CEP97. The procentrioles continue to grow throughout G2 and into S phase (Kuriyama and Borisy, 1981). Phosphorylation of rootletin and C-NAP1 by Nek2 kinase stimulates centriolar separation at the G2/M transition (Faragher and Fry, 2003).

9.5 Conclusion and Perspective on Chemosensory Primary Cilia

Other than regulating centrosomal duplication and cell division (AbouAlaiwi et al., 2009a, 2011), the physiological role of chemosensory functions of cilia are not well understood. It has, however, been hypothesized that primary cilia regulate centrosomal duplication to control planar cell polarity (PCP). PCP refers to the positioning of cells along a plane and their organization with respect to one another. These cells arrange their organelles and biomolecules in a very particular way. The cellular morphology is then consistent among neighboring cells.

The exact molecular mechanism of PCP signaling is not clear. Nonetheless, Fz and Dsh seem to be the critical components of PCP while the other players appear to regulate the Fz/Dsh effect. Regular PCP signaling is mediated through Fz/Dsh interactions. In addition, Wnt signaling is also able to initiate PCP through receptor tyrosine kinase-like orphan receptors (Nishita et al., 2010). Taken together, both Hedgehog and Wnt signaling systems are involved in PCP. However, their molecular interactions, in terms of centrosomal duplication and cell division, remain to be explored with regards to sensory primary cilia.

Acknowledgements Due to restricted space, we apologize to those whose work is not described in this review. Authors are grateful for stimulating discussion about primary cilia given by research assistants, graduates, undergraduates and pharmacy students in our laboratory. Authors thank Charisse Montgomery for her editorial review of the manuscript. Work from our laboratory that is cited in this review has been supported by grants from the NIH (DK080640) and the NIH Recovery Act Funds. We are thankful to The University of Toledo research programs.

References

Abdul-Majeed S, Nauli SM (2011) Calcium-mediated mechanisms of cystic expansion. Biochim Biophys Acta 2011 Oct 12. doi:10.1016/j.bbadis.2010.09.016

AbouAlaiwi WA, Lo ST, Nauli SM (2009a) Primary cilia: highly sophisticated biological sensors. Sensors 9:7003–7020

AbouAlaiwi WA, Takahashi M, Mell BR, Jones TJ, Ratnam S, Kolb RJ, Nauli SM (2009b) Ciliary polycystin-2 is a mechanosensitive calcium channel involved in nitric oxide signaling cascades. Circ Res 104:860–869

AbouAlaiwi WA, Ratnam S, Booth RL, Shah JV, Nauli SM (2011) Endothelial cells from humans and mice with polycystic kidney disease are characterized by polyploidy and chromosome segregation defects through survivin down-regulation. Hum Mol Genet 20:354–367

Ahumada A, Slusarski DC, Liu X, Moon RT, Malbon CC, Wang HY (2002) Signaling of rat Frizzled-2 through phosphodiesterase and cyclic GMP. Science 298:2006–2010

Andersen JS, Wilkinson CJ, Mayor T, Mortensen P, Nigg EA, Mann M (2003) Proteomic characterization of the human centrosome by protein correlation profiling. Nature 426: 570–574

Bahe S, Stierhof YD, Wilkinson CJ, Leiss F, Nigg EA (2005) Rootletin forms centriole-associated filaments and functions in centrosome cohesion. J Cell Biol 171:27–33

Bettencourt-Dias M, Rodrigues-Martins A, Carpenter L, Riparbelli M, Lehmann L, Gatt MK, Carmo N, Balloux F, Callaini G, Glover DM (2005) SAK/PLK4 is required for centriole duplication and flagella development. Curr Biol 15:2199–2207

Brailov I, Bancila M, Brisorgueil MJ, Miquel MC, Hamon M, Verge D (2000) Localization of 5-HT(6) receptors at the plasma membrane of neuronal cilia in the rat brain. Brain Res 872: 271–275

Brohmann H, Pinnecke S, Hoyer-Fender S (1997) Identification and characterization of new cDNAs encoding outer dense fiber proteins of rat sperm. J Biol Chem 272:10327–10332

Bumcrot DA, Takada R, McMahon AP (1995) Proteolytic processing yields two secreted forms of sonic hedgehog. Mol Cell Biol 15:2294–2303

Caro I, Low JA (2010) The role of the hedgehog signaling pathway in the development of basal cell carcinoma and opportunities for treatment. Clin Cancer Res 16:3335–3339

Cavalier-Smith T (1974) Basal body and flagellar development during the vegetative cell cycle and the sexual cycle of *Chlamydomonas reinhardii*. J Cell Sci 16:529–556

Chang P, Giddings TH Jr, Winey M, Stearns T (2003) Epsilon-tubulin is required for centriole duplication and microtubule organization. Nat Cell Biol 5:71–76

Chauvet V, Tian X, Husson H, Grimm DH, Wang T, Hiesberger T, Igarashi P, Bennett AM, Ibraghimov-Beskrovnaya O, Somlo S, Caplan MJ (2004) Mechanical stimuli induce cleavage and nuclear translocation of the polycystin-1 C terminus. J Clin Invest 114:1433–1443

Chen Z, Indjeian VB, McManus M, Wang L, Dynlacht BD (2002) CP110, a cell cycle-dependent CDK substrate, regulates centrosome duplication in human cells. Dev Cell 3:339–350

Chretien D, Buendia B, Fuller SD, Karsenti E (1997) Reconstruction of the centrosome cycle from cryoelectron micrographs. J Struct Biol 120:117–133

Corbit KC, Aanstad P, Singla V, Norman AR, Stainier DY, Reiter JF (2005) Vertebrate Smoothened functions at the primary cilium. Nature 437:1018–1021

Crasta K, Surana U (2006) Disjunction of conjoined twins: Cdk1, Cdh1 and separation of centrosomes. Cell Div 1:12

Dahmane N, Lee J, Robins P, Heller P, Ruiz i Altaba A (1997) Activation of the transcription factor Gli1 and the Sonic hedgehog signalling pathway in skin tumours. Nature 389:876–881

Dawe HR, Farr H, Gull K (2007) Centriole/basal body morphogenesis and migration during ciliogenesis in animal cells. J Cell Sci 120:7–15

Dejmek J, Safholm A, Kamp Nielsen C, Andersson T, Leandersson K (2006) Wnt-5a/Ca2+-induced NFAT activity is counteracted by Wnt-5a/Yes-Cdc42-casein kinase 1alpha signaling in human mammary epithelial cells. Mol Cell Biol 26:6024–6036

Dierker T, Dreier R, Migone M, Hamer S, Grobe K (2009) Heparan sulfate and transglutaminase activity are required for the formation of covalently cross-linked hedgehog oligomers. J Biol Chem 284:32562–32571

Dix CI, Raff JW (2007) Drosophila Spd-2 recruits PCM to the sperm centriole, but is dispensable for centriole duplication. Curr Biol 17:1759–1764

Dupuis-Williams P, Fleury-Aubusson A, de Loubresse NG, Geoffroy H, Vayssie L, Galvani A, Espigat A, Rossier J (2002) Functional role of epsilon-tubulin in the assembly of the centriolar microtubule scaffold. J Cell Biol 158:1183–1193

Evangelista M, Tian H, de Sauvage FJ (2006) The hedgehog signaling pathway in cancer. Clin Cancer Res 12:5924–5928

Evans L, Mitchison T, Kirschner M (1985) Influence of the centrosome on the structure of nucleated microtubules. J Cell Biol 100:1185–1191

Faragher AJ, Fry AM (2003) Nek2A kinase stimulates centrosome disjunction and is required for formation of bipolar mitotic spindles. Mol Biol Cell 14:2876–2889

Fisk HA, Mattison CP, Winey M (2003) Human Mps1 protein kinase is required for centrosome duplication and normal mitotic progression. Proc Natl Acad Sci USA 100:14875–14880

Fong KW, Choi YK, Rattner JB, Qi RZ (2008) CDK5RAP2 is a pericentriolar protein that functions in centrosomal attachment of the gamma-tubulin ring complex. Mol Biol Cell 19:115–125

Fry AM, Mayor T, Meraldi P, Stierhof YD, Tanaka K, Nigg EA (1998) C-Nap1, a novel centrosomal coiled-coil protein and candidate substrate of the cell cycle-regulated protein kinase Nek2. J Cell Biol 141:1563–1574

Gilula NB, Satir P (1972) The ciliary necklace. A ciliary membrane specialization. J Cell Biol 53:494–509

Graser S, Stierhof YD, Nigg EA (2007) Cep68 and Cep215 (Cdk5rap2) are required for centrosome cohesion. J Cell Sci 120:4321–4331

Guarguaglini G, Duncan PI, Stierhof YD, Holmstrom T, Duensing S, Nigg EA (2005) The forkhead-associated domain protein Cep170 interacts with Polo-like kinase 1 and serves as a marker for mature centrioles. Mol Biol Cell 16:1095–1107

Handel M, Schulz S, Stanarius A, Schreff M, Erdtmann-Vourliotis M, Schmidt H, Wolf G, Hollt V (1999) Selective targeting of somatostatin receptor 3 to neuronal cilia. Neuroscience 89: 909–926

Haycraft CJ, Banizs B, Aydin-Son Y, Zhang Q, Michaud EJ, Yoder BK (2005) Gli2 and Gli3 localize to cilia and require the intraflagellar transport protein polaris for processing and function. PLoS Genet 1:e53

Hinchcliffe EH, Sluder G (2001) Centrosome duplication: three kinases come up a winner! Curr Biol 11:R698–R701

Hiraki M, Nakazawa Y, Kamiya R, Hirono M (2007) Bld10p constitutes the cartwheel-spoke tip and stabilizes the 9-fold symmetry of the centriole. Curr Biol 17:1778–1783

Hou B, Kolpakova-Hart E, Fukai N, Wu K, Olsen BR (2009) The polycystic kidney disease 1 (Pkd1) gene is required for the responses of osteochondroprogenitor cells to midpalatal suture expansion in mice. Bone 44:1121–1133

Hovater MB, Olteanu D, Hanson EL, Cheng NL, Siroky B, Fintha A, Komlosi P, Liu W, Satlin LM, Bell PD, Yoder BK, Schwiebert EM (2008) Loss of apical monocilia on collecting duct principal cells impairs ATP secretion across the apical cell surface and ATP-dependent and flow-induced calcium signals. Purinergic Signal 4:155–170

Hsu WB, Hung LY, Tang CJ, Su CL, Chang Y, Tang TK (2008) Functional characterization of the microtubule-binding and -destabilizing domains of CPAP and d-SAS-4. Exp Cell Res 314:2591–2602

Huangfu D, Anderson KV (2005) Cilia and Hedgehog responsiveness in the mouse. Proc Natl Acad Sci USA 102:11325–11330

Huangfu D, Liu A, Rakeman AS, Murcia NS, Niswander L, Anderson KV (2003) Hedgehog signalling in the mouse requires intraflagellar transport proteins. Nature 426:83–87

Hung LY, Tang CJ, Tang TK (2000) Protein 4.1 R-135 interacts with a novel centrosomal protein (CPAP) which is associated with the gamma-tubulin complex. Mol Cell Biol 20:7813–7825

Hung LY, Chen HL, Chang CW, Li BR, Tang TK (2004) Identification of a novel microtubule-destabilizing motif in CPAP that binds to tubulin heterodimers and inhibits microtubule assembly. Mol Biol Cell 15:2697–2706

Ibrahim R, Messaoudi C, Chichon FJ, Celati C, Marco S (2009) Electron tomography study of isolated human centrioles. Microsc Res Tech 72:42–48

Ishikawa H, Kubo A, Tsukita S, Tsukita S (2005) Odf2-deficient mother centrioles lack distal/subdistal appendages and the ability to generate primary cilia. Nat Cell Biol 7:517–524

Ishitani T, Ninomiya-Tsuji J, Nagai S, Nishita M, Meneghini M, Barker N, Waterman M, Bowerman B, Clevers H, Shibuya H, Matsumoto K (1999) The TAK1-NLK-MAPK-related pathway antagonizes signalling between beta-catenin and transcription factor TCF. Nature 399:798–802

Ishitani T, Kishida S, Hyodo-Miura J, Ueno N, Yasuda J, Waterman M, Shibuya H, Moon RT, Ninomiya-Tsuji J, Matsumoto K (2003) The TAK1-NLK mitogen-activated protein kinase cascade functions in the Wnt-5a/Ca^{2+} pathway to antagonize Wnt/beta-catenin signaling. Mol Cell Biol 23:131–139

Jerka-Dziadosz M, Gogendeau D, Klotz C, Cohen J, Beisson J, Koll F (2010) Basal body duplication in Paramecium: the key role of Bld10 in assembly and stability of the cartwheel. Cytoskeleton (Hoboken) 67:161–171

Jia J, Jiang J (2006) Decoding the Hedgehog signal in animal development. Cell Mol Life Sci 63:1249–1265

Johnson RL, Rothman AL, Xie J, Goodrich LV, Bare JW, Bonifas JM, Quinn AG, Myers RM, Cox DR, Epstein EH Jr, Scott MP (1996) Human homolog of patched, a candidate gene for the basal cell nevus syndrome. Science 272:1668–1671

Kasbek C, Yang CH, Yusof AM, Chapman HM, Winey M, Fisk HA (2007) Preventing the degradation of mps1 at centrosomes is sufficient to cause centrosome reduplication in human cells. Mol Biol Cell 18:4457–4469

Kellogg DR, Moritz M, Alberts BM (1994) The centrosome and cellular organization. Annu Rev Biochem 63:639–674

Kemp CA, Kopish KR, Zipperlen P, Ahringer J, O'Connell KF (2004) Centrosome maturation and duplication in *C. elegans* require the coiled-coil protein SPD-2. Dev Cell 6:511–523

Kenney J, Karsenti E, Gowen B, Fuller SD (1997) Three-dimensional reconstruction of the mammalian centriole from cryoelectron micrographs: the use of common lines for orientation and alignment. J Struct Biol 120:320–328

Keryer G, Ris H, Borisy GG (1984) Centriole distribution during tripolar mitosis in Chinese hamster ovary cells. J Cell Biol 98:2222–2229

Khodjakov A, Rieder CL (1999) The sudden recruitment of gamma-tubulin to the centrosome at the onset of mitosis and its dynamic exchange throughout the cell cycle, do not require microtubules. J Cell Biol 146:585–596

Kimelman D, Xu W (2006) beta-catenin destruction complex: insights and questions from a structural perspective. Oncogene 25:7482–7491

Kitagawa D, Vakonakis I, Olieric N, Hilbert M, Keller D, Olieric V, Bortfeld M, Erat MC, Fluckiger I, Gonczy P, Steinmetz MO (2011) Structural basis of the 9-fold symmetry of centrioles. Cell 144:364–375

Kleylein-Sohn J, Westendorf J, Le Clech M, Habedanck R, Stierhof YD, Nigg EA (2007) Plk4-induced centriole biogenesis in human cells. Dev Cell 13:190–202

Kohlmaier G, Loncarek J, Meng X, McEwen BF, Mogensen MM, Spektor A, Dynlacht BD, Khodjakov A, Gonczy P (2009) Overly long centrioles and defective cell division upon excess of the SAS-4-related protein CPAP. Curr Biol 19:1012–1018

Kolb RJ, Nauli SM (2008) Ciliary dysfunction in polycystic kidney disease: an emerging model with polarizing potential. Front Biosci 13:4451–4466

Kuhl M, Sheldahl LC, Malbon CC, Moon RT (2000) Ca^{2+}/calmodulin-dependent protein kinase II is stimulated by Wnt and Frizzled homologs and promotes ventral cell fates in *Xenopus*. J Biol Chem 275:12701–12711

Kuriyama R, Borisy GG (1981) Microtubule-nucleating activity of centrosomes in Chinese hamster ovary cells is independent of the centriole cycle but coupled to the mitotic cycle. J Cell Biol 91:822–826

Lange BM, Gull K (1995) A molecular marker for centriole maturation in the mammalian cell cycle. J Cell Biol 130:919–927

Lee S, Rhee K (2010) CEP215 is involved in the dynein-dependent accumulation of pericentriolar matrix proteins for spindle pole formation. Cell Cycle 9:774–783

Leidel S, Delattre M, Cerutti L, Baumer K, Gonczy P (2005) SAS-6 defines a protein family required for centrosome duplication in C. elegans and in human cells. Nat Cell Biol 7:115–125

Liu A, Wang B, Niswander LA (2005) Mouse intraflagellar transport proteins regulate both the activator and repressor functions of Gli transcription factors. Development 132:3103–3111

Loncarek J, Khodjakov A (2009) Ab ovo or de novo? Mechanisms of centriole duplication. Mol Cells 27:135–142

Loncarek J, Hergert P, Magidson V, Khodjakov A (2008) Control of daughter centriole formation by the pericentriolar material. Nat Cell Biol 10:322–328

Ma L, Wang HY (2006) Suppression of cyclic GMP-dependent protein kinase is essential to the Wnt/cGMP/Ca^{2+} pathway. J Biol Chem 281:30990–31001

Ma L, Wang HY (2007) Mitogen-activated protein kinase p38 regulates the Wnt/cyclic GMP/Ca2+ non-canonical pathway. J Biol Chem 282:28980–28990

Marti E, Takada R, Bumcrot DA, Sasaki H, McMahon AP (1995) Distribution of Sonic hedgehog peptides in the developing chick and mouse embryo. Development 121:2537–2547

Masyuk AI, Masyuk TV, Splinter PL, Huang BQ, Stroope AJ, LaRusso NF (2006) Cholangiocyte cilia detect changes in luminal fluid flow and transmit them into intracellular Ca2+ and cAMP signaling. Gastroenterology 131:911–920

Masyuk AI, Gradilone SA, Banales JM, Huang BQ, Masyuk TV, Lee SO, Splinter PL, Stroope AJ, Larusso NF (2008) Cholangiocyte primary cilia are chemosensory organelles that detect biliary nucleotides via P2Y12 purinergic receptors. Am J Physiol Gastrointest Liver Physiol 295:G725–G734

Matsumoto Y, Maller JL (2004) A centrosomal localization signal in cyclin E required for Cdk2-independent S phase entry. Science 306:885–888

Matsuura K, Lefebvre PA, Kamiya R, Hirono M (2004) Bld10p, a novel protein essential for basal body assembly in *Chlamydomonas*: localization to the cartwheel, the first ninefold symmetrical structure appearing during assembly. J Cell Biol 165:663–671

Mayor T, Stierhof YD, Tanaka K, Fry AM, Nigg EA (2000) The centrosomal protein C-Nap1 is required for cell cycle-regulated centrosome cohesion. J Cell Biol 151:837–846

McGrath J, Somlo S, Makova S, Tian X, Brueckner M (2003) Two populations of node monocilia initiate left-right asymmetry in the mouse. Cell 114:61–73

Meraldi P, Lukas J, Fry AM, Bartek J, Nigg EA (1999) Centrosome duplication in mammalian somatic cells requires E2F and Cdk2-cyclin A. Nat Cell Biol 1:88–93

Mogensen MM, Malik A, Piel M, Bouckson-Castaing V, Bornens M (2000) Microtubule minus-end anchorage at centrosomal and non-centrosomal sites: the role of ninein. J Cell Sci 113(Pt 17):3013–3023

Moser JJ, Fritzler MJ, Rattner JB (2009) Primary ciliogenesis defects are associated with human astrocytoma/glioblastoma cells. BMC Cancer 9:448

Moudjou M, Bordes N, Paintrand M, Bornens M (1996) gamma-Tubulin in mammalian cells: the centrosomal and the cytosolic forms. J Cell Sci 109(Pt 4):875–887

Nakagawa Y, Yamane Y, Okanoue T, Tsukita S, Tsukita S (2001) Outer dense fiber 2 is a widespread centrosome scaffold component preferentially associated with mother centrioles: its identification from isolated centrosomes. Mol Biol Cell 12:1687–1697

Nauli SM, Zhou J (2004) Polycystins and mechanosensation in renal and nodal cilia. Bioessays 26:844–856

Nauli SM, Rossetti S, Kolb RJ, Alenghat FJ, Consugar MB, Harris PC, Ingber DE, Loghman-Adham M, Zhou J (2006) Loss of polycystin-1 in human cyst-lining epithelia leads to ciliary dysfunction. J Am Soc Nephrol 17:1015–1025

Nauli SM, Kawanabe Y, Kaminski JJ, Pearce WJ, Ingber DE, Zhou J (2008) Endothelial cilia are fluid shear sensors that regulate calcium signaling and nitric oxide production through polycystin-1. Circulation 117:1161–1171

Nishita M, Enomoto M, Yamagata K, Minami Y (2010) Cell/tissue-tropic functions of Wnt5a signaling in normal and cancer cells. Trends Cell Biol 20:346–354

Okuda M, Horn HF, Tarapore P, Tokuyama Y, Smulian AG, Chan PK, Knudsen ES, Hofmann IA, Snyder JD, Bove KE, Fukasawa K (2000) Nucleophosmin/B23 is a target of CDK2/cyclin E in centrosome duplication. Cell 103:127–140

Ou YY, Mack GJ, Zhang M, Rattner JB (2002) CEP110 and ninein are located in a specific domain of the centrosome associated with centrosome maturation. J Cell Sci 115:1825–1835

O'Toole ET, Giddings TH, McIntosh JR, Dutcher SK (2003) Three-dimensional organization of basal bodies from wild-type and delta-tubulin deletion strains of *Chlamydomonas reinhardtii*. Mol Biol Cell 14:2999–3012

Paintrand M, Moudjou M, Delacroix H, Bornens M (1992) Centrosome organization and centriole architecture: their sensitivity to divalent cations. J Struct Biol 108:107–128

Peel N, Stevens NR, Basto R, Raff JW (2007) Overexpressing centriole-replication proteins in vivo induces centriole overduplication and de novo formation. Curr Biol 17:834–843

Pelletier L, Ozlu N, Hannak E, Cowan C, Habermann B, Ruer M, Muller-Reichert T, Hyman AA (2004) The *Caenorhabditis elegans* centrosomal protein SPD-2 is required for both pericentriolar material recruitment and centriole duplication. Curr Biol 14:863–873

Pelletier L, O'Toole E, Schwager A, Hyman AA, Muller-Reichert T (2006) Centriole assembly in *Caenorhabditis elegans*. Nature 444:619–623

Piehl M, Tulu US, Wadsworth P, Cassimeris L (2004) Centrosome maturation: measurement of microtubule nucleation throughout the cell cycle by using GFP-tagged EB1. Proc Natl Acad Sci USA 101:1584–1588

Piel M, Meyer P, Khodjakov A, Rieder CL, Bornens M (2000) The respective contributions of the mother and daughter centrioles to centrosome activity and behavior in vertebrate cells. J Cell Biol 149:317–330

Piel M, Nordberg J, Euteneuer U, Bornens M (2001) Centrosome-dependent exit of cytokinesis in animal cells. Science 291:1550–1553

Porter JA, von Kessler DP, Ekker SC, Young KE, Lee JJ, Moses K, Beachy PA (1995) The product of hedgehog autoproteolytic cleavage active in local and long-range signalling. Nature 374:363–366

Praetorius HA, Leipziger J (2009) Released nucleotides amplify the cilium-dependent, flow-induced [Ca2+]i response in MDCK cells. Acta Physiol (Oxf) 197:241–251

Ratnam S, Nauli SM (2010) Hypertension in autosomal dominant polycystic kidney disease: a clinical and basic science perspective. Int J Nephrol Urol 2:294–308

Rohatgi R, Snell WJ (2010) The ciliary membrane. Curr Opin Cell Biol 22:541–546

Rohatgi R, Milenkovic L, Scott MP (2007) Patched1 regulates hedgehog signaling at the primary cilium. Science 317:372–376

Rusan NM, Rogers GC (2009) Centrosome function: sometimes less is more. Traffic 10:472–481

Saneyoshi T, Kume S, Amasaki Y, Mikoshiba K (2002) The Wnt/calcium pathway activates NF-AT and promotes ventral cell fate in *Xenopus* embryos. Nature 417:295–299

Saqui-Salces M, Merchant JL (2010) Hedgehog signaling and gastrointestinal cancer. Biochim Biophys Acta 1803:786–795

Schmidt TI, Kleylein-Sohn J, Westendorf J, Le Clech M, Lavoie SB, Stierhof YD, Nigg EA (2009) Control of centriole length by CPAP and CP110. Curr Biol 19:1005–1011

Sheldahl LC, Park M, Malbon CC, Moon RT (1999) Protein kinase C is differentially stimulated by Wnt and Frizzled homologs in a G-protein-dependent manner. Curr Biol 9:695–698

Sluder G, Begg DA (1985) Experimental analysis of the reproduction of spindle poles. J Cell Sci 76:35–51

Slusarski DC, Corces VG, Moon RT (1997a) Interaction of Wnt and a Frizzled homologue triggers G-protein-linked phosphatidylinositol signalling. Nature 390:410–413

Slusarski DC, Yang-Snyder J, Busa WB, Moon RT (1997b) Modulation of embryonic intracellular Ca2+ signaling by Wnt-5A. Dev Biol 182:114–120

Snyder JA, McIntosh JR (1975) Initiation and growth of microtubules from mitotic centers in lysed mammalian cells. J Cell Biol 67:744–760

Sorokin S (1962) Centrioles and the formation of rudimentary cilia by fibroblasts and smooth muscle cells. J Cell Biol 15:363–377

Soung NK, Kang YH, Kim K, Kamijo K, Yoon H, Seong YS, Kuo YL, Miki T, Kim SR, Kuriyama R, Giam CZ, Ahn CH, Lee KS (2006) Requirement of hCenexin for proper mitotic functions of polo-like kinase 1 at the centrosomes. Mol Cell Biol 26:8316–8335

Soung NK, Park JE, Yu LR, Lee KH, Lee JM, Bang JK, Veenstra TD, Rhee K, Lee KS (2009) Plk1-dependent and -independent roles of an ODF2 splice variant, hCenexin1, at the centrosome of somatic cells. Dev Cell 16:539–550

Spektor A, Tsang WY, Khoo D, Dynlacht BD (2007) Cep97 and CP110 suppress a cilia assembly program. Cell 130:678–690

Stanton BZ, Peng LF (2010) Small-molecule modulators of the Sonic Hedgehog signaling pathway. Mol Biosyst 6:44–54

Strnad P, Gonczy P (2008) Mechanisms of procentriole formation. Trends Cell Biol 18:389–396

Stucke VM, Sillje HH, Arnaud L, Nigg EA (2002) Human Mps1 kinase is required for the spindle assembly checkpoint but not for centrosome duplication. EMBO J 21:1723–1732

Tang CJ, Fu RH, Wu KS, Hsu WB, Tang TK (2009) CPAP is a cell-cycle regulated protein that controls centriole length. Nat Cell Biol 11:825–831

Taylor MD, Liu L, Raffel C, Hui CC, Mainprize TG, Zhang X, Agatep R, Chiappa S, Gao L, Lowrance A, Hao A, Goldstein AM, Stavrou T, Scherer SW, Dura WT, Wainwright B, Squire JA, Rutka JT, Hogg D (2002) Mutations in SUFU predispose to medulloblastoma. Nat Genet 31:306–310

Tsou MF, Stearns T (2006) Mechanism limiting centrosome duplication to once per cell cycle. Nature 442:947–951

Turner KJ, Sharpe RM, Gaughan J, Millar MR, Foster PM, Saunders PT (1997) Expression cloning of a rat testicular transcript abundant in germ cells, which contains two leucine zipper motifs. Biol Reprod 57:1223–1232

Vorobjev IA, Chentsov Yu S (1982) Centrioles in the cell cycle. I. Epithelial cells. J Cell Biol 93:938–949

Wang H, Lee Y, Malbon CC (2004) PDE6 is an effector for the Wnt/Ca2+/cGMP-signalling pathway in development. Biochem Soc Trans 32:792–796

Xiao Z, Zhang S, Mahlios J, Zhou G, Magenheimer BS, Guo D, Dallas SL, Maser R, Calvet JP, Bonewald L, Quarles LD (2006) Cilia-like structures and polycystin-1 in osteoblasts/osteocytes and associated abnormalities in skeletogenesis and Runx2 expression. J Biol Chem 281: 30884–30895

Xu C, Rossetti S, Jiang L, Harris PC, Brown-Glaberman U, Wandinger-Ness A, Bacallao R, Alper SL (2007) Human ADPKD primary cyst epithelial cells with a novel, single codon deletion in the PKD1 gene exhibit defective ciliary polycystin localization and loss of flow-induced Ca2+ signaling. Am J Physiol Renal Physiol 292:F930–F945

Xu C, Shmukler BE, Nishimura K, Kaczmarek E, Rossetti S, Harris PC, Wandinger-Ness A, Bacallao RL, Alper SL (2009) Attenuated, flow-induced ATP release contributes to absence of flow-sensitive, purinergic Ca^{2+} signaling in human ADPKD cyst epithelial cells. Am J Physiol Renal Physiol 296:F1464–F1476

Yang-Snyder J, Miller JR, Brown JD, Lai CJ, Moon RT (1996) A frizzled homolog functions in a vertebrate Wnt signaling pathway. Curr Biol 6:1302–1306

Zhao L, Jin C, Chu Y, Varghese C, Hua S, Yan F, Miao Y, Liu J, Mann D, Ding X, Zhang J, Wang Z, Dou Z, Yao X (2010) Dimerization of CPAP orchestrates centrosome cohesion plasticity. J Biol Chem 285:2488–2497

Zheng Y, Wong ML, Alberts B, Mitchison T (1995) Nucleation of microtubule assembly by a gamma-tubulin-containing ring complex. Nature 378:578–583

Zhu F, Lawo S, Bird A, Pinchev D, Ralph A, Richter C, Muller-Reichert T, Kittler R, Hyman AA, Pelletier L (2008) The mammalian SPD-2 ortholog Cep192 regulates centrosome biogenesis. Curr Biol 18:136–141

Chapter 10
Identification of Mechanosensitive Genes in Chondrocytes and Osteoblasts and Their Role in OA Pathogenesis

Yves Henrotin, Laurence Pesesse, and Christelle Sanchez

Abstract Osteoarthritis is a pathological condition resulting of the deregulation of the homeostatic mechanisms of the whole joint, triggered by mechanical and biochemical factors. Mechanical factors play a key role in this process by directly breaking the matrix scaffold or by altering the production of matrix components and anabolic/catabolic factors. However, recent studies have shown that only some, but not all genes and regulatory pathways are sensitive to mechanical stimuli. It is also emphasized that cell responses to mechanical stimuli are dependent of their physical characteristics including the magnitude and the frequency. This chapter proposes a review of the responses of chondrocytes and osteoblasts to mechanical stimuli described in the literature recently.

Keywords Osteoarthritis · Cartilage · Bone · Mechanical stimuli

10.1 Introduction

Osteoarthritis (OA), the most common form of arthritis, is a complex disease characterized by a progressive loss of cartilage, an abnormal subchondral bone remodeling and a chronic inflammation of synovial membrane. These metabolic and structural modifications are driven by biochemical and mechanical factors which act in concert to deregulate the homeostasis of joint tissues. Numerous biochemical mediators including metalloproteinases, cytokines, reactive oxygen and nitrogen species, prostanoids and cytokines are directly involved in OA onset and progression. Secreted cytokines, mainly those classified as pro-inflammatory, therefore, are critical mediators of metabolic and structural changes occurring in OA joint tissues. IL-1β, TNF and IL-6 have been identified as the most active cytokines in OA,

Y. Henrotin (✉)
Bone and Cartilage Research Unit, Institute of Pathology Level 5, University of Liège, 4000 Liège, Belgium
e-mail: yhenrotin@ulg.ac.be

A. Kamkin, I. Kiseleva (eds.), *Mechanical Stretch and Cytokines*, Mechanosensitivity in Cells and Tissues 5, DOI 10.1007/978-94-007-2004-6_10,

although others, such as IL-8, IL-15, IL-17, IL-18 and IL-21, leukemia inhibitory factor (LIF) and chemokines have been implicated (see details in Kapoor et al., 2011).

IL-1β, IL-6 and TNF-α are produced by the major cells of the joint, including synovial cells, chondrocytes and osteoblasts. They stimulate the production of a number of inflammatory and catabolic factors. IL-1β and TNF-α can act independently or in synergy with other cytokines, such as IL-6 which is considered as a potent cofactor of IL-1 activity. A large number of studies have demonstrated that IL-1β, TNF-α and IL-6 down-regulate the synthesis of the major extracellular components and stimulate the release of several proteolytic enzymes, among which are the matrix metalloproteinases MMP-1 (collagenase-1), MMP-3 (stromelysin-1) and MMP-13 (collagenase-3) (see details in Goldring et al., 2011). These three proteases are key actors of the cartilage degradation because they cleave the two major components of cartilage extracellular matrix, type II collagen and aggrecan molecules. IL-1β, TNF-α and IL-6 also stimulate the production of prostanoids (e.g. PGE_2) and reactive oxygen and nitrogen species (e.g. nitric oxide). These mediators are directly involved in the development of chronic inflammation and cartilage degradation. Further, IL-6 is a critical cytokine of the metabolic changes of subchondral bone occurring in OA. It triggers osteoclast differentiation and bone resorption. This effect of IL-6 is indirect and mediated through the production by IL-6-stimulated osteoblasts of receptor activator of NF-kB ligand (RANKL), IL-1β, parathyroid (PTH)-related protein (PTH-rP) and PGE_2, all of which trigger the production of IL-6 by osteoblasts.

Most forms of OA are secondary and results from superimposed risks factors affecting distribution and severity of loading forces acting on specific joints, such as injury, overuse to overexercises, malalignment, joint instability from loss ligamental or meniscal support or underloading. In regard to underloading, disuse immobilization from chronic inanition, neurologic disorders, or postoperative casting can lead to cartilage atrophy, with increased vulnerability to cartilage injury, unless a progressive program of rehabilitation is conducted. This underlines the fact that mechanical stimuli are essential to maintain the homeostatic balance between the catabolic and anabolic events in chondrocytes. Indeed, many studies have demonstrated that the mechanical conditions of the joint is important and regulate the biological activity of the chondrocytes, but also osteoblasts in vivo. Considerable effort has been directed toward understanding the process by which physical signals are converted to biochemical signals by the chondrocytes and the osteoblasts. Currently, there is significant evidence that these cells may transduce mechanical signals into biochemical responses through various intracellular and intercellular signaling pathways involving integrin, cyclic AMP, etc. (see Liedert et al., 2006).

In this chapter, we review in detail the genes involved in OA pathogenesis which are sensitive to mechanical stimuli. By this way, we attempt to make the connection between mechanical and biochemical factors involved in OA pathophysiology.

10.2 Mechanosensible Genes in Chondrocytes

Non-physiological mechanical strains are known to induce cartilage degradation (Sun, 2010). The intracellular events induced by mechanical overload are similar to those that are initiated by proinflammatory cytokines in arthritis. Magnitude-dependent signals of mechanical strain utilize the NF-κB transcription factors as common elements to abrogate or aggravate proinflammatory responses (Agarwal et al., 2004). Furthermore, mitogen-activated protein kinases (MAPK), activated by IL-1β and TNF-α, are activated by static loading in chondrocytes as well (Fitzgerald et al., 2008).

PGE_2 is also a major catabolic mediator involved in cartilage degradation and chondrocyte apoptosis (Miwa et al., 2000; Hardy et al., 2002). Higher levels of PGE_2 are observed in OA cartilage when compared to normal cartilage (Jacques et al., 1999). Mechanical compression of cartilage explants (0.5 Hz, 1 MPa) significantly increased PGE_2 production, resulting from an increase in COX-2 and mPGES1 expression in chondrocytes (Gosset et al., 2006). Shear stress is known to induce IL-6 production by chondrocytes (Mohtai et al., 1996) and high magnitude cyclic tensile strain increased expression of IL-1β and TNF-α in these cells (Honda et al., 2000).

While the pathways involved in the destruction of cartilage by mechanical overloading are being elucidated, less is known concerning the mechanisms initiated by underloading. Joint immobilization in animal models show that cartilage degradation is mediated by MMP-8 and MMP-13, the main enzymes responsible for the degradation of type II collagen, which are also involved in hypertrophic differentiation of chondrocytes (Ando et al., 2009). In these models, immobilization also increases expression of HIF-1α, VEGF, and decreases expression of chondromodulin-1 (Sakamoto et al., 2009).

While overloading or reduced loading cause cartilage degradation, moderate levels of activity maintain normal cartilage integrity. Physiologic mechanical stimulation of the articular cartilage generates biochemical signals which increase the anabolic activity of the chondrocytes (15% compressive strain amplitude at 1 Hz) (Lee and Bader, 1997).

Studies demonstrate that biomechanical signals within a physiologic range of intensity, duration, and frequency are potent antiinflammatory signals that attenuate proinflammatory genes induction in chondrocytes. Agarwal group showed that cyclic tensile strain with a magnitude of 10% or higher is proinflammatory in nature, whereas 3–6% cyclic tensile strain is anti-inflammatory and inhibits IL-1-induced proinflammatory gene induction (Gassner et al., 1999; Xu et al., 2000). At low magnitudes in vitro, biomechanical signals inhibit IL-1β- or TNF-α-induced transcriptional activation of COX-2, MMPs, IL-1β, and other proinflammatory molecules (Chowdhury et al., 2003, 2006; Agarwal et al., 2004; Ferretti et al., 2005). Furthermore, a recent study shows that chondrocytes exposed to moderate levels of dynamic cyclic load (0.2–0.5 MPa) inhibit IL-1β-induced matrix degradation (Torzilli et al., 2010). In vivo, motion-based therapies have been demonstrated to mitigate joint inflammation in rabbit with antigen-induced arthritis. Mechanical

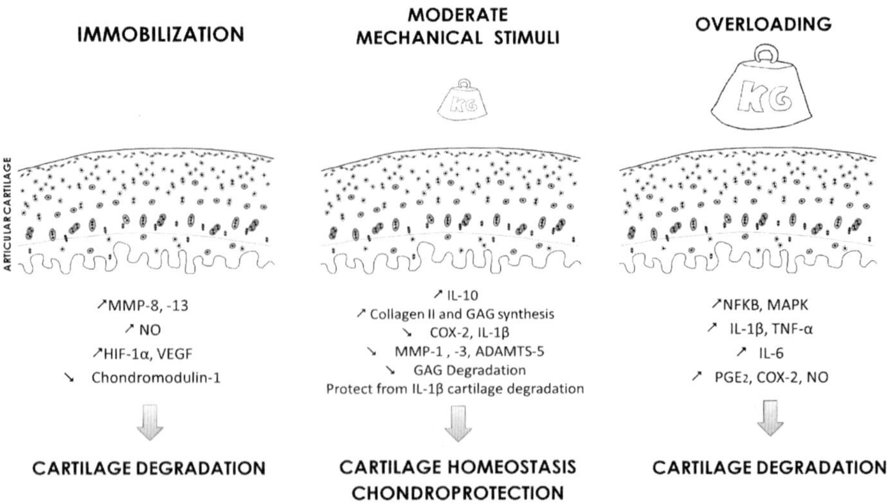

Fig. 10.1 Effect of mechanical stimuli on cartilage

signals generated from these passive joint motion therapies were reported to suppress expression of proinflammatory catabolic mediators, such as IL-1β, COX-2, and MMP-1 as well as inducing expression of the antiinflammatory cytokine, IL-10 (Ferretti et al., 2005; 2006) (Fig. 10.1).

10.3 Mechanosensitive Genes in Osteoblasts

Bone remodeling results from the coordinated and balanced activities of osteoblasts and osteoclasts. Osteoblasts functions are intimately linked to osteoclast activity via the osteoblastic production of cytokines, growth factors and prostaglandins. The production of some of these factors is controlled by mechanical strains (Liedert et al., 2006). Recently, a number of in vitro models have attempted to screen genes and signaling pathways involved in this mechanism, mainly by stretching osteoblasts (Mikuni-Takagaki, 1999; Ignatius et al., 2005; Kusumi et al., 2005; Liu et al., 2006a; Siddhivarn et al., 2006; Tang et al., 2006; Kanno et al., 2007; Sasaki et al., 2007) or by submitting them to a fluid shear stress (Chen et al., 2003; Inoue et al., 2004; Tanaka et al., 2005). Osteoblasts possess mechanosensors activating intracellular signals including ion channels, integrins, calveolar membrane structure and cytoskeleton (Rubin et al., 2006). Nevertheless, response to physical signals may be quite different according to the type of mechanical stimulus applied. Fluid shear stress have been shown to elicit multiple intracellular signaling pathways involving intracellular calcium rise, extracellular signal-regulated kinase (ERK)1/2 activation of c-Fos and nuclear factor (NF)-κB translocation (Chen et al., 2003; Inoue et al., 2004). Downstream of such signaling events is induction of various gene expression including type I collagen (COL1A1), osteopontin (OPN), insulin-like growth factor-I (IGF-1) and cyclooxygenase (COX)-2 (Nomura and Takano-Yamamoto, 2000; Chen et al., 2003).

Cyclic tensile strains were also potent activator of the signaling cascade formed by ERK/c-fos/NF-κB (Liu et al., 2006a). Stretching increased the production of vascular endothelial growth factor (VEGF), transforming growth factor (TGF)-β1 (Singh et al., 2007), alkaline phosphatase activity (AP), osteocalcin (OC), osteoprotegerin (OPG), MMP-1 and -3 (Kanno et al., 2007), COX-1 and -2, PGD_2 synthase, peroxisome proliferator-activated receptor (PPAR)γ-1 synthesis, but decreased the soluble receptor activator of nuclear factor ligand (sRANKL) released (Tang et al., 2006) by osteoblasts. In contrast, no significant effect was reported on MMP-2, tissue inhibitor of metalloproteinases (TIMP)-1 and -2, and PPARγ-2 synthesis (Siddhivarn et al., 2006; Sasaki et al., 2007). More relevant of the in vivo loading stress, little is known on the effect of compression on bone cells. In fact, compression associates both stimuli, including stretching and fluid flow. Current experimental approaches in this field have included animal studies with bones biopsy culture after external loading (Fox et al., 1996; Noble et al., 2003; Hoffler et al., 2006; Mann et al., 2006). Compressive strain induced an early ERK activation and c-fos mRNA and protein production (Jackson et al., 2006) and subsequently an increase of AP, COL1A1, PGE_2 and PGI_2 synthesis, and mRNA expression levels of Runt-related gene-2 (RUNX)2, OC, OPN, and COX-2 (Jackson et al., 2006). However, mineralization was inhibited, despite an increase in calcium accumulation, suggesting that loading may inhibit mineralization in order to increase matrix deposition (Jackson et al., 2006). Recently, Koyama et al. have demonstrated that IL-1, IL-6 and IL-8 gene expression in the osteosarcoma cell line Saos-2 is increased by very low compression (up to 3 g/cm^2, static) (Koyama et al., 2008).

To better understand subchondral bone sclerosis, we recently developed an original osteoblast compression model to investigate the influence of mechanical stimuli on bone metabolism (Sanchez et al., 2009). This original model of 3D osteoblast culture allowed studying the compression of osteoblasts/osteocytes embedded in their newly synthesized extracellular matrix. Moreover, the model conserves cell/matrix interactions and allows fluid flow through a three-dimensional extracellular matrix, permitting the study of compression, tension, and fluid shear stimuli.

In our study, high amplitude compression (6 and 10% corresponding to 1 and 1.67 MPa) was applied at the frequency of 1 Hz. These loading conditions are included in the range of mechanical strains applied on bone during locomotion. In vivo bone strain measurements taken during locomotion demonstrated large amplitude strains between 1 and 1.6 MPa at frequencies ranging from 0.5 to 2 Hz (Turner et al., 1995).

In our experimental conditions, we observed a strong accumulation of PGE_2 in the culture medium of loaded 3D-osteoblasts. This confirms previous studies demonstrating that fluid flow, compression and stretching stimulate PGE_2 production by osteoblasts (Bakker et al., 2003, 2006; Reilly et al., 2003; Grimston et al., 2006). Further, we demonstrated that PGE_2 accumulation probably results from an imbalance between PGE_2 synthesis and degradation. Indeed, in our experimental conditions, compression increased COX-2 expression but decreased 15-PGDH expression. 15-PGDH is a cytosolic enzyme which catalyzes the first step in the catabolic pathway of prostaglandins, and is believed to be the key enzyme responsible for the biological inactivation of this biologically potent eicosanoid (Cho et al., 2006).

Another important finding was that compression had no significant effect on COX-1 and mPGES gene expression. This contrast with previously reported data showing that compression of cartilage explants increases mPGES1 expression by chondrocytes (Gosset et al., 2006). This observation is important since it was suggested that PGE_2 can act as a potent stimulator of both bone resorption and formation (Harada et al., 1995; Suzawa et al., 2000; Mitsui et al., 2006). Therefore, our results indicate that compression induces PGE_2 accumulation in bone through the COX-2 and 15PGDH pathways.

PGE_2 is also a mediator involved in IL-6-induced osteoclast formation and bone resorption (Liu et al., 2006b). In this study, we showed that IL-6 is a highly mechano-sensitive gene. IL-6 gene expression is early increased (1 h) and IL-6 protein secretion is highly stimulated (up to 32-fold) by compression. The role of IL-6 on bone physiology is complex. IL-6 clearly stimulates osteoclasts and bone resorption in vivo and in vitro (Palmqvist et al., 2002; De Benedetti et al., 2006). Data on in vivo and in vitro effects of IL-6 on osteoblasts are still conflicting, and several models have shown contradictory results (Franchimont et al., 2005). In a recent model of transgenic mice overexpressing IL-6, a marked decrease in osteoblast activity, proliferation and expression of gene of bone matrix protein was found (De Benedetti et al., 2006). This last result suggests that IL-6 could downregulates bone formation.

Using specific inhibitors, we have identified some signaling pathways involved in the compression-induced IL-6 and PGE_2 production. It clearly appeared that alpha5beta1 integrin, intracellular Ca^{2+}, NF-κB and ERK1/2 were involved. In contrast, these effects of compression were not affected by JNK and p38 inhibitors. These data confirm the involvement of Ca^{2+}, ERK1/2 and NF-κB in the transduction of mechanical signals (Chen et al., 2003; Inoue et al., 2004; Jansen et al., 2004). We have also investigated the role of IL-1 in the compression-stimulated IL-6 and PGE_2 production. IL-1 is a well known PGE_2 and IL-6 inducer suggesting that the effects of compression on these mediators could be mediated by IL-1. This hypothesis is supported by the work of Koyama and collaborators, who have recently demonstrated that compression induced the production of IL-6 by osteoblasts subsequently to IL-1β production (Koyama et al., 2008). However, in our model, IL-1ra did not significantly modify IL-6 and PGE_2 production indicating that IL-1 is not involved in this response. This discrepancy can be explained by differences in the source of osteoblasts (osteosarcoma cell line vs primary osteoblasts) and in the compression model used (monolayer vs 3D). Finally, using an IL-6 neutralizing antibody and piroxicam, a COXs inhibitor, we have investigated the potential interaction between IL-6 and PGE_2 during compression. Interestingly, IL-6 neutralization significantly reduced compression-stimulated PGE_2 production and the full inhibition of PGE_2 synthesis by piroxicam drastically inhibited IL-6 production. These findings provide evidence for an interaction between PGE_2 and IL-6

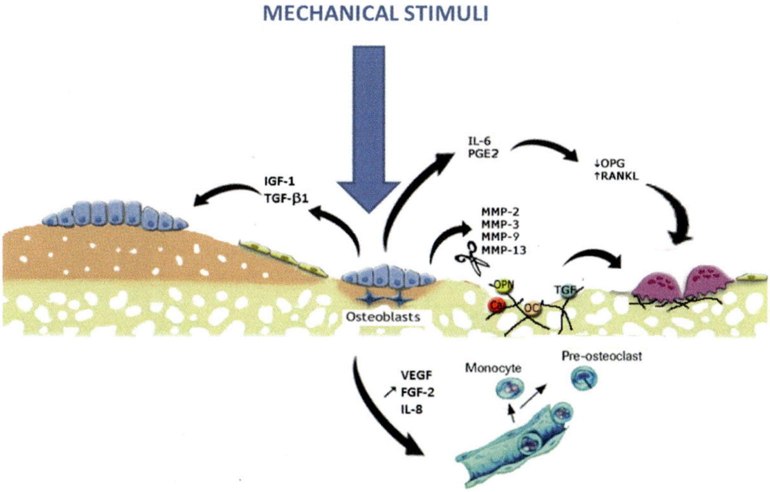

Fig. 10.2 Hypothetical model of mechanical stress impact on bone

to amplify the response of osteoblasts to compression. As shown in Fig. 10.2, this increased IL-6 and PGE_2 production by osteoblasts could increase osteoclasts differentiation and activation via effect on the RANK/RANKL/OPG system (Liu et al., 2005). Indeed, we observed that compression decreased OPG gene expression, but did not modify RANKL. OPG is a decoy receptor for RANKL which inhibits osteoclast activation. Therefore, we can conclude that compression by decreasing OPG/RANKL ratio, could promote osteoclasts activity and bone remodeling. Since it was reported that IL-6 is the mediator of PGE_2-induced suppression of OPG production by osteoblasts (Liu et al., 2005), we speculate that IL-6 and PGE_2 are major mediators of the compression-induced bone remodeling.

We also observed an increase of MMPs expression and production, particularly MMP-3. These MMPs could degrade, in parallel to osteoclasts, the bone matrix and release some growth factor and cytokine, i.e. TGF-β, which then enhance bone matrix synthesis (Fig. 10.2).

10.4 Conclusion and Perspectives

In the absence of mechanical strains, cartilage atrophy occurs indicating that mechanical stimuli play a critical role in the control of cartilage homeostasis. Moderate levels of mechanical stimuli have anabolic and anti-inflammatory effects, suggesting that mechanical stimuli exert a beneficial effect on cartilage. In contrast, abnormal mechanical strains play a critical role in OA pathogenesis, not only by inducing cartilage matrix breakdown and cells apoptosis, but also stimulating the production of proinflammatory mediators. Unfortunately, in the literature the words "abnormal" or "moderate" are not clearly defined and the physical characteristic of the stimuli are often missing. There is no consensus on the definition of a "physiological" and a "pathological" mechanical stimuli in vitro. Different frequency, magnitude and duration are reported in the literature. Therefore, the interpretation of the effects of mechanical stimuli is confused. A systematic research of the impact of frequency, magnitude and type of mechanical stimulus on chondrocytes response should be performed.

References

Agarwal S, Deschner J, Long P, Verma A, Hofman C, Evans CH, Piesco N (2004) Role of NF-kappaB transcription factors in antiinflammatory and proinflammatory actions of mechanical signals. Arthritis Rheum 50:3541–3548

Ando A, Hagiwara Y, Tsuchiya M, Onoda Y, Suda H, Chimoto E, Itoi E (2009) Increased expression of metalloproteinase-8 and -13 on articular cartilage in a rat immobilized knee model. Tohoku J Exp Med 217:271–278

Bakker AD, Joldersma M, Klein-Nulend J, Burger EH (2003) Interactive effects of PTH and mechanical stress on nitric oxide and PGE2 production by primary mouse osteoblastic cells. Am J Physiol Endocrinol Metab 285:E608–613

Bakker AD, Klein-Nulend J, Tanck E, Heyligers IC, Albers GH, Lips P, Burger EH (2006) Different responsiveness to mechanical stress of bone cells from osteoporotic versus osteoarthritic donors. Osteoporos Int 17:827–833

Chen NX, Geist DJ, Genetos DC, Pavalko FM, Duncan RL (2003) Fluid shear-induced NFkappaB translocation in osteoblasts is mediated by intracellular calcium release. Bone 33:399–410

Cho H, Huang L, Hamza A, Gao D, Zhan CG, Tai HH (2006) Role of glutamine 148 of human 15-hydroxyprostaglandin dehydrogenase in catalytic oxidation of prostaglandin E2. Bioorg Med Chem 14:6486–6491

Chowdhury TT, Bader DL, Lee DA (2003) Dynamic compression counteracts IL-1 beta-induced release of nitric oxide and PGE2 by superficial zone chondrocytes cultured in agarose constructs. Osteoarthritis Cartilage 11:688–696

Chowdhury TT, Bader DL, Lee DA (2006) Anti-inflammatory effects of IL-4 and dynamic compression in IL-1beta stimulated chondrocytes. Biochem Biophys Res Commun 339:241–247

De Benedetti F, Rucci N, Del Fattore A, Peruzzi B, Paro R, Longo M, Vivarelli M, Muratori F, Berni S, Ballanti P, Ferrari S, Teti A (2006) Impaired skeletal development in interleukin-6-transgenic mice: a model for the impact of chronic inflammation on the growing skeletal system. Arthritis Rheum 54:3551–3563

Ferretti M, Srinivasan A, Deschner J, Gassner R, Baliko F, Piesco N, Salter R, Agarwal S (2005) Anti-inflammatory effects of continuous passive motion on meniscal fibrocartilage. J Orthop Res 23:1165–1171

Ferretti M, Gassner R, Wang Z, Perera P, Deschner J, Sowa G, Salter RB, Agarwal S (2006) Biomechanical signals suppress proinflammatory responses in cartilage: early events in experimental antigen-induced arthritis. J Immunol 177:8757–8766

Fitzgerald JB, Jin M, Chai DH, Siparsky P, Fanning P, Grodzinsky AJ (2008) Shear- and compression-induced chondrocyte transcription requires MAPK activation in cartilage explants. J Biol Chem 283:6735–6743

Fox SW, Chambers TJ, Chow JW (1996) Nitric oxide is an early mediator of the increase in bone formation by mechanical stimulation. Am J Physiol 270:E955–E960

Franchimont N, Wertz S, Malaise M (2005) Interleukin-6: an osteotropic factor influencing bone formation? Bone

Gassner R, Buckley MJ, Georgescu H, Studer R, Stefanovich-Racic M, Piesco NP, Evans CH, Agarwal S (1999) Cyclic tensile stress exerts antiinflammatory actions on chondrocytes by inhibiting inducible nitric oxide synthase. J Immunol 163:2187–2192

Goldring MB, Otero M, Plumb DA, Dragomir C, Favero M, El Hachem H, Hashimoto K, Roach HI, Olivotto E, Borzi RM, Marcu KB (2011) Roles of inflammatory and anabolic cytokines in cartilage metabolism: signals and multiple effectors converge upon MMP-13 regulation in osteoarthritis. Eur Cell Mater 21:202–220

Gosset M, Berenbaum F, Levy A, Pigenet A, Thirion S, Saffar JL, Jacques C (2006) Prostaglandin E2 synthesis in cartilage explants under compression: mPGES-1 is a mechanosensitive gene. Arthritis Res Ther 8:R135

Grimston SK, Screen J, Haskell JH, Chung DJ, Brodt MD, Silva MJ, Civitelli R (2006) Role of connexin43 in osteoblast response to physical load. Ann N Y Acad Sci 1068:214–224

Harada SI, Balena R, Rodan GA, Rodan SB (1995) The role of prostaglandins in bone formation. Connect Tissue Res 31:279–282

Hardy MM, Seibert K, Manning PT, Currie MG, Woerner BM, Edwards D, Koki A, Tripp CS (2002) Cyclooxygenase 2-dependent prostaglandin E2 modulates cartilage proteoglycan degradation in human osteoarthritis explants. Arthritis Rheum 46:1789–1803

Hoffler CE, Hankenson KD, Miller JD, Bilkhu SK, Goldstein SA (2006) Novel explant model to study mechanotransduction and cell-cell communication. J Orthop Res 24:1687–1698

Honda K, Ohno S, Tanimoto K, Ijuin C, Tanaka N, Doi T, Kato Y, Tanne K (2000) The effects of high magnitude cyclic tensile load on cartilage matrix metabolism in cultured chondrocytes. Eur J Cell Biol 79:601–609

Ignatius A, Blessing H, Liedert A, Schmidt C, Neidlinger-Wilke C, Kaspar D, Friemert B, Claes L (2005) Tissue engineering of bone: effects of mechanical strain on osteoblastic cells in type I collagen matrices. Biomaterials 26:311–318

Inoue D, Kido S, Matsumoto T (2004) Transcriptional induction of FosB/DeltaFosB gene by mechanical stress in osteoblasts. J Biol Chem 279:49795–49803

Jackson RA, Kumarasuriyar A, Nurcombe V, Cool SM (2006) Long-term loading inhibits ERK1/2 phosphorylation and increases FGFR3 expression in MC3T3-E1 osteoblast cells. J Cell Physiol 209:894–904

Jacques C, Sautet A, Moldovan M, Thomas B, Humbert L, Berenbaum F (1999) Cyclooxygenase activity in chondrocytes from osteoarthritic and healthy cartilage. Rev Rhum Engl Ed 66:701–704

Jansen JH, Weyts FA, Westbroek I, Jahr H, Chiba H, Pols HA, Verhaar JA, van Leeuwen JP, Weinans H (2004) Stretch-induced phosphorylation of ERK1/2 depends on differentiation stage of osteoblasts. J Cell Biochem 93:542–551

Kanno T, Takahashi T, Tsujisawa T, Ariyoshi W, Nishihara T (2007) Mechanical stress-mediated Runx2 activation is dependent on Ras/ERK1/2 MAPK signaling in osteoblasts. J Cell Biochem 101:1266–1277

Kapoor M, Martel-Pelletier J, Lajeunesse D, Pelletier JP, Fahmi H (2011) Role of proinflammatory cytokines in the pathophysiology of osteoarthritis. Nat Rev Rheumatol 7:33–42

Koyama Y, Mitsui N, Suzuki N, Yanagisawa M, Sanuki R, Isokawa K, Shimizu N, Maeno M (2008) Effect of compressive force on the expression of inflammatory cytokines and their receptors in osteoblastic Saos-2 cells. Arch Oral Biol 53:488–496

Kusumi A, Sakaki H, Kusumi T, Oda M, Narita K, Nakagawa H, Kubota K, Satoh H, Kimura H (2005) Regulation of synthesis of osteoprotegerin and soluble receptor activator of nuclear factor-kappaB ligand in normal human osteoblasts via the p38 mitogen-activated protein kinase pathway by the application of cyclic tensile strain. J Bone Miner Metab 23:373–381

Lee DA, Bader DL (1997) Compressive strains at physiological frequencies influence the metabolism of chondrocytes seeded in agarose. J Orthop Res 15:181–188

Liedert A, Kaspar D, Blakytny R, Claes L, Ignatius A (2006) Signal transduction pathways involved in mechanotransduction in bone cells. Biochem Biophys Res Commun 349:1–5

Liu XH, Kirschenbaum A, Yao S, Levine AC (2005) Cross-talk between the interleukin-6 and prostaglandin E(2) signaling systems results in enhancement of osteoclastogenesis through effects on the osteoprotegerin/receptor activator of nuclear factor-{kappa}B (RANK) ligand/RANK system. Endocrinology 146:1991–1998

Liu J, Liu T, Zheng Y, Zhao Z, Liu Y, Cheng H, Luo S, Chen Y (2006a) Early responses of osteoblast-like cells to different mechanical signals through various signaling pathways. Biochem Biophys Res Commun 348:1167–1173

Liu XH, Kirschenbaum A, Yao S, Levine AC (2006b) Interactive effect of interleukin-6 and prostaglandin E2 on osteoclastogenesis via the OPG/RANKL/RANK system. Ann N Y Acad Sci 1068:225–233

Mann V, Huber C, Kogianni G, Jones D, Noble B (2006) The influence of mechanical stimulation on osteocyte apoptosis and bone viability in human trabecular bone. J Musculoskelet Neuronal Interact 6:408–417

Mikuni-Takagaki Y (1999) Mechanical responses and signal transduction pathways in stretched osteocytes. J Bone Miner Metab 17:57–60

Mitsui N, Suzuki N, Maeno M, Yanagisawa M, Koyama Y, Otsuka K, Shimizu N (2006) Optimal compressive force induces bone formation via increasing bone morphogenetic proteins production and decreasing their antagonists production by Saos-2 cells. Life Sci 78: 2697–2706

Miwa M, Saura R, Hirata S, Hayashi Y, Mizuno K, Itoh H (2000) Induction of apoptosis in bovine articular chondrocyte by prostaglandin E(2) through cAMP-dependent pathway. Osteoarthritis Cartilage 8:17–24

Mohtai M, Gupta MK, Donlon B, Ellison B, Cooke J, Gibbons G, Schurman DJ, Smith RL (1996) Expression of interleukin-6 in osteoarthritic chondrocytes and effects of fluid-induced shear on this expression in normal human chondrocytes in vitro. J Orthop Res 14:67–73

Noble BS, Peet N, Stevens HY, Brabbs A, Mosley JR, Reilly GC, Reeve J, Skerry TM, Lanyon LE (2003) Mechanical loading: biphasic osteocyte survival and targeting of osteoclasts for bone destruction in rat cortical bone. Am J Physiol Cell Physiol 284:C934–C943

Nomura S, Takano-Yamamoto T (2000) Molecular events caused by mechanical stress in bone. Matrix Biol 19:91–96

Palmqvist P, Persson E, Conaway HH, Lerner UH (2002) IL-6, leukemia inhibitory factor, and oncostatin M stimulate bone resorption and regulate the expression of receptor activator of NF-kappa B ligand, osteoprotegerin, and receptor activator of NF-kappa B in mouse calvariae. J Immunol 169:3353–3362

Reilly GC, Haut TR, Yellowley CE, Donahue HJ, Jacobs CR (2003) Fluid flow induced PGE2 release by bone cells is reduced by glycocalyx degradation whereas calcium signals are not. Biorheology 40:591–603

Rubin J, Rubin C, Jacobs CR (2006) Molecular pathways mediating mechanical signaling in bone. Gene 367:1–16

Sakamoto J, Origuchi T, Okita M, Nakano J, Kato K, Yoshimura T, Izumi S, Komori T, Nakamura H, Ida H, Kawakami A, Eguchi K (2009) Immobilization-induced cartilage degeneration mediated through expression of hypoxia-inducible factor-1alpha, vascular endothelial growth factor, and chondromodulin-I. Connect Tissue Res 50:37–45

Sanchez C, Gabay O, Salvat C, Henrotin YE, Berenbaum F (2009) Mechanical loading highly increases IL-6 production and decreases OPG expression by osteoblasts. Osteoarthritis Cartilage 17:473–481

Sasaki K, Takagi M, Konttinen YT, Sasaki A, Tamaki Y, Ogino T, Santavirta S, Salo J (2007) Upregulation of matrix metalloproteinase (MMP)-1 and its activator MMP-3 of human osteoblast by uniaxial cyclic stimulation. J Biomed Mater Res B Appl Biomater 80:491–498

Siddhivarn C, Banes A, Champagne C, Riche EL, Weerapradist W, Offenbacher S (2006) Prostaglandin D2 pathway and peroxisome proliferator-activated receptor gamma-1 expression are induced by mechanical loading in an osteoblastic cell line. J Periodontal Res 41:92–100

Singh SP, Chang EI, Gossain AK, Mehara BJ, Galiano RD, Jensen J, Longaker MT, Gurtner GC, Saadeh PB (2007) Cyclic mechanical strain increases production of regulators of bone healing in cultured murine osteoblasts. J Am Coll Surg 204:426–434

Sun HB (2010) Mechanical loading, cartilage degradation, and arthritis. Ann N Y Acad Sci 1211:37–50

Suzawa T, Miyaura C, Inada M, Maruyama T, Sugimoto Y, Ushikubi F, Ichikawa A, Narumiya S, Suda T (2000) The role of prostaglandin E receptor subtypes (EP1, EP2, EP3, and EP4) in bone resorption: an analysis using specific agonists for the respective EPs. Endocrinology 141:1554–1559

Tanaka SM, Sun HB, Roeder RK, Burr DB, Turner CH, Yokota H (2005) Osteoblast responses one hour after load-induced fluid flow in a three-dimensional porous matrix. Calcif Tissue Int 76:261–271

Tang L, Lin Z, Li YM (2006) Effects of different magnitudes of mechanical strain on Osteoblasts in vitro. Biochem Biophys Res Commun 344:122–128

Torzilli PA, Bhargava M, Park S, Chen CT (2010) Mechanical load inhibits IL-1 induced matrix degradation in articular cartilage. Osteoarthritis Cartilage 18:97–105

Turner CH, Yoshikawa T, Forwood MR, Sun TC, Burr DB (1995) High frequency components of bone strain in dogs measured during various activities. J Biomech 28:39–44

Xu Z, Buckley MJ, Evans CH, Agarwal S (2000) Cyclic tensile strain acts as an antagonist of IL-1 beta actions in chondrocytes. J Immunol 165:453–460

Subject Index

A. Kamkin, I. Kiseleva (eds.), *Mechanical Stretch and Cytokines*, Mechanosensitivity in Cells and Tissues 5, DOI 10.1007/978-94-007-2004-6,